LIVING IN THE LABYRINTH OF TECHNOLOGY

From the very beginnings of their existence, human beings distinguished themselves from other animals by not taking immediate experience for granted. Everything was symbolized according to its meaning and value: a fallen branch from a tree became a lever; a tree trunk floating in the river became a canoe. *Homo logos* created communities based on cultures: humanity's first megaproject. Further symbolization of the human community and its relation to nature led to the creation of societies and civilizations. *Homo societas* created ways of life able to give meaning, direction, and purpose to many groups by means of very different cultures: humanity's second megaproject. Western civilization may well be creating humanity's third megaproject, based not on symbolization for making sense of and living in the world, but on highly specialized desymbolized knowing stripped of all peripheral understanding.

Willem H. Vanderburg's *Living in the Labyrinth of Technology* looks critically at contemporary society and attempts to create an understanding of the world that is not shackled to overspecialized scientific knowing and technical doing. In his narrative, Vanderburg focuses on two interdependent forces, namely, people changing technology and technology changing people. The latter aspect has proved to be the more critical one for understanding the spectacular successes and failures of contemporary ways of life. As technology continues to change the social and physical world, the experiences of this world 'grow' people's minds and society's cultures, thereby recreating human life in the image of technology. *Living in the Labyrinth of Technology* argues that the twenty-first century will be dominated by this pattern unless society intervenes on human – as opposed to technical – terms.

WILLEM H. VANDERBURG is the director of the Centre for Technology and Social Development and a professor in the Department of Civil Engineering, the Institute for Environmental Studies, and the Department of Sociology at the University of Toronto.

Living in the Labyrinth of Technology

Willem H. Vanderburg

UNIVERSITY OF TORONTO PRESS
Toronto Buffalo London

© University of Toronto Press Incorporated 2005
Toronto Buffalo London
Printed in Canada

ISBN 0-8020-4432-8 (cloth)
ISBN 0-8020-4879-x (paper)

Printed on acid-free paper

Library and Archives Canada Cataloguing in Publication

Vanderburg, Willem H.
   The labyrinth of technology / Willem H. Vanderburg.

   Includes bibliographical references and index.
   ISBN 0-8020-4432-8 (bound).   ISBN 0-8020-4879-x (pbk.)

   1. Technology and civilization.   2. Technology – Social aspects.
   3. Industrialization – Social aspects.   I. Title.

   CB478.V383 2005   303.48'3   C2004-907409-1

This book has been published with the help of a grant from the Humanities
and Social Sciences Federation of Canada, using funds provided by the
Social Sciences and Humanities Research Council of Canada.

University of Toronto Press acknowledges the financial assistance to its
publishing program of the Canada Council for the Arts and the Ontario
Arts Council.

University of Toronto Press acknowledges the financial support for its
publishing activities of the Government of Canada through the Book
Publishing Industry Development Program (BPIDP).

To the memory of the life and work of Jacques Ellul,
my postdoctoral mentor from 1973 to 1978,
and my colleague until his death in 1994,
and to Rita, my life's companion.

Contents

Preface   xi

Introduction: Where Are We Going with Technology?   3

**Part One: Disconnecting from and Reconnecting to the Earth
and the Gods**

1 Industrialization as 'People Changing Technology': Disconnecting
from and Reconnecting to the Earth   17
1.1 Revisiting the Process of Industrialization   17
1.2 The Technology-Based Connectedness of Society   21
1.3 Living with Materials   34
1.4 Living with the Economy   39
1.5 Living Together Socially   51
1.6 Living Together Politically   56
1.7 Living with the Law   60
1.8 Disconnecting from and Reconnecting to the Earth   63
1.9 Some Implications   67

2 Industrialization as 'Technology Changing People': Disconnecting
from and Reconnecting to the Gods   71
2.1 Symbolization and Cultural Moorings   71
2.2 Symbolization and the Life-Milieu   81
2.3 Culture as the Symbolic Basis for Individual Life   83
2.4 Culture as the Symbolic Basis for Collective Life   93
2.5 Industrialization as Cultural Unfolding   107
2.6 New Cultural Moorings   110

2.7 Religion, Morality, and Art   125
2.8 The First Generation of Industrial Societies   129

3 Living with New Moorings to the Earth and the Gods   133
3.1 Serving Technology   133
3.2 On Becoming Human Resources   147
3.3 Technology and the Human Journey   159
3.4 No Detached Observers   164

Part Two: Disconnecting from and Reconnecting to Experience and Culture

4 People Changing Technology: Severing the Cultural Moorings of Traditional Technological Knowing and Doing   173
4.1 Transcending the Limits of Technological Traditions   173
4.2 The Destruction of Technological Traditions   176
4.3 Parallel Modes of Knowing   185
4.4 The Technological Knowledge of a Society   191
4.5 A Discontinuous Change in Technological Knowing and Doing   195

5 Scientific and Technological Knowledge in Human Life   208
5.1 Scientific Education and Culture   208
5.2 Contemporary Technological Doing Embedded in Culture   221
5.3 Contemporary Technological Knowing and Doing in Relation to Culture   228

6 Adapting to the New Technological Knowing and Doing   236
6.1 The Emergence of Universal Technology   236
6.2 Living with a New Economy   259
6.3 Living in a Mass Society   274
6.4 Living with a Limitless Politics   287
6.5 The Intellectual and Professional Division of Labour and the Poverty of Nations   299

Part Three: Our Third Megaproject?

7 Technique and Culture   311
7.1 The Disenchantment of the World Revisited   311

7.2 The Invention of Universal Knowledge   315

7.3 Rationality and Industrialization   325

7.4 Logic, Artificial Intelligence, and Culture   328

7.5 On Creating a New Concept   335

8 Human Life Out of Context   338

8.1 The Technical Approach to Life   338

8.2 Sport   340

8.3 Education   343

8.4 War   352

8.5 Commercial and Political Advertising   354

8.6 Organization   356

8.7 Agriculture   357

8.8 Living with the Technical Approach to Life   361

9 From Experience to Information   376

9.1 The Roots of the Information Explosion   376

9.2 *Homo Informaticus* and the Information Society   378

9.3 Technique and Industry   391

9.4 The Price to Be Paid   412

9.5 Living with Information   421

10 Remaking Ourselves in the Image of Technique:
    Culture within Technique   426

10.1 Technique as Phenomenon   426

10.2 Technique as Life-Milieu   433

10.3 Technique as Consciousness   438

10.4 Possessed by Technique?   444

10.5 Technique as System of Non-sense   461

10.6 Technique as Collective Person   467

10.7 Living with Non-sense   469

Epilogue   483

Notes   489

Index   531

Preface

The twentieth century was an epoch during which the means at our disposal increased on an unprecedented scale, both in diversity and power. However, at the present, early in the twenty-first century, there appear to be very few people who feel that these means have given them more influence over their lives and that consequently things are pretty much turning out as they had hoped. This is probably even more true when we contemplate the lives of our children and grandchildren, born or unborn. For example, although GDP keeps rising, net (i.e., available) wealth is declining. Although we fill our lives with more and more time-saving gadgets, people have less and less time to themselves. Those who have full-time jobs face demands that progressively mine their physical and mental resources, leading to an epidemic dependence on drugs of all kinds to maintain bodily, sexual, and mental functions. The world community has even had to acknowledge that our lives together are unsustainable. What we thought was within reach thanks to science and technology seems to be unattainable. I believe we can all acknowledge this regardless of our values and commitments. A distinction will have to be made between where we thought we were going with science and technology and where science and technology are taking us. Does this mean that we are no longer sure of where we are going? If this is a real possibility, it is time to promote a genuine conversation among ourselves to sort things out. What better way to prepare for this than to retrace our steps, to see how and why our journey of the past two hundred years has turned out so different from our aspirations and commitments?

This book is a narrative of where we are taking science and technology, and where science and technology are taking us. Although indissociably linked, the two are not the same, and our individual and collective journeys combine these two influences. In order to reveal this, the narrative must examine science and technology not as being 'out there,' but instead as being integral to the fabric of relationships woven by human life in the world. All of us participating in this journey are both individuals and part of the fabric, since we help to constitute a society for others, as they do for us. Accordingly, during the last two hundred years, first technology and later science have strengthened some relationships, weakened others, created entirely new ones, and severed others; and in the process they have affected the entire fabric. Two aspects of these relationships make this particularly evident. Because no individual or collective human activity can create or destroy the matter and energy on which it relies, the fabric of relationships embodies networks of flows of matter and energy temporarily borrowed from local ecosystems and the biosphere. Also, each relationship within the fabric is lived, and is thus integral to the lives of one or more individuals and the life of their society. We will examine industrialization in terms of the evolution of the fabric of relationships of human life in the world. These relationships at first severely constrained industrialization, but gradually these constraints were loosened to create an entirely new and universal technology and a very different kind of civilization made up of mass societies. So great was the transformation of the fabric of relationships woven by human life in the world that it will be treated as humanity's third megaproject. It will undoubtedly keep us busy for much of the present century.

The vantage point from which this narrative will be told is rooted in my own intellectual journey. Briefly put, after three degrees in engineering I still did not understand how the consequences of my work were woven together with those of many others to create the social issues and environmental concerns everyone was talking about. Hence, I decided to continue my studies of technology via the social sciences and humanities. I found myself caught between the incommensurate intellectual worlds of various disciplines, some full of technology and little else, and others full of everything else and little or no technology. Sorting out how all of this belonged to one world gradually revealed a very different world from the one I thought I lived in as an engineer and a citizen. After five years of postdoctoral work in the social sciences and humanities, I began a dialogue with students as I taught sociology, engineering, and later, environmental studies. I tried to engage them in the narrative of our collective journey. Since there were so few courses dealing with this subject matter, students from many different disciplines participated, including sociology, economics, political science, anthropology, religious studies, history, nursing, social work, and engineering. This book would never have been written without these students' confirmation of my own impressions that our lives and our world were really very different from what we believed them to be. Theoretically, we had all encountered at one point or another the concept of alienation, but I tried very hard to help my students become aware of it in their own lives.

Connecting this narrative to our lives involves a recognition that social science theories cannot be confirmed or rejected on the basis of experiments, but that they can be validated in our daily lives. If the world is the way we think it is, and if we engage ourselves in this or that action, will the outcome be what our understanding would lead us to expect? If this happens to be the case, then our understanding of the world is confirmed in a small way. If it is not, we will have to revise it. Many of us like to believe that our daily activities, which help to constitute the fabric of relationships of human life in the world, have a beneficial effect: improving this or that technology, strengthening the economy, alleviating social problems, rehabilitating ecosystems, and much else. Nevertheless, our interventions produce many results that no one intended or would value as positive. Scientific theories and professional codes of ethics cannot account for so much of our world being contrary to our expectations, theories, and beliefs, almost regardless of the political or religious commitments we hold. I wish to point out that this narrative is in no way designed to convert anyone to my own particular point of view, although the beliefs and commitments integral to my outlook have evolved with it. Instead, I hope that my perceptions will help my readers to better understand why they hold the beliefs and commitments they do. Our dialogue with this narrative should take into account the obvious fact that the overwhelming majority of us hold the beliefs and commitments we do because of the time, place, and culture we were born into and raised in. To think that we can leave all that behind and participate as 'objective observers' would be as naive as thinking that it is possible to take a picture of the world without standing with a camera in a particular place at a particular time. We are all people of our time, place, and culture, and our narrative cannot exclude any of that. Who would we be, how would we live, and what would our world look like without science and technology? It is simply unimaginable.

At this point I must admit the possibility that even if we agree with this narrative, it could just be dead wrong. No one can possibly know enough, or effectively validate the knowledge received from others, to posit a faultless account of reality. After twenty-five years of dialogue with generations of gifted students at a very good university, I remain keenly aware of the possibility that I may have it wrong, and that as a consequence I have wasted much of my life and a part of the lives of my students. This would be an unsettling situation to look back on, particularly after retirement. Nevertheless, if we are to understand our lives and our world, we have no choice but to interpolate and extrapolate the findings of a great many disciplines and to connect these to our experiences in order to create a narrative.

We must be aware of some responses to this narrative that exclude the possibility of communication. Science and technology, the subjects of this narrative, have been attributed a very high value by contemporary cultures. As a result, it becomes very difficult to treat them like any other human creation, that is, as good for certain things, harmful for others, and simply irrelevant to still others. To accept their limitations is to invite an accusation of science or technology 'bashing.' This is a common misunderstanding. Would we accuse the thermostat on our living room wall of 'bashing' our heating system because it endlessly 'criticizes' its influence on the temperature of the room by comparing it to the set point as an expression of our sense of comfort? Without such 'bashing' the temperature of the room could not be regulated. In the same vein, we cannot exercise responsibility for our most powerful creations without constantly comparing their effects on human and all other life with our values and aspirations. Another common reaction is equally problematic. It holds that this narrative is of little or no use in obtaining or holding a job, and without a job it is impossible to have much of a life. Such sentiments in part explain why the 'system' that we have created evolves with so little decisive human intervention. This narrative may well be 'blowing in the wind,' but without it we give up what in many traditions is considered fundamental to what it is to be human.

Finally, a brief note on the relationship between this and my previous works. In *The Growth of Minds and Cultures* I set out my theory of culture, which as an approach to living in the world is diametrically opposite to those arising from science and technology. In *The Labyrinth of Technology* I set out what the Canada Foundation for Innovation has recognized as one of twenty-five leading recent innovations. That book explores the possibility that preventive approaches for technological and economic development could produce a superior ratio of desired to undesired effects than their conventional alternatives; but it also concludes that because of our cultural beliefs and values there is little chance of their being adopted. The present work is a narrative of our journey with science and technology during the past two hundred years, and complements my previous two works. It is my 'blowing in the wind,' but from where I stand I can see no other way. It is inefficient, and by that very fact offers a possibility of genuine change towards a future that affirms all life.

In addition to acknowledging the important contribution of many generations of students, I also wish to acknowledge the decisive role my postdoctoral mentor, Jacques Ellul, has played in forming this narrative. I can think of very few authors of the twentieth century whose analyses of that time are even more true today than when they were written; and the nearly five years I spent with one of them is an intellectual gift that continues to have great value for my own work. In the same breath, I wish to acknowledge the support of Rita, my life's companion, who has spent so much time discussing and editing the way I have expressed this narrative. Writing has never come naturally to me, and were it not for my teaching, I am not sure I could have justified adding yet another book to a civilization that already has far too many books that say too little. I also wish to thank Pauline Brooks, who over the years has patiently word-processed its many iterations, which have constituted the course notes for our students.

My thanks go to the many volunteers of the Ontario Audio Library Service and PAL Reading Services for making my research materials accessible to me as a blind scholar.

Last but not least, I wish to acknowledge the support of the Social Sciences and Humanities Research Council of Canada, which has generously funded this and related research over many years. I trust that this investment of public funds will serve the public good.

LIVING IN THE LABYRINTH OF TECHNOLOGY

Introduction:
Where Are We Going with Technology?

Our collective story of the last two hundred years has been characterized in many different ways, as industrialization, rationalization, modernization, secularization, and most recently, computerization. Some accounts celebrate the many spectacular successes and extrapolate these to predict a brilliant future. Others emphasize the equally spectacular failures, arguing that it is high time we rethought our steps. A few may still proclaim that, despite the injustices and enormous human suffering, the 'laws of history' will soon see a new world rise from the ashes of the present one, but many have turned their backs on this secular gospel of salvation.

In one way or another, all such accounts of the past two hundred years involve the story of where we are taking our most influential creations, namely science and technology, or where these creations are taking us. Surely any satisfactory account will have to explain how our journey has simultaneously led to success and failure, hope and fear, wealth and poverty, and much more. It will have to account for the fact that, despite an exponentially increasing outpouring of goods and services, net wealth production may actually be declining. It will have to explain why, contrary to all religious and political traditions and the human values on which they are based, we have created ways of life that, as a world community, we have declared to be unsustainable. It will have to explain why the nations that took a leading role in the journey also produced the great secular political religions, namely, fascism, communism, and 'hard-line' democracy. It will have to explain why globalization and free trade may well be harming a great many more people and nations than they benefit.[1] It will have to explain why the cities we created for ourselves are not only unhealthy and unsustainable but generate many forces that undermine the lives of their inhabitants. Perhaps the most troubling and threatening of all these contradictions, it will have to account for the fact that so many young people feel that an enquiry like this one is a waste of time, because unless you conform to the 'system,' it will simply roll over you. Perhaps attempting yet another account of our common story for the last two hundred years is 'blowing in the wind,' but how else can we contribute to our urgent need to better understand why we appear not to be going in the direction we thought we were heading? Without a better understanding of how we got here and where we are going, we are condemned to muddle along. We need a better diagnosis if there is to be an adequate prescription for creating a democratic, just, and sustainable future.

The present attempt at telling this story will reveal what may turn out to be the third megaproject undertaken by humanity in the course of its history. Each of these involved nothing short of a complete transformation of the world, and because this world has a profound influence on human beings, a parallel transformation of who they were. The first such megaproject involved the creation of *logos*, transforming immediate experience into a symbolic universe endowed with names, meanings, and values, thereby creating *homo logos* within the natural family of all life. The second involved the creation of societies that interposed themselves between the human groups and nature. This involved the transformation of both the social and natural worlds as well as human awareness, to create *homo societas* within a social whole that became distanced from the natural world. The third involved the creation and growing use of a universal science and technology, in tension with the local *logos* and cultures, as an alternative way of making sense of and dealing with the world. Societies were transformed into mass societies, which, despite their name, have little in common with traditional societies. For more and more people, nature (which was already much transformed by agriculture) was replaced by an urban habitat; and ecological footprints of mass societies and their urban habitats grew to the point of creating an environmental crisis that is transforming our relationship with nature. The influence of this transformation of the world produced first *homo economicus*, symbolized within an economy that was relatively distinct from society, and in turn situated within a separate environment. In the latter part of the twentieth century, *homo economicus* was replaced by *homo informaticus*, symbolized as depending more on scientific and technical information than on *logos* and culture.

The possibility that humanity has embarked on a third megaproject will gradually become clearer as we examine how technology makes, breaks, and transforms relations between people, between people and their society, and between that society and the biosphere. The transformation of these relations in particular involves the separation of three kinds of connections that were previously intertwined. These are culture-based connectedness, technology-based connectedness, and biology-based connectedness. By means of a culture,[2] people symbolize their participation in immediate experience as a moment of their lives, the participation of others in this immediate experience as moments of their lives lived in a group or society, and the participation of everything else in their immediate surroundings as part of a world. For example, the symbolization of what is visually detected from our immediate experience by the retina results in our experiencing its meaning for our lives. No 'edges' are seen around our field of vision, and the world remains upright and stable when we tilt our heads or go running. People symbolize everything in their lives in relation to everything else by means of a culture. All symbolized experiences contribute to the culture-based integrality of people's lives and their world by means of interrelated sets of meanings and values. Before industrialization, this integrality encompassed two other kinds of connections. According to the first law of thermodynamics, no human activities, nor the way of life to which such activities contribute, can create or destroy the matter and energy they depend on. They exchange matter and energy, thereby constituting a network of flows of matter and a network of flows of energy. As a consequence of industrialization, these helped to build a technology-based connectedness that became increasingly distinct from a society's culture-based connectedness. Recently, technology has begun to be used to modify constituents of the DNA pool, thereby affecting the biology-based connectedness of all life. From this perspective, technology alters the fabric of relations that make up individual and collective human life.

The culture-based connectedness of a society results from the cultural approach to making sense of and living in the world. Until the onset of humanity's third megaproject, human groups and societies evolved exclusively on the basis of this cultural approach. It symbolized everything in the life of a group or society, and in this way incorporated the other modes of connectedness to endow human life and the world with meanings and values.

The cultural approach to life is based on the central role *logos* and culture have played in human life as far back as we can go. _Homo logos_ precedes _homo faber_. A tree branch must first be symbolized as something human before it can become a bow or a paddle. A flat stone must first be symbolized as the blade of an axe or the head of a hammer before it can be worked. Plants and animals must first be symbolized as more than mere constituents of an ecosystem before they can be domesticated. Death must first be symbolized as something else before ritual burial becomes essential. More importantly, such instances of symbolization were not piecemeal but were integral to a systematic attempt to make sense of and live in the world by symbolizing immediate experience. Symbolization thus transformed our natural niche in the ecosystems of the biosphere into a symbolic universe inhabited by people with a non-natural, that is a cultural, awareness of themselves, others, and the world. Naming everything in this symbolic universe signified the place and importance of everything relative to everything else in an individual human life – a life lived as a member of the group by means of a way of life in a world endowed with these meanings and values. Although we know next to nothing about how the higher symbolic functions are performed by the brain-mind, it is evident that each experience modifies something of its organization by synaptic and neural changes, thereby symbolically relating that experience to all others and thus placing it in a person's life. As a result, the meaning and value of that experience for that person's life as lived in a community in the world become apparent.

The cultural approach to life gives rise to the culture-based connectedness of people's lives, commonly recognized by expressions such as 'living a life.' This connectedness of human life is usually taken for granted unless it is disrupted by a condition such as a short-term memory disorder or Alzheimer's disease. These conditions appear to interfere with the ability to connect new experiences to the vast structure of neural connections built up in the brain in the course of living a life. New experiences can no longer be lived as moments of a life, instead approaching something like the immediate experiences of separate moments of existence. The separation of afflicted people's being – in time, space, and the social domain – cuts them off not only from their own lives but also from those of others and from their surroundings. Suppose a person suffering from such a condition enters a building for the first time. After turning several corners in the hallways, he will be unable to retrace his steps to the front door because he will not remember the experiences that got him lost, which would enable him to recognize the corners where he turned. Asking for directions is almost impossible because he will immediately be lost in any conversation, being unable to recall what he or someone else just said. For these reasons it becomes very difficult for him to sustain social relationships with others, especially intimate ones with family members and friends. He no longer lives a life in a world, but inhabits a sequence of micro-worlds connected only by the life lived prior to the onset of the condition.

A similar loss of connectedness can occur at the level of collective human life. This is much more difficult to diagnose, and the findings will almost certainly be controversial. Nevertheless, the rise and fall of civilizations involve changes in the connectedness of collective human life. For example, a civilization risks collapse when it is no longer able to give meaning, direction, and purpose to the lives of its members, which can result in various groups going off in their own direction. The response of some ancient Greeks to such a situation has profoundly marked Western civilization. Socrates and Plato detected a weakening in the connectedness of Greek life due to the relativization of its culture through contact with others, and in an attempt to establish a logical foundation for their culture, they sought to discover rational rules underlying daily-life experiences.[3] Much more recently, Karl Marx raised the question of alienation and the problem of a false consciousness as well as their distortion of the fabric of human life.[4] Emile Durkheim was concerned with what he perceived as a fundamental weakening in the connectedness of industrializing societies and with the resulting anomie.[5] Max Weber observed the growth of rationality and its influence on human life, including what he called the disenchantment of the world.[6] Arnold Toynbee sought to explain the rise and fall of civilizations in terms of the constant need to adapt the connectedness of collective human life to new circumstances.[7] Jacques Ellul warned against the reification of human life under the influence of what he called technique.[8] Most recently, artificial intelligence researchers have attempted to describe the connectedness of daily-life experience in terms of the rules, algorithms, micro-worlds, scripts, frames, and other entities that were supposed to be its building blocks.[9] In the course of the twentieth century, our symbolic connection with reality, which earlier had often been taken for granted, became a subject for enquiry by the social sciences and humanities.

When we later examine the cultural approach in some detail, it will become evident that symbolizing everything in relation to everything else respects as much as possible the way everything interacts, depends on, and evolves with everything else. Every living entity helps to constitute the biosphere for all others as they help to constitute its biosphere. Something of this biological whole is thus enfolded into each constituent entity via the DNA in every cell. Similarly, each person helps to constitute the society for other people as they help to constitute the society for that person. Through the internalization of experiences, something of the social whole is enfolded into each member. Cultures symbolize this by everything in human life and the world being integral to that life and to that world. In this sense, it is analogous to the way the constituents of any ecosystem, as well as all the ecosystems within the biosphere, have evolved over an extremely long period of time, thereby helping to constitute the varied conditions essential for supporting life on earth. In a similar way, the cultural approach symbolizes and guides the evolution of a society as a whole as well as its science, technology, economy, social structure, political framework, legal institutions, morality, religion, and aesthetic creations as integral to a way of life. There is no separate economy or social structure in daily life. These exist only as such in scientific disciplines.

From cultural anthropology, depth psychology, and the sociology of religion we know that symbolizing everything in relation to everything else includes the unknown. As I will show later, myths convert the unknown into 'interpolations' and 'extrapolations' of the known yet to be lived and discovered. The central myth represents the most valuable element in a community's life, beyond which nothing more valuable can be lived or imagined. In this way, interrelated sets of names, meanings, and values are absolutized and institutionalized as relations to the 'gods.' Hence, the culture-based connectedness of a society includes its religious moorings. When modern technology began to change this connectedness, it also changed these moorings.

Prior to industrialization, the network of flows of matter and the network of flows of energy usually overlapped with the culture-based connectedness of a society, since metabolic exchanges between activities followed for example, kinship patterns or feudal obligations. Obvious exceptions resulted from trade within and between empires. Whatever the case, any activity must obtain matter and energy either directly from the biosphere or indirectly from a chain of activities connecting that activity to the biosphere. Similarly, the matter and energy transformed by an activity are either returned directly to the biosphere, or returned indirectly via another chain of activities. Hence, all the activities constituting the way of life of a society are connected by a network of flows of matter and another network of flows of energy. These two networks occasionally overlap, as is the case for fuels and the production of goods. What is true for an individual activity is also true for these two networks: they temporarily borrow matter and energy from the biosphere. This includes solar energy, which is prepared for service by the ozone layer. Although the culture of a society is able to symbolize these networks (or portions thereof) in different ways, their contribution to the technology-based connectedness nevertheless constrains the culture-based connectedness. Such is also the case for the biology-based connectedness of society.

The temporary borrowing by a society of all its required matter and energy reveals some of the services the biosphere renders. The ecological footprint refers to the 'amount' of biosphere needed to provide these services. It represents a society's moorings in the 'earth.' When modern technology began to create a technology-based connectedness from the metabolic exchanges between activities that jointly sustain a way of life, it transformed the moorings of that society in the earth. Hence, industrialization includes a change in the moorings of a society to the earth and to 'the gods.'

Out of this transformation of the moorings of industrializing societies gradually emerged the need for humanity's third megaproject. This change introduced an entirely new kind of connectedness based on a universal (i.e., non-cultural) science and technology. At first, their emergence and growth occurred within the culture-based connectedness of Western societies. This situation did not endure very long, given the growing tensions between those activities of a way of life organized and reorganized on the basis of universal knowing and doing and those activities that continued to be evolved on the basis of a culture-based (i.e., local) tradition. A new technical order began to impose itself on the cultural order. It represented a dramatic reversal of the usual relationship since, in the past, each culture had its unique science and technology, which diffused along with it. The intellectual and professional division of labour of modern science and technology no longer represents an elaboration of the culture-based connectedness of a society. High levels of specialization cannot simply be understood in terms of specialists wearing blinkers delimiting their professional fields of vision and action. The following slightly elaborated historical example illustrates the problem.

A group of experts under the auspices of the United Nations were examining the causes of hunger in a Colombian valley and ran into the following problem. The nutritionist made an inventory of all the foodstuffs grown in the valley in order to determine the most nutritious diet possible, supplemented it as required, and made the appropriate recommendations. The specialist in community health suggested that these recommendations would not solve the problem because the inhabitants of the valley suffered from intestinal parasites resulting in diarrhoea and their inability to absorb a nutritious diet. Instead, this specialist recommended that the water supply be improved, sewage treatment be started, and basic health care provided. The economist smiled politely, suggesting that the inhabitants did not have these things because they lacked the resources, and that these could only be created by economic development based on cottage industries and some inhabitants working outside of the valley to send money to their families. The agronomist on the team recommended instead that the inhabitants be taught modern agricultural methods to enable them to grow enough food for themselves and to sell the surplus to generate income enabling them to procure the above necessities of life. The political scientist firmly disagreed. All these things were not happening because the hungry people in the valley had no political voice, and this would not change until they were empowered by forming a political party. I could go on with the diagnoses and recommendations of the sociologist, the demographer, and an expert in systems. However, the point is obvious: Each expert paints a picture of the situation by putting those aspects that correspond to his or her specialty in the foreground and everything else in the background, thereby creating incommensurate diagnoses and recommendations. Each and every expert has the answer but clearly does not know what the question is. In the absence of a science of the sciences, it is impossible to *scientifically* integrate the findings of different specialities to arrive at a comprehensive interpretation of the situation. Hence, science has no access to or basis for evaluating as a whole the culture-based connectedness of a society.

The following story may further illustrate the predicament in which the inhabitants of the Colombian valley found themselves. Suppose three artists each paint a landscape that includes a beautiful old farmhouse, a stand of trees in fall colours, and a picturesque wetlands with a meandering creek. Also suppose that each one, because of his or her interests, decides to put one of these three constituents of the landscape into the foreground of the picture, and the other two in the background. If we had never been to the place depicted in the three paintings, it would be impossible for us to describe the original landscape. Nor would we be able to advise the owners of this property as to where best to build their new home. If we nevertheless did offer our advice, the homeowners would likely question as to whether we really understood the terrain. Similarly, the inhabitants of the Colombian valley would likely have found it next to impossible to incorporate any of the advice into their world of sense, created by the interrelated meanings and values of their culture. To them, the 'paintings' of their situation made by the experts would have appeared as non-sense, that is, as not belonging to their world of sense. Similarly, it would be highly unlikely that the team of experts would have been able to make professional sense of what would probably have appeared to them as the non-sense advanced by the inhabitants. All too often, experts do not solicit the opinions of non-expert 'ordinary' people because in their world of sense the culture has devalued all forms of knowing and doing that are not based on highly specialized scientific and technical knowledge. For the inhabitants of the Colombian valley, the entire issue would likely have disappeared with the team of experts unless their advice was imposed from the outside. These might have left with the conviction that it was not hard to understand why the inhabitants of the valley were in such a predicament, because their way of making sense of and living with hunger made no sense to them at all.

The same kind of collision between the world of sense created by means of a culture and universal (i.e., non-adapted to a particular time, place, and culture) scientific knowing and technical doing has occurred in any society beginning to make extensive use of this highly specialized knowledge to evolve its way of life. In order to prevent a kind of cultural schizophrenia, either highly specialized scientific knowing and technical doing would have been rejected as 'foreign bodies' to the world of sense, or the culture would have had to undergo an entirely new kind of mutation unprecedented in human history. It turns out, as we shall see, that the latter best fits what happened in many nations, particularly following the Second World War. It will also become apparent that this, more than anything else, describes the spectacular successes of contemporary civilization and its equally spectacular failures. It gets right to the heart of our current predicament of succeeding brilliantly at extracting the greatest possible outputs from requisite inputs of matter, energy, labour, capital, and knowledge as measured by performance values such as efficiency, productivity, cost-effectiveness, and profitability. As output-input ratios, these 'values' are entirely moot on the question of whether the enormous gains in output were partly or wholly realized by undermining human life, society, and the biosphere, which has led to equally spectacular failures. Contemporary cultures appear to have lost the capability of evolving their ways of life based on the recognition that everything depends on everything else. We talk about this because we can no longer take it for granted. Instead, contemporary cultures appear to give the following advice to those individuals facing unbearable psychological tensions: 'Let them eat Prozac.'[10] To mass societies facing a growing number of tensions, the advice appears to be: 'Wait until the next election and then everything can be changed.' To those concerned about the environment and future generations: 'Let them dream of sustainable development, globalization, or free trade, as long as they do not insist on social and environmental standards.' All this and much more is directly associated with the dilemmas created by our third megaproject and the way it has permitted a universal world of non-sense to overwhelm worlds of sense, turning them inside out in the process. The symbolic basis by which groups and societies have made sense of and lived in the world is under siege, yet we have found no alternative for sustaining human life. Hence, it is urgent that we understand how through a series of developments we have (largely unintentionally) become caught up in the creation of humanity's third megaproject. A recognition of this fact can help us better understand our current predicament.

The development of humanity's third megaproject initially occurred in three phases. Each one of these will be examined in the three parts of this work. The first phase in humanity's third megaproject involved, but was not limited to, the industrialization of western European countries, roughly between 1750 and 1930. This industrialization was but a small part of these nations completely changing the fabric of relations that made up their ways of life and how these ways of life depended on the biosphere. This change in their culture-based connectedness and technology-based connectedness constituted a disconnecting from the earth and their traditional gods and a reconnecting to the earth and new secular gods. The disconnecting and reconnecting to the earth involved a significant change in the ecological footprints of these societies, from primarily depending on local ecosystems to depending on ecosystems all across the world. As Gandhi put it, 'It took England the exploitation of half the globe to be what it is today. How many globes will it take India?'[11] The disconnecting and reconnecting to their gods involved a decline in the role of Christianity and its eventual replacement by secular religions such as communism, fascism, and hard-line democracy. *Homo societas,* living in the world by means of the cultural approach to life, was transformed into *homo economicus,* living in the world by means of an economic approach to life.

A second phase in the development of humanity's third megaproject began when these societies ran into developmental bottlenecks related to the limitations of knowing and doing based on experience and culture. Such bottlenecks were overcome by these societies gradually disconnecting themselves from such knowing and doing and reconnecting themselves to a knowing and doing based on universal knowledge no longer unique to a particular time, place, and culture. This breaking and remaking of connections in the domain of knowing and doing involved, among others, the transition from appropriate technologies to a single modern universal technology, and from mostly sustainable ways of life to an environmental crisis and growing concerns over potential resource limits. It also contributed to the birth of mass societies and the necessity for mass education. This second phase occurred in western Europe and North America between the 1930s and 1970s, accelerating greatly following the Second World War.

During the third phase of the development of humanity's third megaproject, entire ways of life were transformed. Nations disconnected themselves to a large extent from their cultures, to reconnect themselves to scientific and technical information. Increasingly, every area of their ways of life ceased to be developed according to tradition and culture; instead, they were developed with a view to enhancing technology-based connectedness at the expense of the culture-based connectedness. This first led to what Max Weber, at the beginning of the twentieth century, called the phenomenon of rationality and what Jacques Ellul, decades later, examined as the phenomenon of technique. The gradual disconnection from experience and culture culminated in an explosion of information, necessitating the computer revolution. *Homo economicus* was replaced by *homo informaticus,* living in the world by means of the technical approach to life.

We may now be witnessing the beginning of a fourth phase in the development of humanity's third megaproject, involving the remaking of some physical and biological connections by means of biotechnology and nanotechnology. It is too early to tell what will be involved, apart from a growing genetic pollution of the DNA pool. Nevertheless, it is clear that humanity's third megaproject means a complete remaking of the fabric of human life and that of the world. It should be noted that each of these four phases of humanity's third megaproject has interpenetrated with prior and later developments, and that this interpenetration has been much greater within societies that industrialized in the twentieth century.

PART ONE

Disconnecting from and Reconnecting to the Earth and the Gods

Industrialization as 'People Changing Technology': Disconnecting from and Reconnecting to the Earth

1.1 Revisiting the Process of Industrialization

The Industrial Revolution conjures up images of great inventors, machines, factories, filthy slums, and widespread poverty. We are less likely to associate with it major changes in language, artistic and literary expression, religious and political beliefs, morality, and much else that added up to a fundamental change in the way human beings thought of themselves and of their relations with one another and the world. It began a process of remaking human life in the image of technology, and it has contributed much to who and what we are today. It is for this reason that the process also involved rationalization, modernization, and secularization.

A silly story may illustrate the conceptual difficulties the Industrial Revolution introduced. Once upon a time, an engineering student inherited a Ferrari engine from his Italian grandfather. He decided to mount it in his Volkswagen Beetle. The necessary space was created by removing the back seat and the existing engine, and the new engine was mounted and connected to the rear wheels. When he took the car for a test drive, he promptly lost control in negotiating the first turn, since the vehicle was imbalanced and its suspension not up to the task. He sheepishly took it home in order to modify the suspension and reinforce the understructure. The car then handled much better, but at highway speeds and in windy weather the aerodynamics proved to be problematic, and in cold weather the defrosting system was inadequate. By the time he had a vehicle capable of making full use of the power of his new engine, he had modified just about everything on his car. In the same vein, building and installing an industrialized production system into a traditional society required so many adjustments as to transform both its technology-based connectedness and its culture-based connectedness. By the time such a society could make full use of such a system, it had completely transformed the relationships between its people, the land, and their gods.

We must recall that *homo logos* precedes *homo faber*. The creation and evolution of the human world occurs according to a pattern I will call the cultural cycle. It includes several phases.[1] Externalization refers to something of our life being expressed in our behaviour. Objectivization refers to behaviour that becomes a part of our world in terms of constituents such as customs, sayings, values, or physical objects. Internalization refers to our ability to live a life by creating new neural connections within the structure of the brain-mind. These connections symbolically relate each experience to all others, thereby revealing its meaning and its value for our life. The organization of the brain-mind constitutes the metalanguage for *logos* and culture as well as the 'metaconscious' of our being a person of a particular time, place and culture.[2] Something of all of this is externalized in subsequent behaviour, and so on. Life never quite repeats itself. Hence, creativity and ongoing adaptations are essential.[3] This cultural cycle is also central to the creation and evolution of a human group. When people awaken in the morning, they must re-establish the fabric of their relations with others and the world and adapt it to new circumstances. When at the end of the day they fall asleep, this evolved fabric becomes symbolically represented in the organization of their brain-minds, where it is maintained by restorative sleep functions and readied for the next day. On the collective level, we find the same daily-life cycle as its members together re-establish, evolve, and adapt their way of life. These cultural cycles open up the possibility for technology to participate in the influence that nature and nurture have on people.

Homo logos comes before *homo faber* even in the creation and use of ever-larger numbers of machines. Every phase of the cultural cycle is affected. As people change the world through technology, the world also changes people. Hence, the story of industrialization is as much the story of 'technology changing people' as it is the story of 'people changing technology.' The one is indissociably linked to the other even though it is rarely recognized that in this aspect of a culture, too, the relationship between people and their surroundings is always a reciprocal one. Each and every experience modifies people's organization of the brain-mind and hence their life.

Wherever industrialization has occurred, the relationships between technology, society, and the biosphere (in short, much of the world in which people live) have been entirely transformed. As a result, the cultural cycle becomes gradually permeated by 'people changing technology' and by 'technology changing people.' From this perspective, industrialization, rationalization, modernization, and secularization will gradually appear as facets of a remaking of human life and society. Industrialization's origin in Great Britain and its rapid spread to western Europe marked the transition from one historical epoch to another, with consequences that were felt around the globe. As industrialization spread further, all of humanity was ushered into a new life and a new world. Technology and the economic activities based on it ceased to be unique to a particular time, place, and culture, gradually becoming universal for the first time in human history. Every human society and culture had to adjust in order to accommodate these universal elements.

In order to understand the far-reaching consequences of industrialization as an important facet of how we and our world came into being, I will show how it involved a fundamental change in the way human life was connected to the earth and to the gods, symbolizing our dependence on matter and energy and on meaning, purpose and direction for life, respectively. I will examine the connectedness of human life to the earth by adopting the perspective of thermodynamics and ecological economics,[4] and the connectedness to the gods by adopting a perspective based on a previously-developed theory of culture derived from various disciplines including cultural anthropology, sociology, and psychology.[5]

'People changing technology' refers to a diversity of activities that result in changes in the technology of a society. As such, they disturb the dynamic local equilibria in the network of flows of matter and the network of flows of energy. In order to restore these, adjustments must be made to these local structures, which in turn affect adjacent structures, and consequently the technology-based connectedness of that society. These changes in turn modify how its way of life is rooted in the biosphere by its ecological footprint.[6] At the same time, this diversity of activities modifies the way of life of which they are a part and hence the culture-based connectedness. A complete transformation of the latter would mean a rebuilding of the 'cultural niche' in reality, including its moorings in the 'heavens.'

'Technology changing people' is integral to how the world in general, and a person's social and physical surroundings in particular, influences that person's life. When the density of technologies and their products becomes such that they help to constitute this world, these will contribute to 'technology changing people.' Until industrialization is well under way, this influence of technology via the world it helps to constitute could be, and generally has been, all but neglected. From that point on, it becomes an essential and important part of how we and our world have become who and what we are.

The present chapter will examine industrialization as a fundamental change in the technology-based connectedness of society, concentrating on 'people changing technology' but anticipating 'technology changing people.' The second chapter will examine industrialization as a transformation of the culture-based connectedness of society, primarily focusing on 'technology changing people' and how this in turn affects 'people changing technology.' For some social groups, 'people changing technology' was the story of pursuing a new prosperity through a new way of life; for others it meant the opposite, namely, the loss of a way of life and extreme poverty. 'Technology changing people' meant that the latter groups were eventually ushered into the new way of life since they could not help but be unwilling participants in the remoulding of the way of life and culture of their society. For a limited time, society was culturally divided as some groups clung to a disappearing way of life while others pioneered a new one. This created the illusion of social classes as 'societies within societies.' However, it is extremely difficult to find an indisputable sociological basis for the concept of a social class.[7] For a limited time, political and social struggles focused on the kind of society that was to emerge, but as the cultural reunification of society progressed, the focus shifted to appropriating the benefits bestowed by the new way of life. In other words, industrialization was one facet of changing the cultural niche in reality that the members of a society occupied according to their social roles. In this way, the bourgeois mind and spirit was the forerunner of the spirit of an age to come.[8]

In this development, we recognize the roots of the environmental crisis. The limited capacity of local ecosystems to support the ever-increasing throughput of matter and energy associated with the new technology-based connectedness of industrializing societies made the expansion of the ecological footprint by means of colonization and emigration a necessity. To make this emerging world liveable and justifiable, this necessity had to be symbolized as integral to a limitless human progress, which would eventually benefit all people. We also see the roots of the contemporary cultural crisis. Technology-based connectedness required ever more resources, human effort, and ingenuity, as well as a moral justification that could not be provided by the traditional culture-based connectedness of societies. The former began to dominate the latter, with the result that the role of culture in individual and collective human life came under growing pressure, to the point that it had to be technically reinforced. Until then, culture was without doubt the greatest of human creations, defining us as symbolic beings set apart from all other animals.

1.2 The Technology-Based Connectedness of Society

The process of 'people changing technology' is greatly constrained by the fact that the activities that maintain and evolve a way of life are connected by a network of flows of matter and a partly overlapping network of flows of energy. Industrialization includes these in a technology-based connectedness, which is constrained by its dependence on the biosphere and on the social division of labour of a way of life. The former may be understood as follows.[9] Because the activities that maintain and evolve a way of life can neither create nor destroy matter and energy but only transform them, they must receive all requisite inputs from the two networks, transform these inputs, and return the outputs from these transformations back to the networks. What is true for these activities is equally the case for the entire network of flows of matter and the network of flows of energy. Each one must temporarily borrow matter or energy from a corresponding larger network, of which it is an integral part. These larger networks represent the flows of matter and energy respectively within local ecosystems, and these networks in turn are integral to still larger networks representing these flows within the biosphere. In this way, the biosphere metabolically sustains the way of life of a society.

The first and second laws of thermodynamics can again shed light on the way the biosphere renders these services (what economics partly recognizes in 'land' being a factor of production). All its living and non-living processes involving transformations of matter and energy are connected by a network of flows of matter and another of flows of energy. Since there is virtually no exchange of matter between the biosphere and outer space, there are no inputs and outputs into and out of the network of flows of matter. In the absence of human groups or societies, this network must be built up from overlapping and interdependent closed cycles in which the output of one transformation becomes an input into another, with the result that the biosphere is self-purifying and knows no pollution. In contrast, there is a constant exchange of energy between the biosphere and outer space. Hence, the network of flows of energy is powered by the sun, and the degraded energy given off in the form of low-temperature heat is partly radiated back into space. Because any transformation of energy is irreversible, this network is built up from interdependent linear chains of transformations. These begin by solar energy being captured by plants by means of photosynthesis to produce biomass in the form of plant formations. They in turn support an enormous diversity of life whose interdependencies can be illustrated by a food-web, up which energy passes in the form of biomass as herbivores feed on plants, and carnivores feed on herbivores. Most energy is used for living and transformed to low-temperature heat that is lost to the surroundings.

In order that the ways of life of groups or societies can be sustained by the biosphere, they must create a network of flows of matter and a network of flows of energy that are nested within corresponding networks associated with the living and non-living processes in the biosphere. To better understand the interdependence of these networks, flows across the society-biosphere boundary are frequently distinguished as renewable, non-renewable, or continuing resources, and stock or flow wastes.[10] Resources are renewable when the rate of human consumption is less than or equal to the rate at which the biosphere can regenerate them. They are non-renewable if this is not the case. Continuing resources are not affected by human use, at least not on the present scale. Wastes that can be transformed by natural cycles into resources for human or natural processes are flow wastes. Those that cannot be absorbed in this manner are stored as stock wastes. No solar energy can flow directly into the network of flows of energy associated with a way of life unless it is first 'prepared' by the biosphere, which filters it by means of the ozone layer. Fossil fuels are solar energy stored as a transformed biomass. Hence, the technology-based connectedness of a way of life is anchored in the biosphere, and a fundamental change in the former is likely to change these moorings. For example, when such changes exceed the ability of local ecosystems to supply the required inputs and absorb the resulting wastes, distant ecosystems must be drawn on to supply the deficit.

Next, consider how the technology-based connectedness of a society depends on the social division of labour of its way of life. Each node in the networks of flows of matter and energy represents a human activity involving the transformation of matter and energy. Prior to industrialization, this social division of labour assigned entire activities required to maintain and evolve a way of life to individual people, thereby creating the social roles of farmers, bakers, butchers, craftspeople, doctors, lawyers and so on. As a result, the technology-based connectedness generally corresponded to the social organization of a way of life. Insofar as the requisite inputs and outputs of matter and energy associated with an activity were not directly derived from and returned to local ecosystems, they represented aspects of interdependencies between activities and thus between the people engaged in them and between their families and communities, if these people belonged to different ones. Until the 'great transformation' described by Polanyi,[11] such interdependencies were primarily based on kinship and social ties and very little on impersonal markets. As a result, the technology-based connectedness of a society was integral to, and dominated by, the culture-based connectedness.

The emergence of a technical division of labour gradually distanced the technology-based connectedness from the culture-based connectedness of a way of life. By atomizing the activities of a way of life into sequences of steps assigned to different people, traditional social roles and the corresponding ties between people, families, and communities were shattered. Adam Smith described how pins could be made by a group of people organized by means of a technical division of labour.[12] The number of pins that could be made by such a group was far greater than what they could produce if they each made pins on their own. Adam Smith correctly predicted a new wealth of nations but noted that this would come at a price, as the skills of workers would atrophy.[13] Technically divided work participates indirectly in the culture-based connectedness of a way of life via a sequence of production steps that now produces a good or service previously delivered by a single person occupying a social role. This would have far-reaching consequences for people, tools, and machines.

A technical division of labour converts portions of a technology-based connectedness integral to the culture-based connectedness of a society into a relatively separate socio-technical entity structured around a sequence of endlessly repeated steps. Participation in this entity by people, tools, and machines requires fundamental changes. People can participate only as precursors of automated machines and robots. Tools suitable for widely varying craftwork are ineffective and require adaptation to a single narrow function, thereby becoming the precursors of machine tools. Together these transformations of tools and people pave the way for replacing them with industrial machinery. People, tools, and machines are transformed into elements of socio-technical entities, and it is via this context that they interact with the surrounding technology-based and culture-based connectedness of a society.[14]

From a historical perspective, it is hard to exaggerate the difficulties involved in disconnecting work from the culture-based connectedness of individual and collective human life in order to reinsert it into a technology-based connectedness modified by a technical division of labour. Draconian factory discipline testifies to this.[15] In western Europe, the transition process took some 150 years, while in totalitarian societies it took much less time because the state resorted to massive work-related propaganda campaigns as well as coercion. The creation and expansion of socio-technical entities incorporated human beings on technical and not on cultural terms, thereby strengthening the technology-based connectedness of a society and correspondingly weakening the culture-based connectedness. However, this was not the first step towards the creation of new mega-machine societies, as Lewis Mumford envisaged.[16] If that had been the case, human beings would have fully adjusted to their new role as cogs in a mechanism, and all the difficulties of coping with a highly technical work setting (which are again on the rise as a result of computerization today) would have been avoided.

The technical division of labour strengthened the local technology-based connectedness of a society in the following five ways. First, workers could be assigned narrow, repetitive tasks that they could readily master. No time or effort needed to be devoted to the broader context of the tasks that preceded or followed the one at hand. These constituent elements of work could now be optimized to eliminate wasted motions in order to ensure the best performance. Second, it was now possible to prepare, optimize, implement, and evaluate a production plan in the form of the optimal sequence of the optimized constituent elements. Third, as a consequence of this decomposition and rational recomposition of work, it was possible to assign constituent production steps either to human beings or to machines to obtain the greatest efficiency, productivity, and profitability. This opened the road first to mechanization and later to automation and computerization. None of this was possible with traditional craftwork. Fourth, the tasks assigned to human beings could be performed by those people having just the required skill level as opposed to those able to accomplish the most difficult steps, as in craftwork. This greatly reduced labour costs. Fifth, workers were isolated from one another and group structures eliminated, with the result that a great deal more time was available for production. It follows that the technology-based connectedness was strengthened when desired outputs were delivered from requisite inputs with a greater efficiency, labour productivity, and profitability, which are ratios as opposed to human values.

The strengthening of the technology-based connectedness by means of the technical division of labour simultaneously weakened the culture-based connectedness of individual and collective human life in five ways. First, as noted previously, Adam Smith himself acknowledged that it would make human beings as stupid as they could possibly become.[17] The monotony of carrying out the same trivial operation over and over again atrophied human creativity, skill, problem-solving ability, and a great deal else involved in living. Much of people's humanity was thus separated from their work. Second, insofar as people still used skills, these had no common denominator with craft skills because they did not originate in a person's mind through the accumulation of experience, but with someone else (external to the work process) who determined the one best way to do the job. This arrangement required the suppression of the self in work, necessitating a great deal of mental energy and thus leading to nervous fatigue and a range of psychosocial consequences. This was the well-known hand-brain separation. Normally, even the most routine activities (such as handwriting) express something of a person's life and experience, as shown by the possibility of correlating characteristics of handwriting with personality traits or the use of handwriting as forensic evidence. Third, the more the technical division of labour was developed in conjunction with mechanization, automation, and computerization, the more workers lost control over their work: the methods to be used, the time to be allocated, how work was to be paced, how problems should be looked at and solutions attempted, what deadlines were to be set and met, how to report and when, and how to assess progress. In other words, a separation of knowing, doing, and managing occurred. This not only separated the hand from the brain but also destroyed the self-regulating character of work and replaced it with long, ineffective negative feedback loops that barred workers from effectively dealing with problems. Those who knew and planned the production process were not the same people who did the work, which required supervisors to ensure that the work was done according to plan. Socio-epidemiology has shown that a loss of control is the primary determining variable of healthy or unhealthy work.[18] Fourth, as mechanization, automation, and computerization advanced, greater demands were placed on workers tending and monitoring machines and performing tasks at increasing speeds, with the result that a split second's inattention could lead to serious consequences. At the same time, problem-solving demands increased because of the gap that inevitably occurred between the ideal work plan and how all this worked out in reality, including the problems that originated upstream in the work process. Even if workers could solve some problems, they almost never had the authority to do so. There were other demands as well related to being paced by machines and quotas, and by reporting responsibilities. All these demands coupled with little control created, as we now know, unhealthy work settings. Fifth, the isolation of workers eliminated social support precisely as demands rose and control diminished. In sum, the development of a technical division of labour greatly weakened the culture-based connectedness of individual and collective human life by dehumanizing work as a consequence of excluding much of the expression of the self in externally organized work, allowing people to participate only as a hand or a brain, and reducing social relations to a minimum.[19] Work was transformed into a commodity of 'pure' labour, 'decontaminated' as much as possible of its humanity.

The operations of workers and their social interactions were now primarily technically based and only secondarily founded on experience and culture. In other words, the technical division of labour created a growing gap between the technology-based connectedness of the workplace and the culture-based connectedness of people's working lives. The resulting socio-technical entities may be represented as networks of technically divided production steps assigned to people or machines, as well as technically divided supervisory and managerial functions. The local technology-based connectedness was now built up with flows of matter and energy, machines and commoditized labour. There was no longer any overlap with the corresponding local portion of the culture-based connectedness of individual and collective human life, since it was no longer possible to fully participate in these socio-technical entities as human beings. Long before Frederick Taylor sought to fully separate the hand from the brain, people participated in these networks mostly as 'hand' or 'brain' or as intermediaries between the two, but never in a manner that kept all three together, as was the case with traditional work. In technically divided work, people lived the tension between the growing rational and algorithmic character of the technology-based connectedness of the workplace and the culture-based 'life never repeats itself' nature of their humanity and social world. Hence, they could not become cogs in a mechanism, of which nervous fatigue replacing physical fatigue was a signpost.

The implications for technology were far-reaching. Its reorganization changed the way the elements of a technology were embedded in a way of life. The elements of traditional technologies were designed and evolved to fit into a technology-based connectedness organized according to the meanings and values of the human activities involved, with the result that they were subservient to, and mostly overlapped with, the culture-based connectedness of a way of life. In contrast, the technical division of labour integrated them into a network of technically divided production steps, and it was as elements of these networks that they were now embedded into a way of life. For these technological elements to be effective meant that they must first and foremost be adapted to these networks. The excellence of the design of a machine, or the lack thereof, now depended mainly on its ability to take the outputs from preceding production steps in order to transform them into the inputs required for the next. Everything else became secondary, to the point that considerations such as human factors engineering, occupational health and safety regulations, industrial relations, and quality of working life would later be added out of necessity to compensate for the fact that these new machines, extraordinarily appropriate to their technical context, were equally inappropriate for their human and social context. The relationship between human life and technology was gradually reversed. Traditional technologies were adapted to human life, society, and local ecosystems, but industrialization increasingly necessitated the reverse adaptation.

The socio-technical entities resulting from the technical division of labour inevitably disturbed the local equilibria of the network of flows of matter and the network of flows of energy, thus necessitating additional adjustments, which in their turn had an impact on these networks and so on, thereby creating a kind of chain-reaction-like process. The events in this process cannot be understood apart from their direct and indirect interactions via these two networks. The result was a new dynamic equilibrium in the technology-based connectedness of a way of life. The technical division of labour thus changed the dependence of a way of life on the biosphere via its technology-based connectedness.

The process of industrialization made ever-greater use of the technical division of labour. Consequently, changing the moorings of industrializing societies quickly became a global affair. Since the technical division of labour significantly increased the productivity of human work, the throughput of matter and energy in the chains of activities involved in the production of many goods rose sharply. It also led to a greater concentration of production. Acquiring the requisite inputs of matter and energy for these chains from local ecosystems was often difficult, either because they were simply not available or they were not available in sufficient quantities. This in turn necessitated complementary chains of human activities to provide the required flows of matter and energy for the production chains. On the output side, longer chains were also inevitable because production soon exceeded local needs, thus requiring transportation and distribution. The technical division of labour extended technology-based connectedness within a society and frequently beyond, to colonies where raw materials were procured and finished products sold. All this fundamentally affected how and where people worked and lived in these colonies as well.

Since no production machinery has ever been designed to mimic human work, the technical division of labour preceded mechanization. In turn, mechanization changed the throughput of matter and energy in a productive activity, thereby disturbing the links between it and others in the technology-based connectedness of society. Redressing these disturbances required further mechanization, which created new disturbances, and so on. Hence, mechanization was inseparable from the industrialization of the technology-based connectedness of society. To accommodate this new technology-based connectedness, a society had to make substantial economic, social, political, and legal adjustments in a manner that drove the wedge still deeper between its technology-based connectedness and culture-based connectedness. As industrialization advanced, individual technologies were accommodated to the technology-based connectedness, making them less appropriate to the culture-based connectedness, to the point that in the twentieth century it was recognized that technology had become universal and that an appropriate technology could no longer be taken for granted.[20] Similarly, the new technology-based connectedness could no longer be sustained by local ecosystems, eventually creating the need for sustainable development. The more people changed technology, the more they changed their social and physical surroundings, and the more they changed the kinds of experiences from which they grew their brain-minds and cultures. Consequently, in the course of generations, 'people changing technology' was accompanied by 'technology changing people,' which in turn stimulated, in a self-reinforcing process, the way people changed technology. For example, it created the conditions that made it entirely self-evident to Karl Marx and most subsequent economists that societies have distinct and relatively separate economies. Durkheim interpreted this as a change from what he called an organic solidarity to a mechanical solidarity and was concerned about the cohesiveness of individual and collective human life.[21] There was a risk of anomie: a sense of rootlessness or being cast adrift resulting from the failure of a way of life and culture to give meaning, direction, and purpose to human life. For reasons that will become apparent in the next chapter, the restructuring of human life and society as a result of industrialization had a profound influence on the English language in general and the meaning of words such as culture in particular. The growing distance between the technology-based connectedness and the culture-based connectedness led to the arts being thought of as separate from the mainstream of life. In some cases, they were now to be experienced in separate institutions, such as museums and art galleries, following which visitors returned to 'real life.' As we will examine in Part Two, the growing dominance of technology-based connectedness undermined the role of experience and culture, requiring a growing reliance on reason and rationality, eventually leading to the phenomenon of rationality as examined by Max Weber.[22] His 'iron cage' may be interpreted as technology-based connectedness enclosing the culture-based connectedness of human life and society. As we will see, the growing use of the technical division of labour, followed by mechanization and industrialization, changed the moorings of human life and society in the biosphere as well as those with the culturally created gods.

These and other developments also led to a growing separation of the economy from the remainder of society. The economy became the locus of human activities dominated by technology-based connectedness, and society remained the locus of human activities where culture-based connectedness continued to hold the upper hand. This separation was inconceivable in traditional pre-industrial societies. In these, the economic aspect of their way of life was one among many. As a result, any attempt to explain the interconnectedness of such traditional societies primarily in economic terms plays down and even neglects their strong dependence on other bonds of a different kind, such as kinship, socio-cultural, political, moral, and religious ties. In other words, the culture-based connectedness integrated largely on its own terms the technology-based connectedness. In an industrial society, on the other hand, this interconnectedness may be analysed in a different way by recognizing that the economy has become largely decontextualized. It may also be thought of as the locus of individual technologies being economically appropriate as opposed to culturally appropriate. The economy provides the primary structure of connections on which al-most all other structures appear to (directly or indirectly) depend, at least as seen through the eyes of its members. Consequently, the basic interconnectedness of such a society is often approximated by eco-nomic exchanges represented in monetary terms in input-output mod-els. In these societies, the technology-based connectedness appears to dominate the culture-based connectedness, representing a reversal of the situation prior to industrialization.

The reasons why this approach is not applicable for analysing the connectedness of a traditional society now become evident. In a tradi-tional society, much of what was produced by each family unit and the community of which it was a part was also consumed by them. Any surplus was bartered for something these units could not produce themselves (or not in sufficient quantities). Trade did occur, but on a much smaller scale than in a modern economy. As a result, the interconnectedness of such societies cannot be symbolized by financial flows. Furthermore, the chains of human activities involved in the flows of matter and energy with the biosphere as the ultimate source and sink tended to be short and largely independent from other parallel chains.

Prior to industrialization, the technology-based connectedness of ru-ral English society had been expanded by the putting-out system of textile-making. This system may be thought of as a decentralized textile factory, in the sense that people worked at home but participated on the basis of a technical division of labour. Peasant families carried out a single production step, using hand-powered tools such as spinning wheels or looms. Families and villages were integrated in a network constituted by a flow of goods: raw materials, yarn, and cloth. These could be represented in terms of financial flows, as could the transac-tions resulting from earned wages. When the peasants began to partici-pate in the putting-out system, they worked for money. What they produced for themselves and the 'in-house' services freely provided within families, between friends or neighbours, or within voluntary associations could not be represented by financial flows. This reveals one of the greatest limitations of modern economic models. The totality of these services is incorrectly assumed to be negligibly small in comparison with all those represented by financial flows resulting from market transactions. Similarly, such models disregard the services rendered by the biosphere as the ultimate source and sink of all matter and energy, as habitat, and as life support. Nevertheless, the network of financial flows gives a much better picture of the interconnectedness of a modern society than of a traditional one.

In the remainder of this chapter, I will develop the 'people changing technology' component of a generic model of the process of industrialization. I will begin by examining the introduction of production machinery into the network of flows of goods involved in the way of life of a society, assuming that everything else remains equal. In other words, the productive activities represented as the nodes in this network are first assumed to change only in terms of the desired outputs they produce from the requisite inputs. This assumption will be gradually relaxed by considering the people involved in these productive activities, to reveal the economic, social, political, and legal adjustments that society had to make. It will then become apparent that the network of flows of goods not only integrates the networks of flows of matter and energy but also the networks of flows of labour and capital, at which point it represents much of the technology-based connectedness of society. The technology-based connectedness of this network is based on three principles: the first law of thermodynamics, the second law of thermodynamics, and a cultural equivalent in the form of a continuity principle. This third principle stems from the fact that most people participating in a way of life seek to avoid wasting goods, which would constitute a leakage from the network back to the biosphere before these goods are used. In terms of the network of flows of goods, this means that the output of one activity generally equals the input into another when waste is minimized. This brings in a cultural factor. Societies try to regulate their affairs so that what is produced is consumed, and waste is kept to a minimum. In a traditional barter economy, for example, a person who had planted a hundred cabbages one year, of which forty were consumed by his family and another twenty bartered for something else, is more likely to plant sixty cabbages the next year unless there is reason to believe that circumstances will have changed significantly.

Similarly, in a modern market economy, the mechanism of supply and demand ideally performs the same function. At the risk of stating the obvious, when supply exceeds demand, there is a tendency for prices to fall, since a sale with a lower profit or a slight loss is better than no sale at all. As the price falls, more people may be willing to buy the product, causing the demand to increase. Soon an equilibrium point is reached at which supply equals demand. If the demand for a product exceeds the supply, people may be willing to pay a higher price to get it. As the price rises, fewer people are willing to buy the product, causing demand to fall. Again, an equilibrium point is soon reached where supply equals demand, with the result that no leakage of wasted goods occurs from the network.

After analysing the effects of introducing production machines into the network representing the flows of goods in a society, we need to relax our assumption that everything else remains equal by recognizing that these goods are produced and moved about by people. Their activities are an integral part of a diversity of human activities, woven together into individual lives and a society's way of life. In other words, the analysis should be extended to take into account the economic organization, the social structure, the political structure, and the legal framework of a society. To understand additional changes, particularly the ones related to the morality, religious institutions, and artistic expressions of a society, we must consider how technology changes people – our task for the next chapter.

A network representing the flows of goods in a society can be constructed as follows. Each node of the network represents a process in the production, transportation, consumption, or disposal of some commodity. Each is connected to others by its inputs and outputs. If we assume zero losses, the output of one process equals the input into another to which it is connected. The portion of the network of flows of goods to be considered first is that of the putting-out system of textile-making in England, where industrialization began. Its technical division of labour parcelled out one of the four production steps to each peasant family: the preparation of raw cotton or wool, spinning it into yarn, weaving yarn into cloth, or finishing that cloth. In its simplest form, this portion of the network can be represented by a chain of four interconnected nodes. At one end of this chain we find the inputs from sheep-farming or cotton-growing, and on the other end we find outputs of cloth, which in turn constitute the inputs for other activities, such as the making of clothes and other textile goods. The processes of transportation interpose themselves everywhere. Many other inputs are required to maintain the chain. A distinction can be made between direct inputs, such as bleaches and dyes, which partly or wholly end up in the finished product, and indirect inputs, such as spinning wheels, hand looms, horse carts, roads, ships, and canals, which do not end up in the final product. These inputs flow from other activities, which in turn require still other inputs, and so on. The portion of the network of flows of goods associated with the putting-out system is therefore dependent on the larger network. Ellul describes what happens to the network of flows of goods once one stage in the production process of textiles is mechanized:

> The flying shuttle of 1733 made a greater production of yarn necessary. But production was impossible without a suitable machine. The response to this dilemma was the invention of the spinning jenny by James Hargreaves. But then yarn was produced in much greater quantities than could possibly be used by the weavers. To solve this new problem, Cartwright manufactured his celebrated loom. In this series of events we see in its simplest form the interaction that accelerated the development of machines. Each new machine disturbs the equilibrium of production; the restoration of equilibrium entails the creation of one or more additional machines in other areas of operation.
>
> Production becomes more and more complex. The combination of machines within the same enterprise is a notable characteristic of the nineteenth century. It is impossible, in effect, to have an isolated machine. There must be adjunct machines, if not preparatory ones ... The need for the organization of machines is found even in the textile industry. A large number of looms must be grouped together in order to utilize the prime mover most effectively, since no individual loom consumes very much energy. To obtain maximum yield, machines cannot be disposed in a haphazard way. Nor can production take place irregularly. A plan must be followed in all technical domains.[23]

Landes also recognizes the patterns that draw together many events of the time when he writes:

> In all of this diversity of technological improvement, the unity of the movement is apparent: change begat change. For one thing, many technical improvements were feasible only after advances in associated fields. The steam engine is a classic example of this technological interrelated-

ness: it was impossible to produce an effective condensing engine until better methods of metal working could turn out accurate cylinders. For another, the gains in productivity and output of a given innovation inevitably exerted pressure on related industrial operations. The demand for coal pushed mines deeper until water seepage became a serious hazard; the answer was the creation of a more efficient pump, the atmospheric steam engine. A cheap supply of coal proved a godsend to the iron industry, which was stifling for lack of fuel. In the meantime, the invention and diffusion of machinery in the textile manufacture and other industries created a new demand for energy, hence for coal and steam engines; and these engines, and the machines themselves, had a voracious appetite for iron, which called for further coal and power. Steam also made possible the factory city, which used unheard-of quantities of iron (hence coal) in its many-storied mills and its water and sewage systems. At the same time, the processing of the flow of manufactured commodities required great amounts of chemical substances: alkalis, acids, and dyes, many of them consuming mountains of fuel in the making. And all of these products – iron, textiles, chemicals – depended on large-scale movements of goods on land and on sea, from the sources of raw materials into the factories and out again to near and distant markets. The opportunity thus created and the possibilities of the new technology combined to produce the railroad and steamship, which of course added to the demand for iron and fuel while expanding the market for factory products. And so on, in ever-widening circles.[24]

To sum up, a generic model of the process of industrialization will be built, beginning with a first phase in which the technology-based connectedness of a society is limited to the network representing the flows of goods resulting from its way of life. Initially, the consideration of events will be limited to the way they change this network, and all other things will be assumed as remaining equal. The fact that any flow of goods requires human activities which in turn require communities, institutions, and a way of life will be considered in subsequent stages of building the model. In other words, the assumption of all things remaining equal will be relaxed in stages as the building of the model advances.

1.3 Living with Materials

Consider once again the portion of the network representing the flows of goods in England on the eve of the Industrial Revolution, constituted by the chain of activities involved in the making of cloth. When a war had just ended, so that overseas markets for cloth and for raw materials could once again be safely accessed by merchant ships, entrepreneurs sought to take advantage of the situation by increasing the output of the putting-out system. Their success was rather limited, essentially because they had little control over the work people did at home. Financial incentives caused the output of work to decrease because people preferred to maintain their standard of living as opposed to 'improving' it, which now required less work and left more time for what were regarded as more important things. Reducing their pay frequently led to a reduction in quality standards, as people attempted to make the same output with a slightly lower input, using the difference as compensation. The situation attracted creative attention, and given the conditions that prevailed in England, it was just a matter of time before someone had the idea that machines could provide an answer. If each person now involved in textile-making attended a single machine operating at a speed over which he or she had little or no control, production levels could be increased. Moreover, since these machines were too large to be used in homes, too expensive to be owned by peasants, too energy-intensive to be hand-powered, but not energy-intensive enough to require individual waterwheels, they should be grouped in workshops adjacent to moving water. Work could now be supervised and controlled in ways that were impossible when it was done at home. Given the pool of technological inventions and devices and the technical orientation in the culture, it was not long before the first textile machine was invented – all due to the unique conditions that prevailed in England at the time.[25]

Let us assume it was a spinning machine (although it could just as well have been a machine for any of the other three stages of textile production). The inventor of the spinning machine teamed up with an entrepreneur, and a spinning shop was built adjacent to a river. It was now possible to greatly expand the production of yarn, but this only partially solved the entrepreneur's problem because of the four linked production stages. His portion of the putting-out system was unable to expand the preparation of raw materials in stage one and the weaving of yarn into cloth in stage three, so that the solution of one problem produced two others. Creative attention was focused on the bottlenecks existing between stages one and two, as well as between stages two and three. It again appeared that only machines could provide an answer. Given the prevailing conditions, it was just a matter of time before the preparation of raw materials and the weaving were mechanized. Now two new bottlenecks were created, between the activities that produced raw cotton or wool and the first production stage, and between weaving and the final finishing stage. It is in this way that the sudden and rapid mechanization of one of the oldest human crafts can be explained. It should also be noted that it made no difference which stage of production was mechanized first – the final results would have been the same given the interdependence of the four production stages and their joint dependence on upstream and downstream activities. The spinning shop on the river inevitably evolved into a textile mill. Different entrepreneurs reached this development by means of different paths, but the interplay between individual events and the broad context brought about their convergence and the rise of factories.

Because of the linkages within the network of the flows of goods in a society, no single category of machines was of any value without all kinds of other machines to establish a new dynamic equilibrium between the stages of production. The description is far from complete, however. Each production stage had many more inputs, both direct and indirect, than the one derived from the output of the preceding stage. These inputs connected this portion of the network of the flows of goods to many others. For example, for a putting-out system these included combs, spinning wheels, hand looms, shears, bleaches, and dyes. When production became mechanized, machines, power sources, and factory buildings were among the new inputs. Each one was the output of another chain of activities that in turn required still other inputs all the way back to the biosphere, the ultimate source of all matter and energy. For example, the building of a power loom required a variety of parts that must be manufactured, which in turn demanded materials such as wood, iron, and steel. These again derived from chains of activities such as logging, curing, sawing, and dressing (in the case of wood). If the mechanization of textile-making required substantial increases of any of these inputs, bottlenecks would again be created, which led to the restructuring and possible mechanization of still other activities. This was particularly true for the production of bleaches and dyes. In other words, the repercussions of re-establishing a dynamic equilibrium in the network representing the flow of goods in a society, as a result of mechanizing one single craft, were far-reaching and widespread. For example, to remove a bottleneck in the supply of cotton and obtain more raw cotton for themselves, the English destroyed a well-organized and flourishing system for the making of cotton textiles in India. Many bottlenecks arose from the inability of local ecosystems to supply in sufficient quantities the flows of matter or energy to sustain the chains of activities that furnished necessary inputs into industrial textile-making. This involved an unprecedented extension of the ecological footprint of each and every industrializing nation. In many cases, it necessitated an expanded and more aggressive colonization of other people.

Other bottlenecks occurred in the network of the flows of goods in the mechanization of textile-making. Mechanization began to increase the throughput of matter and energy in the chain of production activities. The result was a bottleneck in transportation. The moving of semi-finished goods between production stages in the putting-out system was eliminated by the factory system, which brought all four stages under one roof. However, as textile production rose, larger quantities of raw cotton and wool as well as inputs such as bleaches and dyes had to be transported to the factories; and greater quantities of wool and cotton cloth were carried to distant markets, since production levels had long exceeded local needs. Transportation associated with indirect inputs also increased. In response to transportation bottlenecks, canals were dug and railroads were built. The implications for the network of the flows of goods were considerable, and may be analysed as before by tracing the inputs to each activity back upstream to their ultimate sources in the earth's crust or the biosphere.

In the case of railroads, for example, some of the inputs were steam engines, cars, track, and buildings. I will limit my attention to one input into most of these, namely, iron and steel. Building railroads required great quantities of iron and steel. Their production in turn demanded iron ore, and coal for the blast furnaces. To obtain sufficient quantities of coal, English mines had to go deeper into the ground, which created the serious problem of water seepage into the mine. This required much better pumps than the ones available at the time. All the developments mentioned above would be threatened unless this bottleneck was resolved, hence a certain urgency arose. Again, creative attention was focused on this matter, and prevailing conditions greatly increased the likelihood of a successful invention. In fact, the steam engine was first invented and used as a pump. The availability of coal increased, as mines could now go deeper. As they did, more steam engines were required, which in turn increased the need for iron, steel, and coal, and this in turn forced coal mines to go still deeper. Together these developments created a self-reinforcing cycle: a greater need for coal led to a greater need for iron and steel, which in turn required more coal.

The mechanization of textile manufacturing also had an enormous impact on the building industry. The relevant indirect inputs into textile making included buildings to house the machines, housing for factory workers, bridges, canals, and mines. Each of these required a whole range of building materials, ultimately extracted from the biosphere via chains of activities yielding the required inputs. One detail is worth mentioning. When textile-making was mechanized, work could no longer be done in the family setting at home and was now carried out in factories. These tended to be concentrated in certain areas, typically in close proximity to clean running water and preferably also to iron and coal mines. The concentration of textile manufacturing in limited areas, as opposed to its distribution throughout the countryside in the putting-out system, required a relocation of the factory workforce so they could live within walking distance of the factories. There was a growing need for new housing, with the required inputs for construction such as lumber, bricks, mortar, roof tiles, glass, and plaster. Each of these can be traced back to its ultimate source via a chain of activities. To sum up: the mechanization of textile-making necessarily had a substantial effect on a large variety of other activities, including agriculture, chemical production, machine-building, transportation, construction, metallurgy, mining, and forestry. These developments in turn influenced each other, as they together helped establish a new dynamic equilibrium in the network constituted by the flows of goods in a society, following the introduction of machines to mechanize human work.

Although this first phase of the building of a model of industrialization is limited to considering the network of the flows of goods in a society, it is already becoming evident that the same results would have been obtained if machines had been introduced in other critically important areas of the network. In fact, a society wishing to industrialize could begin to build a better transportation infrastructure. The generic features of the model are now becoming evident.

As noted, the network constituted by the flows of goods in a society depends on and partly overlaps the network of energy flows necessary to sustain a way of life. Input-output analysis, energy analysis, and the inventory stage of life-cycle analysis examine this network or parts thereof. For example, energy analysis seeks to establish the total energy that must be transformed to directly and indirectly sustain a particular activity.[26] Through this network of energy flows the activities involved in any material, product, or process are linked to many others, thus further contributing to the technology-based connectedness of a society. Once again, the biosphere is the ultimate source and sink of the network of energy flows. I will not include an exploration of this dimension of the technology-based connectedness of a society in this work.[27]

1.4 Living with the Economy

The first step on the road to building a generic model of the process of industrialization examined how the activities of a way of life are connected and sustained by a network of flows of goods, and how this network had to be entirely restructured. The assumption was made that all other things remained equal, thereby disregarding the changes in human life and society that occurred beyond the material focus on the technology-based connectedness of a way of life, as if this larger context participated little in this massive restructuring. By gradually relaxing this assumption one step at a time, it will become evident that nothing beyond the technology-based connectedness of human life and society could remain the same. A first step examines the inevitable changes in people's working lives and the economy. The next step analyses the spillover effects of such changes into the remainder of people's lives, beginning with the transformation of the social fabric, followed by the transformation of the political framework, and concluding with the complete overhaul of the legal system. How all these changes in turn were integral to a complete change in the culture-based connectedness of human life and society requires a different methodological approach, which will be developed in the next chapter.

Whether they were doing their work or organizing it, people's working lives were completely transformed. The introduction of the technical division of labour and the resulting socio-technical entities integrating people and machines focused all efforts on transforming the inputs obtained from the local technology-based connectedness into the desired outputs to be returned to it in the most productive and profitable manner. What the meaning and value were for human life and society became entirely secondary. Gradually, an economic approach to life began to displace the cultural approach to life, ushering in a transformation of which Adam Smith was without doubt the great secular prophet. This approach concentrated on the 'meaning' and 'value' of transforming inputs into outputs for the technology-based connectedness of human life and society, as opposed to the culture-based connectedness. To make this possible, the technology-based connectedness was expanded to include a way of organizing and evolving itself on its own terms. Capital and the Market ruled over an increasingly distinct economy and, through it, over the remainder of society. The wedge that the technical division of labour had driven between the technology-based connectedness and the culture-based connectedness went still deeper. It compelled the latter to transform itself in the image of the former.

Changes in individual economic behaviour could not occur haphazardly, since they were tied together by the technology-based connectedness of society. This poses the following questions: Who or what guided and coordinated the many people making changes in the activities connected by the networks of flows of goods and energy in order to create a new technology-based connectedness for their way of life? What was the basis for this guidance and coordination, and where did it come from? One thing is certain: it did not come from the existing culture-based connectedness. Several reasons are already apparent from the first step in the analysis. The technical division of labour created a distance between the culture-based connectedness and the technology-based connectedness of a society, making it more difficult to use the cultural approach to life to reorganize and evolve the latter. This same technical division of labour also increased the throughput of matter and energy in the technology-based connectedness. Its moorings in the biosphere were strained to the point that this reorganization had to face thermodynamic constraints regardless of their cultural meanings and values. All this further weakened the cultural approach to life. Gradually, by fits and starts, a new economic approach to organizing the activities of a way of life emerged. It built on and accelerated what Polanyi showed constituted one of the great transformations of human history.[28] Nevertheless, the present model will show that this transformation was part of a much more comprehensive one in which an economic approach to life began to undermine the cultural approach, thereby collapsing every traditional culture it touched.

Adam Smith became the prophet of what, two centuries later, would become the secular theology of monetarism, which is virtually unopposed today. For now what concerns us is that, at the time, Adam Smith explained what was happening by means of a theory implying that the Market was taking over much of the role previously played by culture.[29] Briefly put, in markets for goods and services, buyers and sellers were increasingly guided by their self-interest at the expense of their way of life and culture in general and the meaning and value of any market transaction in particular. Such self-interested behaviour was regarded as the essence of human nature, now seen as *homo economicus*. It functioned on the basis of a calculus of optimizing desired outputs obtained from requisite inputs. This is how *homo economicus* allocates time and resources to the pursuit of a particular goal, namely, the maximization of utility and satisfaction. To the extent that people's behaviour began to conform to this model, the cultural approach to life began to make way for an economic approach. A clue to the profundity of this transformation is provided by the fact that, to my knowledge, no religious or moral tradition has ever condoned unrestrained self-interested behaviour – certainly neither Christianity nor Judaism, as the two most important influences on Western civilization. From the perspective of *homo economicus*, not acting in one's own self-interest was irrational and became the greatest of all new 'secular sins.' Lewis Mumford rightly complained that the traditional Christian vices were now converted into virtues and vice versa.[30]

Economic theory claimed that buyers maximized the utility derived from their income while sellers maximized the profits derived from investing their capital. The invisible hand of the Market would ensure the optimal allocation of resources as a way of organizing a society based on performance values (output/input ratios) as opposed to the values of a culture.[31] To make this possible, the role of money in individual and collective human life had to quantitatively and qualitatively change. For producers, this was the consequence of the transformations of activities connected by the networks of flows of goods requiring ever-larger investments; while for factory workers, it resulted from their becoming wage-earners, with money mediating the procurement of all necessities of life. The economic regulation of human activities essentially restricted the context to be taken into account to the technology-based connectedness of these activities, which falls far short of the context taken into account when human activities are guided by a culture. In sum, the technology-based connectedness of the activities of a way of life was expanded to include their economic regulation, which diminished and gradually undermined every traditional culture where this occurred. In order to explain this in some detail, we will first consider how capital became the prime factor of production; and next, how markets organized many of the activities of society, thereby greatly weakening the role of culture.

The factor of production most critical and difficult to procure in the industrialization process as it occurred in England was capital. The large amounts of land required to produce the raw materials had been obtained through colonization. Various agricultural and social transformations, including the enclosure movement, had ensured a large labour pool. Similar agricultural reforms producing a large labour pool are typical for almost all industrializing nations today. Procuring capital was a different story. Industrialization necessitated ever-growing amounts of capital. For example, a comparison of the capital investment required to set up a putting-out system or a textile mill with equivalent production capacity showed the latter to be considerably more capital-intensive. First of all, the machines for a textile mill were much more expensive than the hand-powered devices used in putting-out systems. Machines needed sources of power, while the devices used in putting-out systems did not. Machines had to be housed in factory buildings, which were not needed in putting-out systems. As the concentration and level of production continued to increase, the transportation infrastructure had to be expanded, which, once again, was very capital-intensive. Similarly, other areas in the network of the flows of goods also necessitated increased capital investment.

Bringing together large amounts of capital thus became the greatest bottleneck. So difficult was this challenge that production costs had to be carefully calculated so that market prices exceeded such costs sufficiently to ensure that capital could be accumulated. Capital was the lifeblood of the new 'system,' and nothing essential could be done without ever-larger amounts of it. Factory owners had to ensure that their profit levels were adequate, and that they would be in a position to replace machinery, power sources, or buildings when these became obsolete or worn out. The technologically more advanced machines or power sources were often more expensive. All this had to be taken into account in the calculations. Whether a business was privately owned or owned by the state made no essential difference, at least in this regard. All levels of government had to get involved in helping to build and develop the financial infrastructure. Many changes had to be made to various institutions in order to create an appropriate economic framework within which the capital resources could be created, sustained, and accumulated. England, being a trading nation, had amassed considerable wealth, which constituted the initial pools of capital that financed industrialization. The per capita income was high compared to many modern developing nations, and ways were quickly invented to permit the accumulation of the modest savings of many individuals to create new pools of capital. Governments also became creative in ensuring that the capital investments required for industrialization would be forthcoming. Everything depended on this critical input. The conditions in England on the eve of the Industrial Revolution ensured that all other inputs required for industrialization were either present or could be obtained by socio-political reorganization and by colonization. Hence, the struggle to bring together the huge pools of capital needed to sustain industrialization took centre stage. It was from this feature that the new emerging 'system' derived its name: capitalism. As it developed technologically, it required ever-larger infusions of capital. For example, the creation of a single job required an increasing amount of capital as industrialization progressed. These trends were in evidence everywhere regardless of whether industry was privately or publicly owned.

Some of the implications of capital becoming the prime factor of production come to light when we compare the new 'system' with others where land and labour were the prime factors. In agricultural societies, wealth ultimately depended on land and labour. Large land-owners worked the land using pools of labour organized by means of a variety of 'systems.' Slaves could be used by means of force, but how well they were fed, clothed, and sheltered affected the amount of force required and the long-term viability of this arrangement. Feudal systems obtained the necessary agricultural labour from large numbers of serfs who, in return for their labour, lived on the land and cultivated small plots for their own needs. An extensive set of reciprocal obligations existed between the landowner and the serf, covering a range of situations including illness and war. In still other cases, the required labour was hired by the landowner on a daily basis from the local labour market. Whatever the arrangements, the needs of all the participants had to be met in some fashion (preferably with minimal resources spent on coercion). Additional wealth was created when a surplus occurred. Consequently, the local ecosystem and labour were the ultimate producers of any wealth. Land also produced wealth through a variety of other activities yielding minerals, forest products, or fish. In other words, land as a factor of production refers to the processes of the biosphere used for human purposes by means of labour.

Craft labour producing artefacts needed for a way of life was another source of wealth. Once again, a surplus could occur after the needs of all the participants in this form of production had been met in accordance with their way of life. In economic terms, both the primary and secondary sectors of the economy of a society could support commerce when they produced a surplus, after all the needs and obligations had been taken care of according to custom and tradition. When capital became the prime factor of production during the first phase of industrialization, its owners had an enormous influence on the 'system,' which greatly affected their status in the social hierarchy relative to the traditional elites. In England, the influence of bankers and factory owners quickly rivalled that of large landowners and the nobility.

Science and technology as 'intellectual labour' were not considered to be factors of production. From an economic perspective, this was largely valid for the first phase of industrialization as it occurred in western Europe. Science still played only a minimal role in technological development, for reasons to be discussed in Part Two. The pool of technological inventions and devices accumulated by earlier generations, extensively used and preserved by monasteries, ensured that the required 'technological capital' was available. Actual or potential bottlenecks in the process of transformation associated with industrialization stimulated inventive activity, nurtured by the technical orientation in the culture (more about this in chapter 3). Whenever a new machine or device was required, it was generally invented within a reasonable time, so as not to block or hold up the transformation of society.

The evolution of capital as the prime factor of production was closely associated with a second major change in the economic context of the technology-based connectedness of society, namely, the appearance of a market economy. Several developments made this possible. A relatively distinct economy had been created where the technology-based connectedness dominated the culture-based connectedness. This technology-based connectedness had become a network of flows of goods, labour, technology, and capital, all essentially controlled by the latter. Work had been separated from the family to become concentrated in industry. Working very long hours in a factory and living in the new urban-industrial habitat precluded the possibility of families meeting their basic needs themselves, as had been the case in agriculture supplemented by the putting-out system. In the beginning, because of the long working hours, factory workers essentially sold all their capacity for work in return for a wage. As a consequence, all the necessities of life had to be procured by means of their wage: shelter, food, clothing, and, later, entertainment. In other words, a barter economy, in which extended families produced most of their basic needs and bartered small surpluses to fill other needs or desires, gave way to a market economy in which wage-earners purchased everything. This eventually happened everywhere in the world where industrialization took place.

With two hundred years of hindsight, we are just beginning to appreciate the externalities that are created when a society shifts from the role of culture acting as its organizational principle to the Market acting as such. In the case of the former, everything takes on its meaning and value in relation to everything else; while in the case of the latter, this must be accomplished by the much more limited context of the Market, where the monetary value of everything is its only real meaning and value. Everything in individual and collective human life that cannot be commoditized into a good or service traded in some market is externalized. Whatever living entities of the social or natural worlds are included, these must now be treated as commodities. Glimpses of what this implies appear in the attitudes of indigenous people to the private ownership of land, which for them was not a mere piece of terrain but something that was integral to, and thus inseparable from, living ecosystems.[32] Glimpses also emerge in the current debate over the ethics and practical value of patenting animals, plants, or their constituent 'elements.' Similarly, they appear in the debate over whether there should be tradeable pollution permits at all. For all this and more to be possible, everything had to have a monetary value. Money became the common denominator of all values, and it changed the very nature of traditional values. These had been an expression of the place and significance of something in individual and collective human life. Monetarized values, however, expressed the place and significance of something in the Market. This was impossible in earlier societies, which had no relatively distinct economy, so that the economic dimension of any human activity could not be separated from all the others and valued on its own terms. Taking work and the economy out of the context of human life and society had such a profound influence that it shattered every traditional culture it encountered.[33] The elevation of money to the value of values marked a decisive victory of technology-based connectedness over culture-based connectedness. Money ceased to be merely precious metals or paper, to become something that could procure almost everything. This became its new meaning and value, and people lived as if it was everything.

The economic approach to life resulted in possibly the first transcultural institution, namely, the Market, in which *homo economicus* operates in a manner that is entirely non-cultural and ahistorical. When other societies adopt the Market as an organizing principle, they in effect adopt an economic approach to life as superior to their own traditional cultural approach. This presupposes the prior development of a false awareness of human life in the world based on externalizing what distinguishes all of life from corresponding commoditized counterparts traded in markets.

Initially, the consequences of the qualitative and quantitative changes in the role of money were not very obvious. In a society with a modest standard of living, many people would have to consider very carefully whether to allocate their tiny savings to the purchase of a badly needed pair of shoes or to an equally necessary new overcoat. They would have to consider all aspects of the choice, because once they had bought the one, it would be some time before they would be able to buy the other. Under such circumstances, it was highly unlikely that outside persuasion would have any influence on their decision. It is in this context that concepts such as *homo economicus*, consumer sovereignty, and economic democracy make sense.

The long-term implications of the economic transformation were veiled by an ideology of the Market that overvalued this institution and what it could do for a society, with disastrous results. We continue to believe in the Market and have great difficulty learning that sustainable development will require a more realistic view of what it can and cannot do. In the nineteenth century, the ideology of the Market was persuasive and elegant in its simplicity. It promised hope and salvation through progress as an instrument of economic democracy based on self-interest. People entered the market with their wages or salaries and 'voted' with their money for specific products and services. These choices translated into the demands of a society for goods and services. Producers and providers of these goods and services allocated their capital and competed with one another to satisfy these demands in order to make the greatest possible profit. Those who provided the best goods and services for the lowest prices would succeed. The ones who ignored the demands of society, who offered poor-quality goods and services or demanded exorbitant prices, would soon go out of business. In other words, the Market ensured economic democracy because what a society needed and desired was produced as effectively and as cost-competitively as possible. We will later see that this economic model became less and less tenable in the course of the twentieth century.

The role of the Market as the great invisible hand undermined the culture-based connectedness of individual and collective human life.

The miracle of the Market was the making of a better world from selfish individual behaviour. It goes without saying that the proclamation of the virtue of self-interested behaviour weakened any traditional value system rooted in culture-based connectedness. As industrialization proceeded, it became increasingly necessary for people to behave as *homo economicus*, at a cost to their humanity. Societies were compelled to surrender the regulation of the network of flows of goods associated with their ways of life to markets. Everything that could not be represented by financial flows became a market externality, and to the extent that the Market organized society, became of secondary importance. Goods and services ceased to be means in the service of life (alienated as it was), and their importance was limited to their monetary value within the new 'system.' Acting in the name of any traditional value became irrational to this 'system.' On the level of society, traditional cultures suffered the same fate. Money ceased to be a convenient means to (human) ends, to become an end in itself.[34] Precious metals, and later paper money, originally intended as simple means to facilitate economic exchanges, were transformed into something capable of procuring almost anything, including people's bodies and (cultural) souls. It was this spiritual power over human life and the world against which the religious traditions of the West had warned in vain. People increasingly lived as if money was everything. It altered the form of alienation, but not its effect on human life. This was, as we will see, another step in the process of technology changing people.

Gradually, the weakening of traditional values and the diminishing role of culture undermined what had essentially been a precautionary attitude towards local ecosystems implied in almost every traditional way of life. A tradition embodying the experiences of countless generations was treated as a proven design for living in the world that should be adapted to new circumstances with great caution lest it become non-viable. Initially, this was not very evident because the scale of the network of flows of goods associated with an economy were relatively modest compared to the equivalent networks within local ecosystems. When this began to change as a consequence of industrialization, much of the additional environmental capacity that was required to sustain the rapidly growing technology-based connectedness of society was appropriated from distant ecosystems. Economists of that time assumed that the consequences of all market transactions for the biosphere were negligibly small, with the result that the capacity of the environment to support human economies was taken to be constant. Hence, the diminution of 'natural capital' did not have to be priced. When the flows across the society-biosphere boundary became significant relative to those occurring in natural cycles, their effects were no longer negligible. Nevertheless, if this threatened any service the biosphere rendered to a human economy, it remained unpriced in the Market. In the case of flows of non-naturally occurring materials, even small quantities could have this effect. Recall that the biosphere does not produce any wastes and is self-purifying, because all matter participates in cycles formed by processes in which the 'wastes' of one become the 'resources' for the next, and so on until the loop is closed. Since matter is neither created nor destroyed, any dislocation of these cycles can have serious consequences for all life.[35] In other words, the consequences of market transactions for the biosphere are unpriced by the Market and become externalities. This results in an overexploitation of the biosphere (referred to as the tragedy of the commons),[36] thereby making the productivity of materials and energy insignificant relative to the productivity of labour. This imbalance creates economies in which materials and energy are so cheap that much of these are wasted, while people become so expensive that underemployment and unemployment are a constant threat.

When a service rendered to a human economy by the biosphere has the status of a commons, it cannot be given any economic consideration. By its very nature, it cannot be divided to become private property amenable to transactions in the market. Such so-called externalities to the Market lead to the absurd situation where no distinction is made between economies that grow partly as a result of depleting their 'natural capital' of forests, agricultural land, fish stocks, minerals, or other resources, and those that live only from the 'interest' of this 'capital,' which can be done indefinitely. The choice of the Market over the cultural approach to life amounts to a negation of the experiences of each and every group and society: self-interested behaviour is destructive of people and the land. The environmental crisis is ample evidence of the inability of the great invisible hand of the Market to harness self-interested behaviour into a positive force capable of creating a better world for most people.

The operation of the Market as an invisible hand is equally problematic with respect to the ultimate 'resource' of a society, namely, its culture (sometimes referred to as the human and social capital). Every human being is born into a community and inherits the labours of many generations in the form of a way of life and culture able to give meaning, purpose, and direction to the lives of many people. The pursuit of economic self-interest has as serious an effect on culture as it does on nature. Any market transaction freely entered into by two parties for mutual advantage rarely has consequences limited to these two parties, particularly as the power of technology and the magnitude of its consequences increase. For example, the inhabitants of cities who neither own nor operate cars and whose work does not involve them in the many market transactions associated with cars experience all the effects cars have on modern cities. In the same vein, the effects market transactions have on culture affect everyone in a society.

A more complete understanding of the effect the Market has on culture involves contrasting the economic approach to life with the cultural approach (to be discussed in chapter 2). For now, it is evident that the Market essentially regulates the technology-based connectedness by externalizing the culture-based connectedness of society and the biology-based connectedness of the biosphere. By making the fullest possible use of context, the cultural approach produces no such externalities. The technology-based connectedness of a society is strengthened by increasing the desired outputs human activities obtain from requisite inputs as measured by output/input ratios. Cultural evolution, on the other hand, is evaluated by the meanings and values of these activities for individual and collective human life. In other words, a growing dependence on the economic approach to life at the expense of the cultural approach continues to drive a wedge between the technology-based connectedness of society and its culture-based connectedness, leading to the emergence of an increasingly distinct economy. Its coexistence with the host society necessitates a cultural mutation, which transforms performance ratios into values at the expense of all traditional values. Once a society is split into its economy and everything else, the institution of the Market behaves as an invisible hand with respect to the former and as an 'invisible elbow' (to use Michael Jacobs's phrase[37]) with respect to the latter. For example, land is reduced from an integral part of an ecosystem with all its life-supporting functions to a piece of terrain, labour from an integral part of a human life related to work to the capacity to add monetary value to industrial production, and everything people create and produce to market commodities. Indigenous cultures living in ecosystems that are required to extend the ecological footprint of the industrializing world are at best transformed into 'adopted children' in need of 'cultural upbringing.' In sum, all such changes represent a transfer from what is integral to the natural and cultural orders to a new order with technology-based connectedness at its core. Everything is commoditized in the process, and anything that cannot be commoditized (which includes all life) is excluded. When the technology and way of life of a society primarily evolve on the basis of the economic approach, the former ceases to be appropriate with respect to the culture, and the latter ceases to be sustainable with respect to the biosphere.

The desired consequences of economic decision-making and organizing are inevitably accompanied by unforeseen and possibly undesired consequences. These are what Jacobs referred to as market forces.[38] They produce collective decisions that cannot be derived from the individual decisions of buyers and sellers. Adam Smith identified the best-known positive market force, namely, markets acting as an invisible hand creating the best possible world for most people. He also identified an important negative market force, namely, that although the technical division of labour would increase the wealth of nations, it would also make human beings as stupid as they could possibly become.[39] The environmental crisis provides another example. These are instances of the invisible elbow attached to the invisible hand knocking over a great deal as societies reach for ever more economic growth.[40] Daly and Cobb have shown how this 'elbow' undermines human life, society, and the biosphere.[41]

Markets are blind to culture in general and human values in particular. The basic needs of many human beings cannot be translated into market demands because the poor lack the money to do so. The needs of future generations are all but eliminated by discounting them. Hence, markets cannot ensure a just distribution of resources for present and future generations. They take no account of the depreciation of 'natural capital' or whether there will be enough of this capital left for future generations. Commoditization is another important example of a market force.[42] Everything has to become a commodity separated from its contexts and reduced to a monetary equivalent.[43] Economic growth is focused on what can readily be commoditized and neglects almost everything else, thereby introducing a growing chaos in natural evolution and human history. In sum, markets essentially regulate the technology-based connectedness of a society in terms of flows of capital and of commoditized goods and services, thereby externalizing almost everything else. As a result, the effects of the invisible hand are increasingly undermined by those of the invisible elbow.

Another economic transformation associated with industrialization was the recolouring of the geographical map. With the change in the hierarchy of the factors of production, the economic fortunes of geographical regions were often affected. The most vital regions became those where industry flourished, making all others of secondary or even marginal importance on the world stage. Industries tended to locate in those regions where the inputs required for the mechanization of production were most readily available. These included coal deposits, iron deposits, clean running water, natural harbours, nodes in rail networks, and proximity to large markets. A new hierarchy among nations and regions was created.

1.5 Living Together Socially

The building of a generic model of the process of industrialization was begun by examining the effects of mechanization on the network of flows of goods, with the assumption that the remainder of human life and society were unaffected. It quickly became evident that this assumption was untenable, and it was relaxed to include the required economic changes. In turn, these could not be made without a number of social changes, hence the assumption will be relaxed once again to examine the changes in the structure of the labour force that accompanied changes in the network of the flows of goods, the implications for the social fabric of work separating from the extended family, and the concentration of the population in the new industrial centres.

Each node in the network of flows of goods involved a number of people who had various skills as 'labour inputs.' For the entire network of flows to function properly, a labour force with a diversity of skills was necessary. It is possible to classify all these skills and determine the number of people required for each category. The transformation of the network described above involved substantial changes in these two characteristics of the labour force. Mechanized textile-making and putting-out systems required different skills. Consequently, new skills emerged as old ones disappeared, while still others continued to evolve. At the same time, the ratios of the number of people involved in interdependent categories of activities underwent constant change. Once again, bottlenecks arose when, for example, an inadequate number of people was available who could fix power looms when they malfunctioned or broke down. To avoid such bottlenecks, changes in the network of the flows of goods in a society had to closely match the restructuring of the labour force that ran the network. In some cases, such changes were substantial. In the putting-out systems, spinning was generally done by women while weaving was done by men. The maintenance of a dynamic equilibrium required a ratio of about seven women to one man.[44] When these activities became mechanized, they tended to be carried out mostly by men, and the ratio of spinners to weavers changed constantly as a result of the technologies involved. Thus, the above transformations in the network of the flows of goods caused substantial changes in the structure of the labour force.

The technology-based connectedness of a society helped to determine the structure of the labour force in terms of the categories of skill required to sustain that way of life and the number of people involved in each category. It created the technically subdivided roles in industrial societies. The ongoing changes in these social roles threatened the livelihood of many people as they saw the demand for their skills dwindle or disappear altogether, with profound consequences for their lives, families, and communities. An ever-changing technology-based connectedness thus imposed itself more deeply on the culture-based connectedness of individual and collective life. The size of the labour force required to maintain and evolve industrializing ways of life also changed dramatically. It is estimated that in the eighty years following 1850, nearly fifty million west Europeans emigrated to the colonies, mainly because there was no place for them in the new economy.[45]

A second major change affecting the people involved in the technology-based connectedness was the separation of work from daily life and the extended family. Mechanization made it impossible to continue work in the family setting, and this had profound consequences. It is difficult to exaggerate how deeply the nature of the work changed when it became mechanized. It is common to think of people in earlier societies working much more than we do today. This is an entirely misleading view because work was inseparable from a host of other activities. The closest we can get to it today is the experience of a traditional mother doing housework while her children are playing nearby. She keeps an eye on them and simply drops her work from time to time when approached about something, or when she suspects something is happening that requires her intervention. The telephone may ring, someone may come to the door, a neighbour may drop by for coffee, and so on. A host of household activities are intertwined with bringing up children: sustaining relations with neighbours and friends, arranging for a variety of things by means of the telephone, and possibly some productive activities such as sewing clothes, keeping a vegetable garden, and preserving food for the winter. This is as close as our experiences come to what things were like in earlier societies. The extended family did such things as nursing members who were ill, taking care of a child or an adult with a disability, looking after elderly members, educating children, and much more. In other words, many of the educational, health, and social services provided by state and insurance agencies to modern families were, in those days, taken care of internally. We may say that the extended family had a kind of built-in school, infirmary, unemployment insurance, Workers' Compensation, disability pension, life insurance, and pension plan. It was the locus of a great many enfolded activities, thereby constituting the basic social building block of a society. It is common to think that in earlier times people 'worked long hours,' but it would be more accurate to say that people '*lived* long hours' since work was enfolded into the lives of people. For example, if a neighbour dropped by obviously upset over something, people would put aside their work and sit down to talk things over. Religious holidays were numerous, averaging about one in every three days in some parts of medieval Europe.

Factories fundamentally changed the nature of work. It was now supposed to be nothing but work, with separate breaks and a separate family life. As a consequence, the integrality of people's lives began to be undermined. It is hard to exaggerate the difficulties people experienced in having to split their lives into separate compartments. One window into the situation is provided by the lists of Draconian penalties, such as the heavy fines that were deducted from the wages of workers who 'misbehaved' or in some way were deemed to be 'irresponsible' or 'lazy.'[46] The work discipline in a setting of the extended family was anthropocentric, while the work discipline in a factory was technocentric. People had to learn to deny something of their person and culture, and this dehumanization did not come easily to many. After all, the nature and pace of work was now controlled and dictated by machines. When people went home, they took their work-related experiences with them, and this had profound consequences for their lives. When, after very long working hours, people came home, they still had to attend to everything else. Needless to say, they no longer had what it took to sustain their marriage, bring up their children, have a family life, maintain friendships, be neighbours to one another, and remain part of a community. Comparatively speaking, factory workers were often as financially poor as many peasants; but the lives of the latter were not crushed by their financial poverty as the lives of the factory workers were by their work. The poverty of the factory workers was multidimensional. Their working hours were so long that they essentially sold all the life-energy they had in the labour market. The concept of alienation accurately describes this condition, in which people's lives became possessed by their work in a manner rivalled only occasionally in earlier times, such as the miserable existence of the slaves in the salt mines or on the galleys in antiquity. The Judaeo-Christian tradition warned against this risk, although this warning was rarely heeded. In this tradition, hiring workers was more than simply buying some of their time for a wage. Work, not being a separate part of their lives, affected their whole existence, with the result that there was a risk of enslavement. This made it impossible for people to respond to the love of their God and their neighbour because human freedom is a prerequisite for such relationships. It is impossible to love someone or to be a friend under duress. In this sense, alienation in work mirrored what in the religious traditions of the West was called sin: it jeopardized the ability to live a human life.

The fact that factory work was organized on the basis of a technical division of labour greatly added to the pressures put on the lives of workers, their families, and their communities. All of this would have been unliveable were it not for new political ideologies, which reinterpreted the situation as a small necessary price to pay on the way to the socialist promised land or on the path to progress. Material advances would end the struggle for survival, permitting human ingenuity and energy to be directed to the higher goals of social and moral improvement. Of course, there were the heretics of the day, including the Luddites, who believed that technology should be used only when it directly benefited people and their communities; but they did not understand that industrialization was essentially an all-or-nothing affair.

Finally, the changes in the technology-based connectedness of a society resulting from the mechanization of textile-making also had a substantial effect on the geographical distribution of the population. The putting-out system was in essence a kind of decentralized factory in the sense that it had the beginning of a technical division of labour, but since people still worked at home this productive activity was distributed throughout the countryside. Mechanization caused work to be concentrated in the new industrial areas, necessitating a corresponding redistribution of the working population. In England, this process was accelerated by many peasants being forced off the old estates and by the enclosure movement. It continues to this day in the developing world, when rural people are forced off the land when their services are no longer required because of the high level of mechanization in agrobusiness farming. The massive uprooting of the rural population and the migration towards the industrial areas created an acute shortage of housing. The pressure to provide accommodation left little time for planning, community involvement, or the putting into place of an adequate infrastructure. The result was that in many cases the new urban environments were simply unfit for human habitation – possibly worse than the shanty towns that now surround many cities in the developing world.

The migration of the rural population to the new industrial centres added to the breakdown of the extended family. It greatly weakened the social fabric of the countryside and recreated nothing of its kind in the industrial centres. With everyone working long hours in factories, there was no one at home to take care of people who were ill, hurt in the frequent industrial accidents, or too young or too old to work. In traditional rural societies, there was little need for a police force. Everyone knew and kept an eye on everyone else. Crime was not anonymous: you knew exactly who you were affecting and what it did to their lives. This acted as a powerful deterrent. The downside of this close-knit social fabric was, of course, that it greatly limited individual freedom and self-expression. Because of this breakdown in the social fabric in the new urban industrial environment, a police force became essential. Many of these social problems required solutions that led to further transformations of the social fabric.

Since the overwhelming majority of the inhabitants of the urban industrial centres no longer produced anything for their own needs, all the necessities of life had to be purchased by means of wages. These were usually inadequate. Since it was common for entire families to hire themselves out for the meagre wages, child labour and infant mortality became enormous problems. Eventually, governments had to step in to ensure the continuity of society through legislation to protect children, by restricting the role of the labour market. The availability of drinking water, the removal of sewage and garbage, fire protection, health care, and security also tended to be entirely inadequate because their provision implied significant changes in other social institutions. In sum, there was a complete breakdown of the traditional way of life.

1.6 Living Together Politically

The adjustments described so far could not occur without significant changes in the political framework of society. Since most of us have lived all our lives within a modern nation state, it is very difficult for us to appreciate how different the political sphere was within a traditional way of life. We have grown accustomed to a certain politicization of life. To satisfy ourselves that this is the case, all we need to do is attempt to locate the limits of the political sphere in society by means of the following exercise. Most of us will find it difficult to draw up a list of important issues that are not political in some sense and that therefore have no political solutions. Our dilemma stems from the fact that the state is directly or indirectly involved in almost all aspects of modern life. Its decisions profoundly affect us in areas such as taxation, education, employment, health care, social services, communication, transportation, defence, retirement, the environment, and much more. The influence of the modern state on our way of life is reflected in our attribution of a very high value to it. Citizens look to the state for solutions to a great many problems. If, as a result, many citizens have difficulty imagining what could possibly be outside the political domain and consequently beyond the reach of the modern state, does this mean that they imagine the modern state to be omnipotent and have few, if any, limits? Could this result in secular religious attitudes towards the state? In the extreme, could this be exploited to create secular political religions? Whatever the answers to these questions may be, the politicization of modern life makes it difficult for us to appreciate how limited and relatively well-defined the political sphere was in earlier societies.

Our situation is in sharp contrast with the political sphere in most traditional pre-industrial societies, which was extremely small in scope and primarily local. In pre-industrial societies, the family tended to be the building block, and it met many of its members' material and social needs. What it could not provide was taken care of via the strong ties it had to other extended families within small local communities. Most events beyond these communities had a minimal effect – whatever happened in a distant political capital was of little consequence. Generally speaking, the local level of political decision-making was more important than the regional or national one. Moreover, the political aspect of many decisions was so enfolded into all the others that it was simply one among many.

All this began to change as industrialization transformed society. I have noted that work became separated from the family, and that production was not for local needs but to satisfy the demands of increasingly national and international markets. Hence, economic decision-making had to occur on a higher political level. Similarly, the building of a transportation infrastructure was incompatible with the primacy of local decision-making. Suppose, for example, that some entrepreneurs were planning to serve an emerging industrial area with a new railroad line. It was a controversial issue. Some village councils welcomed the railroad because of the benefits it was expected to bring, while others fiercely opposed it, fearing its disruptions. It quickly became necessary to have a higher political authority deal with such situations. As the fundamental patterns of development became increasingly national and international, local communities were swamped by forces beyond their control, and decision-making in all other spheres also had to shift to higher political levels. Although the process began with the concentration of production pushing up the scope of economic exchanges from the regional to the national and international levels, almost everything else in society was affected. For example, only the state could protect overseas trade routes vital in the extension of the ecological footprint, create the military force necessary to do so, negotiate economic treaties, and decide on economic and social domestic policy. These functions in turn required a whole range of other activities. No government ever had to get so involved in so many matters at the same time. The emergence of the modern nation state appeared to be an inevitable necessity imposed by industrialization.[47] It happened everywhere regardless of the kind of political regime and the espoused ideology.

People's lives could now be disrupted by events far away. A war between two foreign nations could threaten their jobs if local factories were cut off from raw materials or markets for their products. Their own government could get involved in a trade dispute resulting in higher tariffs that made the export of local products more difficult. Financial markets might influence local investment. Gradually, local life became profoundly influenced by distant events against which individuals or local groups were virtually powerless. They looked to government to defend their interests, which further fuelled the kinds of developments described above.

It is easy to understand the growing importance of the role of ideology in the industrializing societies of the nineteenth century. The daily-life activities of their members could be, and frequently were, affected by events very far away. It was not merely that such events were difficult to understand since they were beyond the horizons of their experience and culture. Perhaps more importantly, they and their communities were powerless to do anything about them. Superimposed on all of this were the difficulties of living the tension between the technology-based connectedness and the culture-based connectedness of individual and collective human life. Ideologies came to the rescue. They had to make these events meaningful and liveable where traditional cultures could no longer do so. Marx claimed that such ideologies originated with the ruling class; however, with hindsight it would appear that the most influential and lasting features were also embedded in his own teachings because they did not critique the cultural myths of his time. This permitted communist parties to turn his work into an ideology in the twentieth century. Capitalism was an inhumane system because the technology on which it was built was privately owned and used to advance the interests of the rich and powerful. This was the bad news. The good news was that this situation was only temporary because capitalism was soon to collapse under its own weight. Technology would then be placed at the disposal of the entire community and would propel humanity into a golden age. In other words, technology could do no wrong, and any problems that occurred were the result of private ownership. Alternative ideologies shared what in the long run turned out to be the most important feature, namely, that technology can do no wrong in the liberal state or when it was guided by conservative or liberal values. Another version of this ideology is alive and well in our modern world, namely, that any problems related to modern technology can be resolved by globalization and free trade: liberate markets and humanity will be liberated. It veils the underlying reality and deflects our attention away from genuine solutions to our problems.[48]

In passing, it should be noted that the growth of the state contradicted the religious and political values of almost every important European tradition of that time. For example, the Jewish Scriptures describe the state in terms of human oppression and loss of freedom. In other words, the state was regarded as an institution that wielded the spiritual power of withholding freedom as the prerequisite for any relations based on love. In the Book of Samuel, when the Jewish people ask for a king, they are warned about the terrible consequences for their sons and daughters as well as the heavier tax burdens they will face.

The Christian tradition, which of course had a tremendous influence on the time period we are examining, carries the same negative images of the state. Marx and his followers also generally regarded the state as an institution that was in contradiction with a democratic socialist society. It is simply impossible to have a powerful and influential state in a society in which people have a democratic control over their collective life. In fact, some Marxists, including Lenin, were concerned about the obvious contradiction that, in the Soviet Union, the size and influence of the state kept on growing, which was incompatible with Marx's teaching. Gradually, all these traditions made their peace with a large modern state as long as they could control or effectively influence it. Despite all conservative rhetoric to the contrary, with minor exceptions they have done little to turn the situation around in favour of more grassroots control. The analyses in Parts Two and Three will further show how industrialization, by weakening the culture-based connect-edness of individual and collective human life, compensated for this loss by allowing the state to play an ever-larger organizational role. Today this is once again accompanied by an ideology of 'freeing' markets through globalization and free trade, which is supposed to encourage individual entrepreneurship and the diminution of the role of the state. The fact that, among the one hundred largest economies in the contemporary world, we find a majority of transnational corporations is all but ignored.[49]

If the political changes described above went against the values and aspirations of so many people, and certainly against the religious and political traditions of the West, we would do well to look for deep structural roots. I believe these are to be found in the wedge that industrialization drove between the technology-based connectedness and the culture-based connectedness of human life and society and the corresponding shift from the cultural approach to life to the economic approach. Within the emerging economy, the Market ruled over the technology-based connectedness at the expense of the role of culture. The invisible hand represented the 'rule' of money taking over from the 'rule' of culture. The former was guided by the overriding concern of strengthening the technology-based connectedness in order to increase profits and accumulate capital. If people were to entrust their lives, communities, and future to capital, then capital had better fulfil all their values and aspirations. As we have already seen, this was impossible because the Market externalized the culture-based connectedness of society and the biology-based connectedness of the biosphere. The deficit had to be made up by something else, and the state acted as if it could step into this role. Beyond the economy, the spillover effects of the strengthening of the technology-based connectedness at the expense of the culture-based connectedness so weakened the self-regulating character of culture-based activities that an increasing number had to be externally regulated. Once again, the state stepped into this role. The state and the Market thus sought to take over an ever-larger part of the role culture had traditionally played in a society. Nevertheless, the new 'system' could not function without a role for culture, but this required additional transformations. For example, if people had to surrender their humanity in technically divided work, it had to deliver something of equal value in return, and this had to be assured by the 'system.' Since the latter was focused on delivering ever-larger quantities and varieties of goods, these had to have the potential of doing far more than simply meeting a few human needs. A cultural mutation not only made this self-evident but provided the ultimate legitimation and justification. Hence, the political changes I have mentioned point to the need to further relax our assumption so as to include changes in the role of culture as a consequence of 'technology changing people.'

1.7 Living with the Law

Thus far, I have analysed the process of industrialization and the accompanying transformation of society, beginning with the network of flows of goods and services and putting this in an ever-larger context beginning with the economic context, the social context, and the political framework. These adjustments presented the legal tradition and institutions with a diversity of new situations that could rarely be dealt with by minor extensions of existing laws. This created a considerable legal challenge, compounded by a gradual but fundamental change in the values of society.

The transformations previously described created many situations requiring changes in existing laws, as well as entirely new ones. For example, who was responsible when someone was hurt on the job in a factory? Did employees have any legal claim against their employer in such cases? What were the responsibilities of employers in the domain of occupational health and safety? With the breakup of the extended family, what were the responsibilities of parents for their children? Did the employer or the parents have the greater authority over children working in a factory? When everyone in a family went out to work, who was responsible for those who because of an injury or old age could no longer work?

When peasant society in the countryside broke down as the migration to the urban industrial centres began, the resulting social situation was initially characterized by an extremely weak social fabric. Many social relations became altogether dysfunctional. In this somewhat anarchic situation, who was responsible for social order? When extended families broke down and some members were abandoned, who should take care of them? Should it be the parents, the employer, the parish, or a social agency created by local government?

In the economic sphere, a variety of legal issues also arose. When the role of banks had to be expanded to ensure ever-larger pools of capital, what were the obligations of the banks towards depositors? What were their rights and obligations towards factory owners in the case of a bankruptcy? The creation of the company with joint ownership and limited liability was an extremely important one because it provided larger pools of operating capital. Of course, many details had to be sorted out legally. Under which conditions could a corporation sell shares? What kinds of shares should be permitted? What were the rights of shareholders in a variety of circumstances? What restrictions, if any, should be placed on the trading of shares, and which institutions should be permitted to act as brokers? What should happen in the case of bankruptcy?

Technological development presented its own legal problems. Patent legislation itself was an important legal invention to resolve a social dilemma created by the tension between the rights of the inventor and the potential benefits to society. Given the central role inventions played in keeping the process of industrialization going, it was clear that, given the ever-greater stakes involved, people were willing to fight longer and tougher battles over all kinds of nuances in the current legislation. What rights did employers have over inventions made by employees in their own time, or in an area unrelated to their employment? What rights did an employer have when an employee resigned and promptly applied for a patent? When did one patent infringe on another? The political changes that accompanied industrialization also required a legal foundation. What powers did local or regional governments have to expropriate private property for the building of a railroad or the digging of a canal? What kinds of financial instruments should be available to different levels of government to finance large public projects required by industrialization? When an entrepreneur decided to build a factory in a particular area, thus attracting employees and their families, were they responsible for the necessary infrastructure, the supply of water, and sewage disposal? Should factory owners pay a portion of the social services required in the neighbourhoods where their employees resided?

Examples such as these can be multiplied almost indefinitely. It is clear that during this time of massive and fundamental upheaval, many legal problems arose for which there were no precedents and to which a legal tradition could not easily be applied. However, the new laws and the many revisions of existing ones faced another important challenge. What causes a law to be spontaneously obeyed? In a free and democratic society, a law that fails to be spontaneously obeyed becomes inapplicable. The reasons are obvious: the use of force on a massive scale is open only to a totalitarian regime, and it simply isn't practical to incarcerate a substantial portion of a population. Why, then, are most laws spontaneously obeyed in a free and open society? Under which conditions may they become inapplicable? For now, a simple answer will have to suffice.

When the habitat of a population changes substantially, as in the case of a rural population migrating to an urban industrial centre, their traditional values and way of life are of limited use in the many new situations that need to be symbolized. At the same time, the entrepreneurs and factory owners strongly internalize many new values closely associated with the new urban-industrial milieu and the 'system' they are helping to build. As a result, there is a sort of coexistence between two 'societies': one based on a way of life that is decaying, and the other based on a way of life that is emerging. The situation cannot be understood simply in terms of social classes, or social strata. The differences are much deeper than socio-economic ones. What may appear as reasonable laws in the eyes of one group may not appear to be so in the eyes of the other, because of the fundamental differences between the values underlying the two different ways of life. This presents difficult challenges to lawmakers, law enforcers, and governments and disturbs the very foundations of any legal system, be it a common law or a codified one. Until the emerging way of life is able to give meaning, direction, and purpose to the lives of all members of society, governments and factory owners frequently resort to force in the face of massive disobedience of new laws judged to be unfair by those who work in industry. It must be recalled that the laws of a community are spontaneously obeyed only when they correspond to the metaconscious values implicit in the brain-minds of its members.[50]

As a consequence, fundamental changes took place in the legal systems of all industrializing societies. In the case of a common-law tradition, the state had to intervene and play an ever-growing role in its evolution. Marx's accusation that the law was becoming a means for organizing a society controlled by the state acting in the interests of the rich and powerful was hardly without foundation. At the same time, it entirely overlooked the important constraint lawgivers face in a democratic society: for the law to be spontaneously obeyed, the distance between any new laws and the values embedded in the minds and cultures of the members of a society cannot be too great. Consequently, the interpretation of the law as an instrument of the ruling class is far too simplistic. For societies in which the state had already taken charge of the common-law system and converted it to a codified system, the tremendous legal challenges that accompanied industrialization simply confirmed that the new role of the state in the evolution of the legal system of a society was more than a simple necessity and represented the way of the future.

The way in which contemporary people live with the law also has its roots in the transformation we are describing. The growing pressure exerted by the technology-based connectedness on the culture-based connectedness necessitated an increased reliance on an economic approach to life and a correspondingly weakened reliance on the cultural approach. As we shall see in the next chapter, legal institutions had generally embodied the cultural values of the group or society. The weakening of these values led to a legal crisis, necessitating a change in the role legal institutions played in human life. Wherever the cultural order broke down, the law had to step in and create a measure of order, but this order was of a different kind. The full implications of this development did not become apparent until the second half of the twentieth century, when, in some industrialized societies, there were more lawyers than engineers – a clear symptom of the inability of cultures to play their traditional roles. These legal adjustments were both cause and effect of the political changes described earlier, including the cultural mutation resulting from 'technology changing people.'

1.8 Disconnecting from and Reconnecting to the Earth

Although the analysis of the process of industrialization is far from complete, some of its generic features are already apparent. The activities comprising the way of life of any society are linked together in two interdependent ways. A culture-based connectedness stems from their participation in a coherent way of life that gives meaning, direction, and purpose to the members of society. A technology-based connectedness stems from their participation in chains of activities through which matter and energy are temporarily borrowed from the biosphere. Industrialization has greatly added to the length and interconnectedness of these chains as compared to ways of life based on hunting and gathering or agriculture. There the chains were generally confined to local ecosystems and were only occasionally extended by distant trade. The principal exceptions were the centres of empires, whose ecological footprints frequently included remote hinterlands connected by militarily protected trade routes. In other words, the technology-based connectedness of a society reveals the moorings of its way of life in the biosphere.

What has been accomplished thus far is a generic examination of how the process of industrialization systematically changes the moorings of a way of life in the biosphere. The analysis began with a materials focus that explored how industrialization affects the network of the flows of goods in a way of life. The initial assumption was that all other aspects of society would remain the same. This was gradually relaxed to consider the people and some of the institutions required to produce these flows. The fundamental pattern that was uncovered may be summarized as follows. The mechanization of work produces a local disturbance in the fabric of a society, which can be resolved in one of two ways. Either the attempt is abandoned and things are restored to their original dynamic equilibrium, or additional adjustments must be made. What will happen depends greatly on the ability of prevailing conditions to 'supply' the material, socio-economic, and cultural 'prerequisites' and to further contribute to a favourable climate. The importance of this (to be examined in chapter 3) cannot be overstated because if any adjustments are made new disturbances are created in adjacent areas of the fabric of a society as the original ones are resolved. The process must continue until the entire fabric of a society becomes transformed as all aspects adjust to one another to create new patterns for human life.

Although the technology-based connectedness provides the dynamism for the industrialization process, it fundamentally depends on the culture-based connectedness without which the lives of the members of society would be without meaning, direction, or purpose. For this reason, the model of the process of industrialization constructed thus far is incomplete. At first, industrialization greatly increases the tension between the technology-based connectedness and the culture-based connectedness of a society. However, as the former gains ground, the latter is profoundly affected by what has earlier been referred to as technology changing people, including their cultural moorings in reality. After all, the changes accompanying industrialization are experienced by the people of the time, thus affecting their consciousness and culture. In particular, the influence on the moral, religious, and artistic aspects of a society and on the integrity of its culture remain to be analysed. Examining how industrialization changes the moorings of a society in the biosphere deals with its interaction with natural evolution but not with the way a society makes history.

It is essential to recognize the prejudgments we bring to the task of making sense of industrialization. These stem from our vantage points as people of our time, place, and culture, which are profoundly marked by industrialization. This too will be examined in some detail in a later chapter, but for now it is essential to be aware of the fact that these vantage points put the economy of a society centre stage. Nevertheless, people do not live by bread alone, life is more than work, and history has a meaning. Our vantage points normalize the fact that the economy became relatively distinct from the remainder of individual and collective human life: production steps were separated from the whole of human work; work became separated from the extended family; individual technologies were removed from the mainstream of daily life, to be incorporated in ensembles and structures housed in factories and industrial centres; values became disconnected from human life; and market transactions became divorced from the remainder of society and the biosphere. Gradually, everything had to be adapted to machines: the activities of daily life became synchronized to them and the technical schedules they imposed. Food was reconstituted so that it could be processed and transported by machines; the urban habitat had to be accommodated to machines for transportation; social relations supported by machines were favoured over those that did not; human needs were transformed into market demands that could be satisfied by machines; farms were reorganized to accommodate machines; human values became the expression of necessities imposed by the technology-based connectedness instead of human sentiments; and a great deal more. This is symptomatic of technology-based connectedness impinging on culture-based connectedness, gradually leading to desymbolization and a weakening of meanings and values without supplanting culture altogether. We as modern people have had to learn to make history in new ways in order to make sense of and live in a new world brought about in part by industrialization. It is essential not to impose our cultural vantage points on a time that was very different despite the growing influence of industrialization.

The process of industrialization has some of the same properties as that of a chain reaction. Initially, events may occur somewhat randomly, disturbing the 'molecules' in the structure of a society, triggering further disturbances, and so on. All of these remain linked through the fabric of a society and its dependency on the biosphere and will jointly produce the kind of transformations described in this chapter and the next. The sequence of events in time can vary considerably and yet produce very similar transformations. For example, England had an abundance of highly skilled craftsmen and a shortage of natural resources, while in North America the opposite was the case. It resulted in different symbioses between people and machines in industrial organizations, the American system eventually proving more successful.[51] Germany drew more heavily on a pool of scientifically trained craftsmen, technicians, and engineers than did England, giving it a substantial advantage in those industries whose knowing and doing were limited by a weak correlation between what could be observed and what was happening, thus restricting what could be achieved by a technical knowing and doing dependent on an apprenticeship system.[52] The industrialization of the manufacture of luxury goods sometimes took a different initial path based on machines extending the capabilities of highly skilled craftsmen rather than displacing them as much as possible. However, this depended on networks of cooperating small enterprises which, in the long term, did not survive because they relied too heavily on socio-cultural conditions that were eroded as a result of 'technology changing people.'[53] In North America, industrialization occurred on what was treated as a 'socio-cultural green field,' while in western Europe it occurred on a 'socio-cultural brown field.'

In the twentieth century, industrialization in many countries could not establish a new dynamic equilibrium for a variety of reasons, including the lack of various 'resources' including a supportive cultural framework. Does this mean that the model of industrialization I am building presupposes a technological determinism? Not at all. However, it does confirm that, seen as a joint process of 'people changing technology' and 'technology changing people,' the force of the latter greatly constrained the outcomes, leading to the universalization of

Western technology and values. A society is a multidimensional tissue of relations that is rewoven each and every day by means of the experiences and activities of its members, guided by a culture. The growing technology-based connectedness and its influence on the culture-based connectedness led to a convergence of historical pathways, making possible such concepts as industrial society and our common future. A society can begin the process of industrialization anywhere in its network of the flows of goods involved in its way of life. Because this *is* a network itself embedded in the structure of a society, in turn embedded in that of the biosphere, this should not be surprising. This will be further developed in subsequent chapters.

1.9 Some Implications

The reverse of the process of industrialization can occur when large technological systems break down and disintegrate. Roberto Vacca, an engineer, was preoccupied with this possibility.[54] He argued that our technology-based society is so highly interconnected that we are exposing ourselves to immense risks. He pointed to major breakdowns that have occurred in large systems, concentrated in the north-eastern part of the United States:

1. In 1965, an instability in the network for the distribution of electrical power in New York, New England, and Ontario caused some regions to be without power for up to fourteen hours in an area occupied by thirty million people. In New York alone, six hundred thousand people were stranded in the subway. No vehicles could refuel because the electric motors at the gas pumps did not function.
2. In 1970 a combination of circumstances caused the breakdown of the Penn Central Railroad system, which served, among others, the cities of New York and Philadelphia. Of the 413 trains that normally ran, 117 did not run at all, and the ones that did run suffered major delays.
3. In 1969 an unexpected increase in demand for service in the New York telephone system, combined with maintenance problems, caused an almost total stoppage of one of its automatic switchboards. For several days, it was virtually impossible to obtain a free line in that area, and for months some subscribers encountered long delays or had to abandon the use of their telephones altogether.

Vacca argued that although the probability is extremely small, a coincidence of several such breakdowns could occur. In fact, the breakdown of one system could increase the demand for another and indirectly impair the functioning of still others. He described a plausible sequence of events, in which one event triggers another to produce a major disaster that could plunge society into a complete breakdown of law and order. Since many technological support systems are essential in the modern highly populated urban areas, their breakdown can constitute a fundamental threat to life in such areas. Because of a huge investment of time and resources, the Y2K problem was largely avoided, but it reminded us once again of the fragility of our systems. In any case, a technology-based connectedness is not, as we will see later, self-regulating the way culture-based connectedness was prior to industrialization.

Charles Perrow has shown that complex socio-technical systems have a propensity for what he calls 'normal accidents.'[55] During these, separate human, organizational, and technical factors that may be trivial in themselves link together to alter such a system in a way that is unpredictable to its designers and operators. As a result, some systems create a fundamental conflict with an open, democratic society. His theory explains a new category of risk that can compel societies to surrender some of their values and aspirations in order to control these risks.[56] It is not yet clear whether the events of 11 September 2001 will finally lead to some serious consideration being given to the fragility of modern societies that depend on a high level of technology-based connectedness involving complex socio-technical systems. On that day, the targets were symbolic, the symbols of global military might and trade. Imagine what could happen if religious or political fascist groups sought to create normal accidents or a chain-reaction process in reverse. There would be little choice but to turn democratic societies into police states in the name of protecting the public or of protecting democracy or freedom (what little meaning this can have in a society with an extremely high level of technology-based connectedness). The only alternative is to completely rethink the role technology should play in a society that wishes to remain open and democratic. It would appear that governments have already made up their minds for us.

The warnings of Vacca and Perrow, as well as those of authors such as Amory Lovins, who drew our attention to the risks of relying heavily on plutonium breeder reactors,[57] will perhaps be taken more seriously after the events of 11 September 2001. What will happen if terrorist groups use a strategy of maximizing the number of victims by taking advantage of the fact that contemporary societies have not heeded the warnings of the authors mentioned above? What will happen if the 'engineering' of a number of normal accidents triggers a chain reaction in reverse? Will this lead to rapidly pushing the possibilities of information technology to monitor all people to its limits, thereby endangering what little democracy is left? In Parts Two and Three, we will examine how the technology-based connectedness continues to rapidly expand, with all the risks to Western values that this entails.

From our consideration of the process of industrialization thus far, it follows that the transfer of technology from one society to another can be fraught with difficulties. This is particularly true when the technology is transferred from a so-called industrially advanced society to a developing nation. Our earlier story of the engineering student inheriting the Ferrari engine can once again help us understand the problem. If the student had made only some of the adjustments, his vehicle would not have been able to make full use of the new powerful engine. Depending on which adjustments were made, the car could even be dangerous. When it comes to technology transfer, foreign aid, and interventions by the World Bank or the International Monetary Fund, we continue to show that we have not learned the lesson from the above story. Technology transfer is fraught with risks because the relationship between technology-based connectedness and culture-based connectedness is different in the industrially advanced and the developing worlds.

Our understanding of the process of industrialization can also shed light on the problems faced by the developing nations. Regardless of where it begins in the socio-cultural fabric of a society, it will produce severe local disturbances. When these are resolved by further rationalization and mechanization, new disturbances occur that can be understood and predicted if the structure of the system is known. As these changes gradually transform the daily lives of the members of the society, the newly internalized experiences link up into new metaconscious patterns in the mind. In the course of many generations, these changes become so extensive that the cultural roots of a society begin to shift, further reinforcing the developments that led to that change in the first place. Historians of technology have largely ignored these deeper accompanying cultural changes. This has contributed to a serious lack of understanding of what is happening in the developing world. England was able to reach a new dynamic equilibrium by entirely restructuring its traditional socio-cultural fabric, because of the historically exceptional conditions that prevailed within it on the eve of the Industrial Revolution. Its socio-cultural system was capable of furnishing the massive material and cultural resources required for its total restructuring. In the developing world such conditions generally do not exist. The influx of material and cultural elements during the discovery journeys, the colonial period, and the continuing transfer of technology after independence set off the process of industrialization; but the new dynamic equilibrium typical of an industrial society often cannot be reached because of the lack of some material and cultural resources. These societies no longer have a cultural integrality because their traditional and industrial elements are incompatible. Their situation is fundamentally unstable. Much of what happens in the developing world can be interpreted as the social, economic, political, and religious symptoms of this instability. These countries are stuck somewhere in the chain-reaction-like process of transformation, unable to go forward or backward. The serious erosion of their traditional cultures, which have not been replaced by new ones, undermines the meaning, purpose, and direction that the people need for their lives. As a result, these societies are vulnerable to external manipulation by the transnationals and foreign interests. Much foreign aid, interventions by the World Bank, the dictates of the International Monetary Fund, the actions of the World Trade Organization, and the transfer of technology often aggravate the situation, because these actions are not compatible with the socio-cultural structures of these societies.[58]

chapter 2

Industrialization as 'Technology Changing People': Disconnecting from and Reconnecting to the Gods

2.1 Symbolization and Cultural Moorings

'People changing technology' by means of the technical division of labour, mechanization, and industrialization separated the technology-based connectedness from the culture-based connectedness of human life and society. It was impossible to change this technology-based connectedness, moored as it was in the earth, by respecting the culture-based connectedness, moored in the traditional gods. As a result, on the heels of 'people changing technology' came 'technology changing people.' People internalized the many changes in their world, which gradually led to a mutation in the organization of their brain-minds and, through it, in their culture. In the course of generations, this committed everyone to the new way of life and to new moorings to their gods. This development will now be examined in some detail by means of a theory of culture according to which culture acts as the symbolic way in which the members of a society individually and collectively make sense of and live in the world. This theory examines how babies and children begin to participate in the culture-based connectedness of their society by 'growing' the organization of their brain-minds with symbolized experiences. It also examines how societies evolved their ways of life prior to industrialization. The theory will then be used to study how all this changed as a consequence of 'technology changing people.' This analysis will focus on the culture-based connectedness of human life and the world. (Readers who wish to postpone the detailed review of my theory of culture can skip sections 2.3 and 2.4, and come back to them before reading Part Three.)

As long as the culture-based connectedness remains strong, the possibility of living a life can be taken for granted, much as we could take for granted that human cultures created technologies that were appropriate to them and that these helped create ways of life that were for the most part sustainable by local ecosystems. Each of these developments is part of a similar pattern. When something can no longer be taken for granted because it is weakening, in danger of disappearing, or for some other reason causing difficulties, that 'something' must be named and symbolized in order that we can think about it, discuss it, and act on it. Here, that 'something' is the culture-based connectedness of individual and collective human life that makes living possible. It comes under pressure from a process of desymbolization, which is undoing a great deal of what was accomplished by humanity's first two megaprojects.

We have seen that the culture-based connectedness of a person's life can be seriously weakened by short-term memory disorders or the early phases of Alzheimer's disease. Attempting to understand the lives of people facing these difficulties is possibly as close as we can come to imagining human life without symbolization, that is, human life lived on the level of immediate experience. Contrast this with the way most people live the culture-based connectedness of their lives. Let us try to understand a moment in the life of a student listening to a lecture. To make sense of her experience, we would have to understand how it is embedded in her life as a whole, as well as how that life is interwoven into the lives of others. For example, she is attending the institution with some reluctance. Her parents strongly urged her to go to medical school because, as immigrants, they never had the opportunity. It also offered a chance to get away from home and from further involvement in the family business. It had prevented further straining the relationship with her parents, which began when she went ahead with a friendship in high school of which they had strongly disapproved. Nevertheless, they continue to show their love despite the many clashes, and she respects them for their hard work and their success in building the family business. More specifically, she does not really like this course but, since it is compulsory, she cannot drop it. There is a student she has been trying to meet for some time in her attempts to make friends in the large and impersonal institution, but three seats away is as close as she has been able to get to her today. It is difficult to concentrate because this morning she has received a letter from a friend who is having a hard time and is experimenting with drugs. The seat is uncomfortable and bothers her sore muscles, a result of practices for the hockey team. It is also difficult to concentrate because she is very hungry after spending her lunch hour in the library completing an overdue assignment. Thus, this moment enfolds many threads that together weave the fabric of her life. Its patterns are related to her life as family member, student, lover, friend, and member of the hockey team. If someone were attempting to write her biography, in order to make sense of any one event in her life he or she would have to relate it to various subplots that weave in and out of one another in the continuity and integrity of her life's narrative.

Each of these subplots is shared with other people as they become a part of her life, and she of theirs. She is a big part of the lives of her parents, her friends, and her lover. To understand some of these subplots requires a knowledge of the lives of her parents. Some aspects may be deeply embedded in her family's history, sometimes going back several generations. Some subplots may not coexist harmoniously as they pull her in different directions, creating tensions and contradictions. In the extreme, she may almost be different persons in some of these subplots. Hence, living a life involves symbolically enfolding something of the lives of others and, through them, something of the habits, manners, customs, beliefs, and values of the culture of society.

Examples like this one illustrate that individual lives, as well as the collective existence of a society, have a unique integrity, thanks to the symbolization of each experience as a moment of a person's life shared with others. Each life weaves a tissue of relationships with a multitude of complex patterns that overlap and interpenetrate those of others, since any social relationship involves at least two people. All of us experience the continuity and integrity of our own being and those of other people. Many of our experiences bear this out. Meeting an old friend I have not seen since I moved away ten years ago, I may exclaim: How you have changed! And yet I had no difficulty recognizing this person. Our personality is embedded in the way we approach and do things. This is evident in a situation in which I say of a close friend: I never thought you would do such a thing! All this reflects a certain integrity and continuity of being in our lives. Indeed, we live lives and not a random sequence of only weakly related bits and pieces of happenings. I am not, however, implying that we know exactly what the patterns of our lives are or that such patterns can somehow make life predictable.

Because people live lives with a measure of integrity, it is possible to write biographies. To be sure, any such narrative is complex, with many subplots that weave in and out of the story. Furthermore, the complexity of any life is such that different writers could produce very different accounts as they select different events, subplots, and aspects as particularly significant, characteristic, or fundamental to later events. However, if each one does a thorough job, the narratives are likely to complement one another.

A society's collective being also has an integrity. Despite constant change, there is a continuity that can be narrated as the history of a people. In such a narrative, there are many beginnings and endings as individuals are born, grow up, live their lives, and die. Yet they are a part of a larger story that we recognize when we say things like: He is a real Frenchman! We are all born into a culture's design for living in a society we did not create. It existed before us and will continue after us. We play subordinate parts in the lives of others as they do in ours, and in so doing we change something in that design as we assert ourselves with a measure of freedom and autonomy. The result is that our individual and collective stories are unpredictable. They do not follow a set of laws or models as encountered in nature. Yet our individual and collective lives remain recognizable despite the many changes, because of the continuity of their culture-based connectedness. That is why a historical narrative is possible, although it is lived before it can be told or scientifically examined.

Being born and growing up are integral to the culture-based connectedness of individual and collective human life. Death ends that participation, with the result that some portions of the culture-based connectedness of human life will no longer evolve. It is as if something in the lives of those left behind also dies, and this adds greatly to their sorrow and sense of loss. People's being in the world is enfolded into the culture-based connectedness of individual and collective human life, which precedes and continues after them in ways that may or may not be affected by their lives. In societies with high levels of mobility, this dropping in and out of the lives of others is a frequent occurrence, be it with a lesser finality, since old acquaintances or friendships can be renewed. In all of this, no one can be a detached observer since our awareness of ourselves and the world is integral to this culture-based connectedness of human life.

The culture-based connectedness of individual and collective human life has its own moorings. This may be illustrated by the following metaphor. One expression of the meaning and value the members of a community give to everything in their experience is a dictionary. Suppose a dictionary found its way to another galaxy, to be examined there without any possible reference to the contexts out of which it came. The structure of the dictionary would then appear somewhat arbitrary because it is circular. Each sign is defined in terms of some others, which in turn are defined in terms of still others, eventually including itself. It would thus appear as a relatively arbitrary structure of signs. Although everything is related to everything else in that structure, there is no way of knowing why the meaning of any sign is what it is, having been reduced to a relatively arbitrary position within the structure. Since this structure is circular, the meaning of any sign will not be grounded. This metaphor is not entirely frivolous since it essentially represents the box into which structuralism enclosed itself with regard to human language. In other words, the dictionary ultimately makes sense only in the context of the people who have lived the reality symbolized in it. However, they might begin to experience the relative character of their way of life in the face of the unknown or through contacts with other cultures, were it not for some additional cultural elements that reinforce the culture-based connectedness of individual and collective human life.

Such cultural elements eliminate the threat of the unknown by symbolizing it as an extension of the known. As a result, future discoveries and experiences will simply elaborate a people's culture and way of life without undermining it. At least symbolically, the community is now in touch with all of reality. The cultural world of the society becomes closed to what is 'wholly other' in relation to its way of life, and its history becomes a cyclic or linear process of more of the same. All of this has occurred only symbolically, so that new developments or discoveries that are non-cumulative with respect to the absolutized way of life become destabilizing intrusions. When they become too significant, a process of de-absolutization or desacralization occurs and a new process of absolutization or sacralization begins as the current historical epoch of a society comes to an end and a new one begins, through a process which fundamentally changes the way everything is contextualized in relation to everything else.

It goes without saying that all this raises more questions about the culture-based connectedness of human life than it answers. Is the relationship between each moment of my life and that life as a whole somewhat analogous to the integrality of my physical self, where each cell in my body has enfolded within it the DNA that serves as a biological blueprint for the whole of my physical being? My DNA enfolds something of my parents, and their DNA enfolds something of their parents, creating a tree-like structure of genetic relationships that ultimately links me to all of humanity and perhaps beyond to other life forms. The DNA includes a blueprint of my brain, and, since this brain is in a small way modified by each and every experience, each cell also enfolds something of the interaction between natural evolution and cultural history. My DNA can help produce new cells to replace old and worn-out ones, thus ensuring my physical continuity even though all my body cells are replaced at least every seven years, except for the brain cells. Can the analogy shed light on how my physical continuity enfolds the continuity of my whole being as a person of my time, place, and culture? Can this analogy be extended to account for how the physical integrality of my being is connected to so much else in the world, past and present? How can I extend such an account to include the integrality of my whole being?

Elsewhere, I have attempted an account of the culture-based connectedness in human life within traditional, pre-industrial societies.[1] It showed that we can think of culture as performing a function for the group or society somewhat analogous to DNA. Each moment expresses something of my person and my life, which in turn expresses something of my time, place, and culture. There exists a dialectical tension between how I affect that moment and how it, in turn, affects me. It is as if each experience has embedded within it a 'cultural DNA' as a blueprint of who I was, am, and wish to be, expressed through the culture and way of life of my society. Even a routine activity like handwriting reflects something of my personality and culture.

The dialectical tension between the way I affect a moment of my life and the way that moment affects me may lead to crises in the integrality of my life. Such crises can occur on any one of three interdependent levels related to the reciprocal relationships with my surroundings. On the physical level, I depend on a constant throughput of matter and energy to live and to repair or replace the cells of my body. This borrowing of matter and energy also constitutes a body-load (by containing toxic substances, bacteria, and viruses), which, if it exceeds my immunological resources, can threaten my physical integrality and make me ill and, in the extreme, cause my death. On the social level, my nervous system and psychological well-being depend on a reciprocal exchange of experiences as well as the nurture, love, and support of others. Excessive monotony, sensory deprivation, an upbringing in isolation or by animals, the withholding of nurture and love during infancy, a lack of emotional and social support, a want of love or belonging, dilemmas from which there appear to be no way out – all these and more can impair the development of my social self and consequently my relationships with others, which, in the extreme, may lead to substance abuse, a nervous breakdown, mental illness, or even suicide. On the spiritual level, I belong to a community that cannot know all of reality and hence must live as if the reality it has come to know is reality itself. Moreover, this community does not know 'the good,' and hence must live as if the very best thing in our individual and collective human life were 'the good.' Together with my community, I am thus plunged into a spiritual world of illusion and make-belief elevated to absolutes. Anomalous experiences, contacts with other ways of life, and socio-economic or political crises can, in the extreme, lead to anomie or suicide in my own life, and to a decline or collapse of the group or society to which I belong.

Crises on one level can lead to crises on the other two levels: toxins can affect my personality and social behaviour, disease can make it impossible for my body to respond to me, effectively cutting me off from others and the world, and malfunctions of the brain-mind can reduce my life to little more than isolated moments of existence. In the course of my life, I encounter many situations that produce contradictions between particular moments and the rest of my life, which may not lead to crises but which result in the integrality of my life embodying a constant tension between it and the disorder from within or without. Living my life involves a constant weaving and repairing of relationships in the face of illness and death. All three levels come together in the way I have learned to symbolize my life in the world by means of my community's culture.

The reciprocal relationship between myself and my surroundings compels me to live my life in a dialectical tension between freedom and alienation, depending on whether or not the influence I exert over my surroundings is greater than the influence my experiences of these surroundings 'possess' the organization of my brain-mind and thus my life. In the same way, I live the tensions between health and illness or death, work and life, moral integrity and chaos.

This integrality, inadequately thought of as 'cultural DNA,' can help us understand much of human life that falls beyond science. It can help us understand concepts such as the spirit of an age, an idea whose time has come, and the history of a people. It can help explain why great art is recognizable across the centuries and across cultures, because it encapsulates something of the human condition that resonates within us. It can help explain the diffusion of jazz, because this music embodies something of the condition of slavery and speaks to the members of a society in which rationality has created what Weber called the 'iron cage.'[2] It explains why some poetry and literature endures while some does not, in terms of how well they reflect the cultural soul of a community to whom the writer belonged and how deeply they touch that of the community to which the readers or listeners belong. All this makes sense in terms of experience and culture but falls beyond the reach of scientific explanation. It also falls beyond much of Western philosophy, which, because of its asocial and ahistorical character, took on an anti-life orientation.

Human beings approach the world with a prejudgment, prejudice, or bias about what it is to be human. First, there is a connectedness to natural evolution through the human body in general and the brain in particular. Second, there is a connectedness to cultural evolution and history, which embody the experiences of countless generations and are internalized by individuals as structures of experience that constitute their minds. Third, there is the connectedness between the two, resulting from the interaction between natural and cultural evolution and between the brain and the mind. At birth, this connectedness is embryonic in nature, sufficient to allow babies to get started on their lives by experiencing something of what is mostly a meaningless and valueless blur, and to act by means of a physical embodiment in this blur.[3]

The development of the organization of the brain-mind appears to involve three processes that are an outcome of natural evolution, cultural history, and their interaction.[4] All three processes are dimensions of symbolization. The first is the ability to generalize an experience. Once an experience has been symbolized by synaptic and neural changes, it is easier to live similar experiences. In this sense, it acts as a paradigm or exemplar. When additional experiences suggest fundamental differences, a second process, namely, differentiation, delimits the generalization of any experience by distinguishing it from all previous experiences that turned out to be fundamentally different from one another. Neither process requires any analysis of what the similarities and differences are. Together these two processes permit people to live in a world that is ultimately unknowable by giving everything a place in the culture-based connectedness of individual and collective life, which is experienced as its meaning and value. A third process, integration, generates a great deal of metaconscious knowledge by symbolically mapping all experiences into a structure of experience. It is somewhat analogous to going beyond the experimental data in a scientific experiment by fitting a curve through them. There is a great deal of evidence suggesting that this metaconscious knowledge is implicit in human behaviour. It generates feelings and intuitions about the deeper patterns affecting individual and collective life.

All three processes are metaconscious, in the sense that they involve going beyond individual experiences in ways to which we have no conscious access. Together they develop a symbolic structure of experience from a highly limited organization of the brain-mind at birth in a manner that is analogous to the growth of an embryo, in which cells continually divide and differentiate to form the constituent elements of the body, such as tissues and organs.[5] The organization of the brain-mind does not literally contain memories, models, paradigms, or processes of differentiation and integration. For example, memories are symbolized by synaptic and neural changes that they produce in the organization of the brain-mind. Similar experiences will activate the same parts of this organization. Differentiation occurs when this activation pattern needs to be expanded to symbolize what are now being lived as very different experiences. The process of integration symbolically integrates experiences through the making of synaptic connections, changes in these connections, and changes in the behaviour of neurons. Little or nothing is known about the symbolic functioning of the brain-mind. However, symbolization is integral to the living of a human life.

A structure of experience symbolizes a person's life as lived in a particular time and place via his or her experiences differentiated in a culturally unique way. It may be conceptualized as a tree-like structure connecting various wholes. The smallest wholes contributing to an experience are sensations. These are the elements of experience that correspond to the constituents or processes of the external world within any dimension of experience corresponding to the five senses. Other dimensions contribute sensations from the inner world, such as being hungry or sleepy, feeling sick or anxious, and the things we think or imagine. Many of these dimensions of experience integrate sensations into gestalts, placing some in the foreground and others in the background according to human intentions and decisions. In turn, these gestalts are integrated into a single overall experience of a moment of our life, with a foreground-background distinction according to how we live that moment. (For example, a student sitting in a classroom may focus on the facial features of his professor, what she is lecturing about, a peculiar spot on the wall, the scent of a neighbour's perfume, a letter received that morning, what groceries to buy on the way home, and so on.) Each experience takes on its meaning and value by systematically being placed in relation to all others within the structure of experience, therefore symbolizing a person's life as shared with other people. These people constitute various groups, which contribute to the fabric of a community that is part of a society functioning within a civilization and the biosphere. Since each entity is enfolded in the next smaller and larger wholes, its meaning and value are multidimensional and frequently dialectical. These connections reflect the way of life and culture of the society the person belongs to, and in time make history, which depends on natural evolution. Culture tries to get at the meaning and value of everything lived in individual and collective human life by placing it in the context of everything else.

The living of individual and collective human life requires going beyond each experience of a moment of a person's life, which is made possible by the meta-experience of his or her structure of experience. So fundamental is this meta-experience that it may be regarded as the symbolic commons. It symbolizes the connectedness of an individual's life with the lives of people with whom he or she shares a great deal, and with those of other people via relations based on social roles created and sustained by a way of life. The resulting clusters of relations recreate on a daily basis the customs, traditions, and institutions of a society.

Each relation involved in this connectedness is valued in the context of the others, creating a hierarchy of values anchored in a cultural unity, which, when a society is healthy, minimizes the risk of anomie, relativism, and nihilism. Thus, the members of a community know what is expected of fathers, for example, and what makes for a good or a bad father; they know what is socially acceptable and what is not; what is important for their lives and for their community and what is not; and what constitutes the good life and what does not. In other words, the symbolic commons includes a moral guidance for life. Children learn to spontaneously obey the laws of their society, not by reading them or being instructed about them, but by developing a 'meta-law' in the form of a hierarchy of values within their structures of experience. This marks another important step towards the acquisition of the symbolic commons of their society.

2.2 Symbolization and the Life-Milieu

A human society introduces into nature an artificial or cultural sphere that can neither be absorbed into nature nor detached from it. Human beings do not live in a natural niche. Societies, by means of their cultures, introduce into nature the possibility of an element of choice and free will, which gives rise to contradictory elements in human experience. On the one hand, human beings experience a distance between themselves and the world, obliging them to make decisions regarding their existence in it, which presupposes a certain freedom. On the other hand, they constantly experience how much they are determined by the world. Faced with that contradiction, human beings have rarely exercised their freedom directly, in the face of necessity. Instead, they have done so through symbolization, never accepting things for what they are on the level of immediate experience. In the face of necessity, human beings have surrendered their freedom for a variety of reasons, such as security, comfort, or wealth. Human groups and societies constantly set out to make conscious and metaconscious decisions as if they were free, and they translate these decisions into cultures with liveable ways of life. Were they as spontaneously adapted to their environment as animals, there would be no fundamental decisions to be made and no risk of alienation in the face of necessity.

By symbolizing their surroundings, human groups and societies open up the possibility of distancing themselves from the influence of their milieu. In so doing, they come face to face with the relative character of human life, which is overcome by absolutizing their way of life. The result of sacralization is alienation: to be possessed by someone or something the way a master possessed a slave in antiquity. The concept of alienation is unthinkable without the religious traditions of the West. This condition undermines the ability of anyone to engage in relationships of love and friendship, which can only be entered into freely. Alienation as a threat to the community was a central preoccupation of the Jewish tradition and later the early Christians because being deprived of one's freedom also deprived one of the ability to love one's God and one's neighbour. Alienation is the secular analogue of sin in contemporary social science, even though it implies the value of freedom.[6]

From the cultural cycle, it follows that the culture-based connectedness of individual and collective human life enfolds something of the connectedness of the world in which that life is lived. It is probably impossible to live a life if there is no integral world 'out there.' It is difficult to imagine a connectedness in human life if everything around it is disconnected and random. I have examined this in some detail in *The Growth of Minds and Cultures*. From their experiences, human beings derive a reality as they have come to know it, which constitutes a symbolic home. In other words, out of the reality as it is lived by a community, a true world has to be created. This can be accomplished only by creating an entire symbolic universe that encompasses the unknown as details that remain yet to be discovered and lived, but that cannot call the present into question.

Every culture, through its way of life, encounters one or more phenomena that are so essential and decisive as to permeate most daily-life experiences and thus people's structures of experience. A culture therefore simultaneously reflects not only a major (if not determining) influence on its members' lives, but also the need to exercise some freedom by symbolizing this situation in terms of its meaning and value for life, thereby imposing something human on it. The 'world' in relation to which human life is lived in this way will be referred to as the life-milieu. In prehistory the life-milieu was 'nature,' and for most of human history it was society that interposed itself between the human group and nature.[7] The culture-based connectedness of human life in prehistory was based on cultures that symbolized the entire world as animated by spirits, including human life itself. During history, when people lived in societies, the primary culture-based connectedness was based on cultures that symbolized society as being more or less distinct from but embedded in nature. In this sense, nature, in prehistory, and society, in most of history, have constituted our primary life-milieus in relation to which human life evolved and from which flowed everything necessary to sustain life, but also the most fundamental threats to it.[8] The way the culture-based connectedness of individual and collective human life was symbolized was closely related to the co-evolution of human life and the life-milieu, including how life was rooted in that life-milieu. Because different ways of life were involved, reality filtered into human awareness and cultures differently in prehistory and history. All of this is well known.

Today technology is creating a new life-milieu that includes, but is not limited to, the modern city and the web of technological means used in virtually every daily-life activity, to the point that they interpose themselves between us and others, between us and much of what happens in our society and the world, and between us and nature. It is largely via these technological means that we experience and participate in our world. Their mediation is not neutral. When the density of such technically mediated relationships becomes very high, a new life-milieu is created, through which people now experience the secondary life-milieu of society and the tertiary life-milieu of nature. It is well known how the ways of life, institutions, and cultures of prehistoric groups were profoundly influenced by the natural life-milieu and how this influence diminished and was eventually overshadowed by that of society.[9] There is no reason why the influence of our present life-milieu would be any less significant. Since this life-milieu is unthinkable without modern technology, the latter is bound to have an equally substantial influence on human awareness and cultures. Technology now plays a fundamental role in the cultural cycle, and thus in the co-evolution of human life and our present life-milieu. It is no longer adequate, therefore, to regard technology only in terms of instruments, machines, systems, organization, technocracy, professions, institutions, or as an approach to the world. It is all of these and much more, having become a significant social force that disconnects and reconnects relations in the fabric of human life, society, and the biosphere.

From this perspective, two fundamental changes have taken place in our relationship with technology during the last two centuries. The first was one of scale, in the sense that technology now plays a prominent role in the cultural cycle and thus in the culture-based connectedness of all groups and societies. The second was a qualitative one, as the disconnecting, reconnecting, and mediating of relationships involved a transformation in the nature of these relationships and thus in the culture-based connectedness of individual and collective life, including the way it is embedded into the biosphere.

2.3 Culture as the Symbolic Basis for Individual Life[10]

The members of any society symbolize their experiences and shape the relationships between themselves and the world into a coherent way of life by means of a culture. The considerable diversity of cultures can be interpreted as suggesting that the way human beings are linked to reality is genetically determined to only a limited degree. For a long time it was believed that the development of children was the result of the natural unfolding of universal states of mental and emotional growth. Although the roles of both nature and nurture were recognized, the learning of a culture was not regarded as the primary phenomenon.

Along with cultural anthropologists I argue that by acquiring a culture, babies and children learn to make sense of the world and to act in it in a way that is individually unique yet culturally typical. Culture-based symbolization thus acts as the basis for individual existence in reality.

Culture plays an important role in the way we maintain contact with the external world via the five senses. Consider some features discussed in my earlier study.[11] In an experiment designed to demonstrate this contact for the visual dimension of experience, subjects wore goggles that reversed the world from left to right, and right to left. Initially this led to a great deal of confusion: talking with two friends involved hearing them in one place and seeing them in another; sitting down to dinner led to seeing a knife but feeling a fork and vice versa; and the scent of a passerby and the sound of her heels might come from the left while she was seen passing on the right. Eventually the brain-mind learned to resymbolize the visual dimension of experience to bring it in line with the other dimensions, even though the retinal images remained reversed. When upon completion of the experiment the subjects removed their goggles, the same kind of confusion arose once more, since the brain-mind had to again learn to resymbolize the visual experience in relation to the other dimensions.

From this and other experiments, it would appear that the symbolization of what is received from the optic nerve aims to determine the meaning for a person's life. Most of this symbolization is learned rather than innate. For example, newborns can follow movement but cannot focus their eyes until they learn that there is something to focus on. The visual symbolization by the organization of the brain-mind is further refined when toddlers learn to talk about their experiences, thereby aligning this symbolization with the way the language and culture of their community help them to make sense of and live in the world. This is confirmed by significant cultural differences, as, for example, in the way the 'world' of colour is organized, or by the way children who grew up in the wild make sense of their world. The visual experiences of adults are similarly affected by prior experience, as is evident when medical students learn to make sense of X-rays, which amounts to seeing something meaningful where before they might have seen meaningless blotches, experienced as a kind of 'visual noise.'

Culture also plays an important role in the symbolic integration of the experiences derived from the external world in each of the five sensory dimensions, and from the body and the mind via several additional dimensions of experience. Such experiences are 'stored' in 'memory' as modifications to the organization of the brain-mind, which symbolised them. The significance of this has been obscured in the English language by the use of the same word and concept for human and machine memory, even though a great deal of experimental evidence suggests that these are fundamentally different. Machine memory stores information already separated from any context in a manner unaffected by any previous or subsequent storing of information. It is a contextless memory. The brain-mind does the opposite: each instance is symbolized as a moment of a life lived in a way that is individually unique yet typical of a time, place, and culture by being 'mapped' in a structure of experience. It is not a question of mindlessly storing and retrieving facts, but of mindfully living a life in the world. Although almost nothing is known about the higher symbolic functions, there is considerable evidence confirming that the memory of an experience can be affected by earlier memories and can subsequently be affected by later ones, as would be expected since they evolve with the structure of experience. In this way, human memory appears to make the fullest possible use of context. This characteristic of the brain-mind has evolved to cope with a living world in which nothing is ever repeated in quite the same way. Machine memory, in contrast, copes extremely well with the world of machines, in which everything is based on repetition and algorithms. If human memory had been genetically limited to functioning as machine memory does, humanity could not have survived.

Further confirmation of the above hypothesis regarding human memory comes from the extensive build-up of metaconcious knowledge, which is evident in a great deal of human behaviour. This knowledge cannot be derived from considering experiences one at a time. As in a scientific experiment, what we learn about ourselves, our society, and our physical surroundings is much more than the 'data' taken one at a time would permit. The principal difference between living a life in the world and practising science is that we have no conscious access to how the brain-mind goes beyond our individual experiences and builds up metaconscious knowledge about our life in the world.

In my theory of culture, I have shown that much of individual and collective human life can be explained if we adopt the (admittedly simplistic) analogy that the organization of the brain-mind 'plots' each experience, as if we were plotting the data from an experiment in the sense that each experience may be regarded as a data point in our 'experiment' of living in the world. In science, we go beyond the evidence by interpolating and extrapolating the data by fitting a curve through it. It is not until we have gone beyond the facts that meaningful new information about the behaviour of nature becomes apparent. If no curve could be fitted through the data, we would expect a flaw in the experimental design. It is important to reflect for a moment on what our behaviour means. What is the scientific basis for it? Have we not gone beyond the experimental evidence to leave the domain of science and enter that of speculation? Why does this strengthen our confidence in the data? The answer appears to be that fitting a curve through the experimental data confirms our (non-scientific)) prejudgment of the world's behaviour as being continuous, non-random, and non-chaotic. Our confidence in the data is strengthened precisely because the curve confirms these prejudgments.

Children's playful behaviour reflects an absence of such prejudgments. These form as a consequence of the development of meta-conscious knowledge, which results from symbolizing each experience by modifications to the organization of the brain-mind, thus 'plotting' it along with all prior experiences. Long before this happens, a child's behaviour implies a great deal of metaconscious knowledge that could only have come from the child going beyond his or her individual experiences to fully contextualize them in relation to all other experiences of the child's life. Again, much of what babies and children learn about their social and physical surroundings and themselves is more than the 'data' taken one at a time would permit. Symbolizing each experience by modifying the organization of the brain-mind may be regarded as interpolating and extrapolating the 'data.' This process turns each experience into a moment of their lives, making it possible for them to live those lives. I will briefly recapitulate a few features of their growing up and its dependence on metaconscious knowledge.

Culture acquisition begins with babies learning to symbolize what little they can experience of the world, based on the correspondingly limited organization of their brain-minds.[12] How does a newborn baby begin to relate to the outside world without having any direct experience of it? At first the baby's world is virtually totally undifferentiated. A baby's senses perceive mostly a host of undifferentiated stimuli that have no meaning whatsoever; even a consciousness of the possibility of their having a meaning is absent. Let us try to imagine what the baby's situation is like. As far as her sight is concerned, the situation can possibly be likened to that of an adult who looks at a photograph which is so badly blurred that he cannot make out what it represents. The patches of colour bear no relationship to one another; they form a random distribution without any order and thus without any meaning. The adult may attempt to interpret the photograph because he knows from prior experience that it must represent something real, but until an interpretation is found the photograph yields only retinal stimuli. In the baby's case, however, there is a complete unawareness of potential meaning so that she cannot focus her eyes because there is nothing to focus on. She can only stare blankly. Her other senses are in a similar situation.

We cannot say the baby's existence is meaningless, however, since, emotionally, she senses that there is a loving care directed towards her. This feeling possibly arouses a sense of curiosity about the world that responds to her. In any case, the absence of loving care can greatly retard her development. She will soon discern the emotional tone of her mother's voice from all other sounds and differentiate her responses accordingly. Similarly, she may also begin to note that part of the blur she blankly stares at moves, and that this movement coincides with acts and sounds of loving care. Slowly the baby learns to see the retinal stimuli as a meaningful whole – that is, as a sensation.

The baby establishes additional relationships with her surroundings as soon as the possibility of their having a meaning presents itself. Suppose that she is looking at a uniformly coloured cube suspended against a background of patterned wallpaper. There is no reason why her brain should group the retinal stimuli of the visible faces of the cube together, since they appear to have different shades depending on how the light falls on them. They might just as well be part of the wallpaper. Any grouping of retinal stimuli is equally plausible unless some prior knowledge or further experience with the cube intervenes. When the baby learns to manipulate the cube she experiences it as a whole independent of the patterned wallpaper, thus permitting her to see the cube as a sensation.

In a similar way, an increasing number of visual sensations will emerge out of the background of undifferentiated stimuli as the baby establishes meaningful relationships with her surroundings. This experience is comparable to one we are all familiar with, namely, the puzzle of finding hidden faces in a drawing in which some of the lines belong simultaneously to the picture and to a face. Once we learn to see some of the lines as belonging to a face, say, among the lines representing the leaves of a tree, it is as if the face leaps out from the picture. We have learned to see a new sensation by rearranging some of the lines of the picture. We have learned to see a face where we saw none before. As the baby learns to recognize an increasing number of visual sensations, the background of anything she looks at will increasingly be made up of sensations too. It will enable her to begin to derive clues about whatever she is looking at from the background context such as relative size, spatial perspective, and so on.

In order to have a visual sensation of someone or something, the baby's mind must be able to separate the corresponding retinal stimuli from the mass of undifferentiated stimuli and learn to interpret them as a whole. In other words, the baby must be able to establish a meaningful relationship with that person or object. This permits a foreground-background distinction between the retinal stimuli so that the baby can learn to see the foreground stimuli as a sensation on the basis of the experience afforded by the established relationships. Since all the senses are involved in these relationships, a context is created in the mind in which the stimuli detected by the other senses can also begin to have a meaning. Once a baby has established a relationship with a constituent of her world she can direct her full consciousness towards it, enabling her to differentiate between various kinds of relationships, such as those that are reciprocal and those that are not, those that involve sounds and those that do not. That she does make such distinctions is evident from her facial expressions and the increasing variety of sounds that she utters. The large variety of baby languages, or, more appropriately, the sets of baby signals more or less understood by those who regularly interact with her, show that the distinctions are largely her own.

As the child learns to recognize more and more visual sensations among the undifferentiated stimuli because of the development of the mental structures built up from experience, her ability to communicate also changes significantly. At first the phrases adults utter are nothing more than undifferentiated aural stimuli, and she may only discern their emotional tone. As the child learns to see facial expressions, gestures, and body posture, things change. The situation becomes more like that of an adult trying to communicate with another adult who speaks a foreign language. To communicate anything at all they would have to pay attention to body language (facial expression, gestures, posture, and so on) as well as the emotional tone of the phrases. Any face-to-face conversation between people speaking the same language is in fact facilitated by such cues, but they are normally registered metaconsciously. As the child becomes aware of body language her ability to communicate improves.

Suppose, for example, that the child has been encouraged to say 'mama.' She may then use the word indiscriminately and call all persons 'mama.' The reactions from everyone but her mother give her the impression that something is wrong. Since by that time the child has learned to single out her mother by the unique way she behaves towards her, she may put two and two together and use the word correctly. Otherwise, the child can only be baffled by the response, so that she may temporarily drop the use of the 'mama' altogether since it cannot have any meaning in her world.

In sum, babies gradually learn to symbolize what initially is little more than an undifferentiated blur, out of which emerge 'mama,' other people, animals, and a great deal else. The limited organization of the brain-mind with which they are born expands as a result of their experiences, and at the same time differentiates more and more 'skills' for making sense of and living in the world. This is why artificial intelligence could not succeed. It cannot be built up rule by rule, script by script, microworld by microworld. In learning to make sense of and living in the world the brain-mind develops in a manner analogous to the way an embryo develops. Each 'part' emerges from within the whole. A baby senses that there is a world 'out there,' and from the beginning there is a human being interacting with it. All this happens, therefore, from the vantage point or prejudgment of being embodied in the world as a human being. Elsewhere,[13] I have examined in some detail how this learning to make sense of and live in the world involves the development of a growing awareness of our physical embodiment in the world, our social selves as cultural beings, and a symbolic world derived from our experiences of reality. Later, a child's brain-mind can be 'grown' from very different experiences such as watching television, playing video games, and surfing the Net.

As a first example of metaconscious knowledge, consider the conversation distance that the members of a culture maintain, without being aware of it, when talking to each other. If we suppose that symbolized experiences are directly differentiated from those that most resemble them, then the structure of the cluster of differentiated symbolized experiences derived from these kinds of relationships will imply that if we stand too close to someone we are considered pushy, while if we stand too far away we are seen as unfriendly. The emotional tones associated with different parts of the structure will point to the culturally normal conversation distance. The structure of the cluster will imply a norm we 'learned' without realizing it. The reason for this is that the 'knowledge' we acquired is metaconscious, in the sense that it cannot be derived from any specific experience. It cannot be recalled from memory because it is generated by processes that systematically integrate symbolized experiences into a structure of experience that lies on a level beyond that of consciousness.

In other words, a distinction must be made between the subconscious (repressed experiences as well as the 'knowledge' implied in the genetically determined organization of the brain) and the metaconscious (the 'knowledge' implied in the structure of experience constituting the mind).[14] The latter plays a central role in the way that a culture structures individual and collective existence. It can be shown that, in the course of being socialized into a culture, children build up metaconscious knowledge in their structures of experience. They implicitly 'learn' such things as their culture's conversation distance; eye etiquette, conceptions of time, space, and matter; an image of their social self and the social selves of others; and the values and the way of life of a culture, including its myths and sacred.

I will give one other example. We have probably all at some time remarked about a good friend: 'I never thought she would do such a thing.' When we examine the basis for such a statement, it is immediately clear that we did not recall all our past experiences of the friend one at a time in order to analyse them for typical patterns of behaviour. We did not match such patterns to the personality types described in the literature to discover our friend's personality type and whether the behaviour in question was surprising because it fell outside the usual patterns. This kind of information was not the reason for our surprise. The remark was made because the experience in question could not easily be related to the metaconscious patterns in our structure of experience derived from the many previous experiences shared with our friend. We can intuit this, and these findings are, of course, important in interpersonal relationships. However, we might be hard-pressed to justify our feelings because the conscious mind has no access to the structure of experience. In the same way, we build up metaconscious patterns of our social selves from our encounters with others.

The brain-mind supports the ability of the members of a society to live their lives by means of multiple levels of symbolization. Each moment is symbolized relative to all others by processes of differentiation and integration, thereby revealing its meaning and value for their individual lives. Each symbolic structure of experience further symbolizes each individual life as helping to constitute and evolve the society for others, as they do for that person. Finally, each structure of experience also symbolizes all contacts beyond that society as integral to the world in which the society lives. For example, the intermediary and macro-levels of people's structures of experience symbolize the way of life of a society as lived by each person. When people come from similar positions in the social hierarchy, their micro- and intermediary structures will also be similar.

This is fundamental to human communication. The condition for its occurrence is that both parties must be both similar and different. If they were both the same, each one would know what the other one thought, and there would be nothing to communicate. If both parties were totally different, whatever one person would seek to communicate would have nothing even remotely resembling it in the world of the other person, so that no communication could take place. A conversation must reflect a tension between differences and similarities. The two people must be different enough that communication brings something new and thus enriches each of their lives. For example, the use of the word 'mother' in a conversation represents a set of experiences for each party, none of which may involve the same mother. Nevertheless, because of the metaconscious integration, they will share a sense of the role of mothers in their way of life, whether good or bad mothers, and so on. For each party, the word 'mother' may conjure up rather different associations. Nevertheless, they share enough to be able to talk about mothers together.

Conversations are essential in keeping social relations healthy. They motivate people to spend time together. The drawback is that in the process of sharing much of their lives, two parties will become more and more similar. When they know a great deal about how the other person thinks and feels, conversations may become much more routine and contribute less in terms of enriching each other's lives. In other words, the relationship tends to run down over time unless people learn to constantly recreate and enrich it. This is difficult, and when certain kinds of relationships are fundamental to the organization of a society, as was marriage and the extended family in pre-industrial societies, society may step in and artificially keep relationships together through external pressures, such as by making divorce difficult or even impossible. Similarly, a group is healthy when each member can contribute to the lives of the others while they, in turn, contribute to that member's life. When that ceases to be the case, there is no reason for belonging to the group. The group then weakens and may even disintegrate. On a larger scale, similar processes occur in a society. Through communication, the members of a society simultaneously create individual diversity and delimit that diversity through a unity implied in their way of life. The vitality of a society depends on its members being different enough to share or disagree about things, but not so different that they feel that they have come from totally different social worlds and have nothing in common. Human communication and social relations are based on a dialectical tension between individual differences and a shared unity. It is obvious that this has nothing in common with machine 'communication' based on information.

The structures of experience of the members of a society may be likened to mental maps with each individual's social self (metaconsciously derived from all their social experiences) as the map reader, provided it is clearly understood that both the map and the map reader are symbolically enfolded into each person's brain-mind. These mental maps permit people to orient themselves in their social and physical surroundings. In other words, the structures of experience of the members of a society form a symbolic medium through which they experience and act on reality. A great deal of our routine behaviour is modelled on typical earlier experiences. This, of course, does not mean that we are determined by our past. Our structures of experience include all aspects of our lives, including our hopes and fears for the future, our ambitions and plans, our dreams and fantasies, our convictions, thoughts, ideas, and so on; and at any time the routine usage of our mental maps can be overruled by thinking a situation through. A map differs from an algorithm or program because it requires someone to read and make use of it. This possibility is ensured by the metaconscious image of one's social self implied in a structure of experience. Since life never repeats itself, each paradigm contained in the map must be creatively adapted to the new situation. As a result, a person can cope with much of the living of a human life as a matter of routine, permitting that person to focus attention on those aspects that are particularly unusual, interesting, threatening, or, for some other reason, of particular meaning or value. For example, in a face-to-face conversation, the eye etiquette, conversation distance, body language, and emotional expression require no special attention, thereby permitting people to concentrate on what is most essential. Again, even the most routine activity can be interrupted by a sudden thought that causes a person to rethink that activity, but all this helps to sustain the extraordinary complexity and diversity of human experience.

Creativity is fundamental to life, where nothing is ever exactly the same as before. Each moment of a person's life cannot be lived by executing a kind of program or algorithm based on his or her past. What the mental map allows the person to do is to anticipate a new situation as being similar to one previously experienced so that conscious attention can be focused on those aspects of the situation that are particularly interesting, dangerous, troublesome, or, for some other reason, in need of special attention. By delegating as much as possible to routine, the use of a mental map allows for a much fuller and richer response to each situation than would be possible with an algorithm or program. When a person's mental map begins to behave more like a program or algorithm because of excessive monotony requiring minimal or no adaptation, a whole range of defence mechanisms intervenes. People begin to daydream, get sleepy, or start hearing other things, or the object of their vision simply disappears from sight. Yet despite massive evidence to the contrary, our culture accepts the idea that the brain-mind is essentially similar to an information-processing device, and that human behaviour is essentially information processing. Yet the strengths and weaknesses of human beings and computers are almost opposite. Human beings cope poorly with monotony while machines thrive on it. Human beings can deal with a living universe in which nothing is ever the same, while machines have enormous difficulty with such a universe. Despite all this evidence, the mechanistic world view is alive and well, except that the computer has replaced the clock.

The structures of experience of each new generation, although individually unique, will be culturally typical for their society at a particular point in its historical development. While we constantly change, we also remain ourselves, and our society evolves its unique identity. It is clear, then, that when we are socialized into a particular society we are transformed from natural beings into cultural, that is, non-natural beings who primarily relate to reality on the basis of a symbolic structure of experience. Hence, the process of socialization integrates us into the socio-cultural whole constituted by our society.

2.4 Culture as the Symbolic Basis for Collective Life

We have thus far analysed culture from the perspective of the individual. The characteristics of a society, however, cannot be derived from those of its individual members. If the links between an individual and reality are not genetically determined, neither are the links between a society and reality. In other words, a culture must not only symbolically mediate the relationships of the individual members of a society to reality, but also integrate the behaviour of individuals into a coherent way of life. This way of life must find a kind of symbolic basis in the structures of experience of its members.

If the cultural unity of a society were exclusively based on the similarity of the structures of experience acquired by the members of each new generation as they grow up in that society, no society or civilization would endure for very long. Divergence between the structures of experience would quickly increase because of new ideas, new discoveries, the rethinking of certain experiences, and many other such events. These could rapidly erode the unity of a culture. Since life never repeats itself, an ongoing stream of creative adaptations is required to meet the changes in a society and its social and physical environments. Yet from a historical perspective the cultural unity of a society appears to be extremely strong, allowing millions of people to live together even though, on the surface of things, they agree on very little. Many of the civilizations humanity has created thus far have endured for a very long time – in some cases thousands of years. Hence, their cultural unities must have been able to delimit the many opportunities for individual diversity to grow and undermine the culture-based connectedness. To find this cultural unity, we must take another look at how a society mediates its relationships with reality on the basis of a culture.

The members of any culture do not entirely know the reality in which they live. Modern science produces an ongoing flow of new discoveries about reality, and there is no reason to believe that this flow will ever come to an end. In other words, we must make a distinction between reality as it is lived by a society and reality itself. Yet, in their daily lives, the members of a society act as if reality as they know it is entirely reliable and differs from reality only in some non-essential details that remain to be discovered. At first sight this may not be surprising, because our intellectual heritage has told us for a long time that knowledge is cumulative and that we are, therefore, basically adding additional details to the essentially accurate gestalt of our knowledge of reality. This view, specifically for scientific knowledge, has been radically challenged by Thomas Kuhn,[15] whose arguments for the case of our knowledge of physical reality may be summarized as follows. In the West, we have had very different ways of conceptualizing physical reality. There has been the Aristotelian view, followed by the Newtonian view, which in turn was succeeded by the Einsteinian view. Each of these three 'descriptions' of the physical world was elaborated during a period when that knowledge of reality was essentially cumulative. Such periods came to an end when it became apparent that the basic conception of physical reality was no longer adequate because a newly discovered phenomenon contradicted it. This contradiction ushered in a revolutionary non-cumulative transition period.

Thus, the growth of our knowledge of the physical aspect of reality cannot be regarded as a purely cumulative process. The basic gestalt of this knowledge changes from time to time; and in the absence of a complete knowledge of reality, it is impossible to say whether or not during a non-cumulative period a more accurate picture emerges. All we can do is compare reality as it is known during different historical periods. During the cumulative periods, scientists behave as though reality were exactly as they know it except for missing details and improvements in accuracy. They speak of the laws of nature,[16] for example, which are simply models that explain their experience of reality for a certain time. Subsequent generations of scientists typically discover that these earlier conceptions of reality embody certain implicit assumptions and hypotheses that later on turn out to be incorrect. This is inevitable. Scientists cannot but behave as if reality as they know it were reality itself, thereby implicitly assuming that the unknown has the same 'nature' as the known.

The development of scientific knowledge within a particular discipline cannot be likened to what happens in an art class where students learn to draw a model. The longer the pose, the more time there is to add and refine details. Unless the gestalt is incorrect – such as when a student doesn't get the proportions right – the process is entirely cumulative, unlike the growth of scientific knowledge. Also, the 'pictures' of the world drawn by different scientific disciplines cannot be integrated into a larger picture, because there is no science of the sciences. If the growth of scientific knowledge is not cumulative, neither is the growth of culture-based knowledge, as is evident when we compare the knowledge that different or successive civilizations have acquired about the world. This situation raises some important questions. If we cannot assume that the unknown is simply more of the known yet to be discovered and lived, how reliable is the knowledge we already have? How do we know that some new discovery will not call our existing knowledge into question? How do we know that we know enough about the world, with no possibility of an unknown threatening that knowledge? How do we know that we can trust the world as we have come to know it? How do we know that we are sufficiently in touch with reality so as not to have to question our sanity? Somehow we must be able to trust our knowledge of the world, and this requires that the threat of the unknown be neutralized. It is one thing for philosophers to discuss, as an intellectual exercise, whether we are really here or whether the world is 'out there,' but it is impossible to live that way and remain mentally healthy. Imagine trying to do any activity such as walking, driving, or writing an exam if you had to worry about whether what you were perceiving was really there. It is impossible to live with our human finitude without trusting reality as we know and live it.

All this points to a community's need for a reference point to guide its journey in time and space. How this is accomplished follows directly from our earlier discussion of how experiences are symbolically mapped into a structure of experience, thereby becoming moments of a person's life. I adopted the admittedly simplistic analogy that this amounts to the organization of the brain-mind 'plotting' each experience by means of the processes of differentiation and integration in a manner analogous to plotting the data of a scientific experiment. I suggested that our confidence in the data is strengthened precisely because they confirm our prejudgments expecting a law-like behaviour. In the same vein, all experiences symbolically mapped into a structure of experience have symbolized the unknown as interpolations and extrapolations of the known. Metaconsciously, all the experiences of a human life point in that direction, thereby creating a prejudgment expecting reality to be identical to the reality as it has become known. The threat of the unknown is thus eliminated.

Metaconscious knowledge incorporates all specific experiences into a life lived in a world that now appears to be entirely seamless. The unknown is now metaconsciously symbolized as interpolations and extrapolations of the known, and it is only now that symbolization can reach its full depth. This completes the symbolization of each situation as a moment of a person's life, as an event in the collective life of a society, and as an integral part of the world of that society. It is the metaconscious equivalent of interpolating and extrapolating the experimental data in a scientific experiment to symbolize their full meaning by means of a curve. The metaconscious interpolations and extrapolations of a person's experience correspond to what (in cultural anthropology, the sociology of religion, and depth psychology) are referred to as myths. Myths help gather individual experiences into a life, the lives of many people into a society, and the many contacts beyond that society into a world. It is this binding together that becomes institutionalized as a traditional or secular religion. Such a binding together has nothing in common with the kind of rational foundation for human life that many Western philosophers and researchers in artificial intelligence have searched for in vain.

The practical implications of metaconscious myths are far-reaching. This becomes evident when we compare our daily-life dealings with reality with those of other human beings in earlier societies whose myths were 'absolutely other' than ours. Indeed, that modern cultures also have myths appears to me to be an inescapable fact because modern science and technology cannot fundamentally alter the condition of human finitude in an ultimately unknowable reality. Our knowledge also must be grounded in hidden metaconscious interpolations and extrapolations that amount to hypotheses and assumptions about the nature of reality and our existence within it, which correspond to the myths of a society.[17] Although we know how the ways of life of societies in the past were grounded in myths, we are generally unaware of the myths that underlie our own existence. These will undoubtedly become apparent to future generations, but in the meantime we act as if our lived reality is reality itself. This implies the elimination of the relative character of our life's knowledge by *absolutizing* reality as we know it. It also implies that our culture symbolically dominates reality. In other words, what is unknown and therefore threatening to the stability of our knowledge and our lives is converted into mere extrapolations and interpolations of reality as we know and live it. By absolutizing reality as it is known by a society, a system of myths converts the unknown into missing bits and pieces of the known. It helps convert *a* way of life into *the* way of life by making all alternatives unthinkable and unliveable.

The metaconscious processes that integrate the experiences of a life into a coherent whole close the gap between reality and reality as it is lived, effectively obscuring from consciousness all alternative possibilities of interpreting and living in reality. Nothing that is entirely 'other' can exist in this absolutized reality. The unknown becomes simply a storehouse of missing details to be added to the reality as it is known and lived. The system of myths of a society is therefore an important element in the creation of its cultural unity, because a different awareness of human life in the world becomes existentially impossible. Myths, after all, point to what reality will almost certainly be like based on all available experience. They extrapolate and interpolate between all available experiences to create a coherent picture of our lives, our society, and the universe, otherwise our mental maps would simply be a set of incoherent and only loosely related fragments.

The metaphor of connecting individual experiences into metaconscious patterns can help us understand how important myths are to human life in the world. Each experience, symbolized as an alteration in the organization of the brain-mind, becomes a moment of the larger 'pattern' that symbolizes a person's life. In this way, myths form a life from experiences. Experiences of encounters with others are symbolized as moments of their lives, and experiences of the physical surroundings become those of a meaningful world. Consequently, culture sustains and reinforces the living of individual and collective human life. Distinct and separate contacts with others and the world are thus transformed into a meaningful, purposeful, and liveable world. Myths are not merely the connections between these contacts but the very life that makes these contacts possible.

Our contemporary understanding of myths is the exact opposite of how they were understood in the nineteenth century, namely, as the religious and superstitious remnants of a distant past for which there would soon be no place or role in individual and collective human life. In the absence of a revelation from a transcendent, myths are now regarded as everything that is sure and true about our lives and the world. Myths still act as a spiritual force that alienates individual and collective human life by possessing it the way a master possessed a slave in earlier societies. In industrial societies, it is the content of myths but not their role that changes, a fact that becomes evident when we examine how the religious dimension of these societies became secularized, to produce new political secular religions including communism, National Socialism, and hard-line democracy. Myths continue to be the metaconscious roots of all secular religious expression. In the same way that there can be no science without going beyond the experimental data, there can be no human life in a 'cultural niche' without myths.[18]

On the deepest level, we might say that a society's system of myths acts as a kind of cultural DNA. Each experience of our life in the world is now symbolized in terms of an absolutized reality. Thus, each experience becomes a moment of our life, just as each body cell enfolds something of our biological whole. Our reductionistic scientific heritage has made it very difficult for us to see this dialectically constructed integrality of human life and culture.

A system of myths thus reduces the threat of the unknown and eliminates the relativity of a society's knowledge of reality. Yet this is still not enough to provide societies and civilizations with the kind of stability and endurance seen in history. All human activities made possible by the absolutized lived reality would be equally good or bad, equally useful or useless, equally beautiful or ugly, and so on. In other words, without the system of myths performing additional functions, each moment of our life would be equally meaningful, that is, without any meaning at all. Choices might as well be made randomly because there would be no possibility of meaning. Life would be a random sequence of events, a complete chaos that would be existentially unbearable. This is true for both individual and collective existence. The members of a society must be oriented in the 'space' created by all possible relations and be shown how to act in it. Every culture achieves this by means of a hierarchy of values anchored in its system of myths. The latter also gives the members of a culture a reason for living and motivates them to adopt a way of life that has meaning and value for them. Reality as it is known and lived must become a society's home – what we have called its symbolic universe.

The values of a society reflect the basic vitality of a culture. Generally speaking, the metaconscious structure of experience tends to identify one or more phenomena in the life of a community that so permeate it that its very existence, and thus also the lives of its members, become inconceivable without them. For the prehistoric group such a phenomenon was what we would call nature, and for the societies that began to emerge at the dawn of history it became society itself. In other words, these phenomena correspond to the primary and secondary life-milieus for human life.

The metaconscious recognition of these kinds of phenomena confronts a community with a dilemma. The community could decide that such a phenomenon is so all-determining that it has little or no control in the face of this fate. On the other hand, and this is in fact what happens, the community could sacralize the phenomenon by metaconsciously bestowing an ultimate value upon it.[19] Necessity is thereby transformed into 'the good,' and the social order is the expression of the community's members freely striving for that good. The freedom and cultural vitality thus (metaconsciously) created eventually permit the sacred to be transcended as an all-determining force and to make human history possible, although exacting a heavy price. All this may be put into traditional religious language when we recognize that such phenomena are so central and fundamental to individual and collective human life that it would be unthinkable and unliveable without them. In this sense, they have created that life and the world in which it is lived. They become the creators and sustainers of life and its absolute moral authority. As we will see shortly, this metaconscious religious operation has nothing to do with the possibility of there being a transcendent. At the dawn of history the overwhelming influence of what we could call nature was slowly transcended, although that of society eventually took its place. In turn, this societal influence was eventually replaced by that of technique.

The bestowing of an ultimate value upon whatever is most central and determining in the life of a community metaconsciously orders all other values in the structures of experience of its members. People live as if what is most important in their lives is 'the good.' Nothing more valuable, important, and life-sustaining can be lived or imagined. This is why the sacred is also the central myth. Thus, the absolutization of reality as it is known and lived creates a sacred, a system of myths, and a hierarchy of values, which together constitute a cultural unity. Because it is profoundly metaconscious, this unity gives a great deal of stability to a culture as well as a history distinct from natural evolution.

The cultural unity embedded in the structure of experience of every member helps organize thought, communication, and social behaviour. This unity makes the social order self-evident and natural. It provides the basic models for responding to new situations. Myths are not directly experienced by the members of a culture; rather, it is through myths that the world is experienced. The unity of a culture is a symbolic equivalent of DNA.

When the members of a society intuit what their metaconscious has identified as the sacred (the phenomenon that is attributed the highest value), they do not treat it casually but as something that is very special; that is, they tend to treat it with religious awe as the 'value of values.' Life without this most valuable entity would be unimaginable, unliveable, and unbearable. Who would they be? What would their life be like? What world would they live in? Symbolically, the sacred has made them and their world who and what they are. To put it in religious language, this sacred is the creator and sustainer of themselves and their world. To put them in contact with this sacred in order to influence it, a culture's religion is developed around it. Without a cultural unity the members of a society would not have firm roots in reality and no order or meaning for their lives. This 'reaching for the heavens' reassures them that they are really in touch with the universe and that their lives are meaningful, which is necessary since their relationships with reality are not sufficiently determined by means of innate structures of the brain, as is the case for animals. The establishment of a cultural unity in the metaconscious patterns of people's structures of experience ensures that each moment of their lives is lived in the context of this unity – it permeates all experience.

No longer do the members of a society have to be preoccupied with the ultimate questions. These have been put to rest by the metaconscious creation of a 'symbolic foundation' for a culture. By working out the relationship with the sacred by means of a religion, a society reassures itself that it is not lost in reality, that the universe is no longer something it does not understand and over which it has no control. Hence, life and death become bearable. A society can put itself in contact with the powers of its world by personifying its sacred. The future continues the present, since anything that is 'wholly other' is unthinkable and unliveable. All these and many other functions of religion are well known. We simply need to stress here that we are not debating the existence of a transcendent God who reveals Himself to, and communicates with, human beings. This is a separate matter that cannot be ruled out by the theories of religion as a cultural element, as even Karl Marx implicitly recognized.[20] In such a case, however, another phenomenon quite distinct from religion would arise. This was clearly seen by Karl Barth and Jacques Ellul.[21] They make a distinction between faith and religion, where the former is a human response to a revelation from a transcendent God that is received, interpreted, and responded to by means of a culture; but the revelation itself does not belong to that culture, which distinguishes it from religion. This does not rule out the possibility that faith can be turned into a religion.

Their arguments may be briefly summarized as follows. For Marx, when the exploitation of people and nature would cease with the advent of the socialist phase of human history, religion should disappear since it would no longer have any role to play. If it turned out that people still had religious sentiments, then these would not be the consequence of a false consciousness but the result of something real. Barth's arguments lead to a similar conclusion. Both the Judaic and Christian traditions were preoccupied with the dangers a people faced as a consequence of creating idols because these would inevitably bring about alienation. Such false gods were distinguished from the true God. In other words, for Jews and Christians, the fact that without exception all cultures have created a sacred and a religion based on it must be distinguished from the 'wholly other,' who is not a cultural creation and whose communication with people enters into the symbolic universe of a culture as a revelation that must be distinguished from religion. Since these people had their own cultures, they constantly turned what they regarded as a revelation into a religion to satisfy their need for cultural moorings in reality. In sum, Marx and Ellul, each in their own way, as an agnostic and a Christian believer respectively, recognized that the cultural creations of a sacred and a religion had to be distinguished from the possibility or reality of a revelation.

The structures of experience of the members of a culture are no longer merely a neutral map of their lives as they have lived it in society. They have metaconsciously made a spiritual commitment to the highest good they know as symbolized by their cultural unity. Being human implies this (metaconscious) defining commitment to a time, place, and culture.[22] At the same time, the members of a society are enslaved by that commitment until they become aware of it and begin to struggle against it. To the extent that they do, their cultural moorings weaken, thereby creating a measure of insecurity and doubt. Culture acts as a spiritual force in their lives, in the sense that it 'possesses' the very depth of their being. Moreover, these structures of experience collectively act as a kind of 'social gyroscope' that allows a society to change without losing its coherence. The cultural unity is to each individual life in the body social what DNA is to each cell in the body. Its members tend to approach new discoveries and new experiences in similar ways, and certain alternative possibilities are simply hidden from consciousness by means of its system of myths. The rise and fall of societies and civilizations as described by Toynbee and others are therefore the rise and fall of cultural systems on the basis of which millions of people live together within reality.[23] When these systems become incoherent and are unable to support the lives of these people for whatever reason, they can collapse. When these cultural systems begin to malfunction, that is, when they no longer give meaning to the lives of their people, the latter begin to feel that their lives do not make much sense and they develop a feeling of rootlessness, a lack of purpose, and so on. Phenomena such as anomie, suicide, violence, crime, alcoholism, drug abuse, and mental breakdown tend to increase. This is the tragic situation of so many indigenous people whose myths have been shattered through contact with modern cultures.

It is worth noting that there are three interpenetrating yet distinct phases in the evolution of a culture's unity during an epoch of its history. Each one provides a unique cultural basis for a society's socio-political life. The first begins when an emerging power elite in society organizes a new way of life while the rest continue to live as before. If the new way of life continues to develop, it increasingly threatens the old one, creating tensions and conflicts.

During the second phase, the new way of life is sufficiently devel-oped to reveal clearly its possibilities, and its benefits reach a growing portion of society. New generations begin to know the old way of life only in a state of serious decay and thus no longer as a viable alterna-tive. They identify with the new way of life as the only viable one they know. Over the course of generations the locus of conflict gradually shifts from a dispute over the kind of society people wish to have to tensions over a more adequate and just participation in the new way of life for everyone. The process is reinforced by the retroaction of the new way of life on the minds of the generations growing up within it and thus on the culture. A new cultural unity, comprising a sacred, a system of myths, and a hierarchy of values, is formed by metaconsciously extrapolating and interpolating the tendencies of the new way of life. New metaconscious images and models of the present, past, and future are created. It is frequently forgotten that power elites cannot advance their interests at the expense of a society without resorting to force unless their actions have some legitimacy in the cultural unity of a society. To bring about genuine change is not a question of replacing the people occupying the social niches of power but of changing the cultural foundation on which these niches ultimately rest. This is often forgotten when people take to the streets, which typically results in different people occupying the niches of power without much change either to these niches or to the cultural base.

The necessities imposed by the new way of life tend to be sacralized as they permeate the community, creating a new cultural vitality de-rived from the gradual metaconscious establishment of the new unity. Once this happens, society again constitutes a cultural whole, which characterizes the third phase. It ends when the society withdraws its allegiance via a process of desacralization. A transition to a new histori-cal epoch will then emerge, or a process of cultural disintegration will begin.

The locus of political activity shifts as one phase follows another. At first there is a struggle between the old and new societies incarnated by two or more groups that have the characteristics of social classes as

Marx described them.[24] Assuming that the classes pioneering the new way of life continue to be successful in imposing it (and only the myth of progress can make this self-evident), the gradual establishment of the new cultural unity tends to shift the locus of political debate and conflict to one over how the benefits and costs of the new way of life are to be shared. The power structure is challenged in a less radical way as groups struggle over participation in it. Any fundamental calling into question is pushed to the fringes of what constitutes acceptable political debate, where it remains until the cultural unity begins to break down. The conflict between social classes has now been transformed into conflicts between socio-economic strata of society.

The cultural approach to life has permitted groups and societies to deal with the finitude and relational character of their existence. Their members individually and collectively live as if the reality they know and live is reality itself, and as if the greatest good they know is the good itself. This, however, plunges human life into a spiritual condition of make-believe or what in the religious traditions of the West were called idols. Life is linked to absolutes that are not absolutes, which means that all cultural meanings and values are 'vanities' centred around a 'vanity of vanities.'[25] However, such 'vanities' do much more than simply alienate human life. They first and foremost make a finite and relational existence in reality and time possible. The traditional and secular religions, which were built on this vanity of vanities, bind the members of a group or society together and bring a measure of order where otherwise there would be none. The price tag for this, alienation, is but one side of the coin. The other side is to make human life as we know it possible, and that is not merely vanity. However, any claim of creating a foundation for human knowing and doing by means of a philosophy, or of creating an objective science or technology, is a vain attempt to introduce an element of permanence (secular eternity) into human life. This is what secular political religions are made of.

How then shall we live? How shall we cope with the 'unbearable lightness of being?'[26] Possibly the first step is to close all false escape hatches that create illusions. First and foremost we must *live* these questions. To be iconoclastic towards what is most meaningful, valuable, and certain is a way of delimiting the vanity of our existence without any chance of escaping it. Limiting the excesses of potentially totalitarian ideas, systems, or secular religions may be a vanity, but it is one that can save millions of lives and restore dignity to millions more. Today we live as if science is omnipotent in the domain of knowledge, and technology in the domain of means. By overvaluing these human creations, we expect much more from them than they can possibly deliver, and consequently we will not explore alternative avenues for what science and technology cannot deliver. Iconoclasm could help reveal their real character of being very good for certain things, harmful for others, and irrelevant to still others – in short, the relational character of these human activities. To get on with human life in the twenty-first century will require the wisdom of knowing what these human creations can deliver and what they will not be able to deliver, and thereby delimiting our own alienation.

Human life is also relational in time. As a consequence, the cultural unity of any culture has its time, and then it too passes. Yet we live as if there is nothing radically 'other' beyond our world and as if the future is nothing more than an extension of the present. It is not surprising that many cultures symbolized human life in cyclical terms. The West borrowed a concept of history from the Jewish tradition. Once again, an iconoclastic attitude towards what is most meaningful, valuable, and certain helps to delimit the time allotted to a particular cultural unity by creating anomalous experiences that prepare for a beyond, namely, a human future and a history. Once again, there must be no illusions. Whatever can desacralize a cultural unity will have its time as a new sacred and myths. Reaffirming the relational character of human life in time keeps the door open to what is other than the present, and what lies beyond. It helps to make a cultural unity temporal. All false escape hatches must be closed. For example, there is no progressive incarnation of the Idea, no destiny of a socialist paradise, and no progress by means of science and technology that can lift us out of the human condition. Since we cannot know the future (it being radically 'other'), the contribution our lives and our work will make towards it cannot be known. There can be no meritocracy and no social justice by which everyone is allocated their rightful place, according to their contributions to society and to its future. Hence, nobody and nothing is in its rightful place, and any claim to the contrary is yet another illusion.

It is essential to bear in mind the finitude and relational character of human life as we examine industrialization, rationalization, modernization, and secularization. This will bring us face to face with our own illusions, which help us live but against which we must struggle if there is to be a human future and a history. An inquiry into what nurtures iconoclasm and hence human life falls beyond the scope of this work. But one thing is certain. It does not come from any philosophy or system that is based on asocial and ahistorical constructs which inevitably have within them an anti-life orientation. The three 'masters of suspicion,'[27] namely, Marx, Freud, and Nietzsche, reduced human life to a mechanistic play of forces producing, respectively, the dialectical process of history, the processes of the unconscious, and a will to power negating all reciprocal dependencies. There are mechanisms and plays of forces, but that is not all there is to human life; hence these 'scientific' theories become bases for anti-life ideologies. It must be remembered that to the extent that human life evolves and has a history, it involves a creative adaptation within a fabric of evolving relations. Hence, whatever is spontaneous, imaginative, creative, and thus non-routine is at the centre of human life and beyond the reach of those scientific approaches that depend on experimental verification and replication and which can therefore only discover human life as non-life. Such science is very useful for developing the technology-based connectedness of a society, but it also implies that the tension between this connectedness and human life is bound to grow. The more this entraps human life in the world of machines, information, and algorithms, the more science will provide us with an accurate description of what life is becoming. It is now increasingly evident that an unforeseen and undesired consequence of humanity's third megaproject is the desymbolization of experience and culture. We will have to reinvent how to live with our finitude and relational character. However, I am once more ahead of my story.

It must not be concluded that the alienation which comes from human life being rooted in a cultural unity was the consequence of a wrong turn in the evolution of the human species. The brain does not 'wire' human life for alienation. This becomes evident when we compare the behaviour of children with that of adults. Children behave as if they know that reality as they have come to know it is not reality itself. They accept that their mental maps present them with a world of which they have inadequate knowledge. For example, a child may confidently stride into a big store while holding her parent's hand, but if she breaks away to go and explore and then becomes lost, she quickly becomes anxious and may burst into tears. For children, no longer having access to a parent to show them how to live in the world and to rescue them if they are stuck calls everything into question. In other words, children do not act as if their mental map is complete and reliable. They are open to outside guidance. Adults, on the other hand, choose to live as if their mental maps are the final word on everything, and this makes it impossible to live in the world any longer with a certain playfulness, as children do. In other words, they live as if their structures of experience are algorithms, thus robbing themselves of the freedom an incomplete mental map could bring. If we rebel against being possessed by our cultural unity, we must engage in a lifelong struggle that is iconoclastic with respect to our culture. Our structures of experience are nothing more than an expression of our human finitude, and there is nothing absolute about them. Western civilization has been profoundly affected by a belief in revelation, which has plunged it into endless tensions with its religions.

2.5 Industrialization as Cultural Unfolding

Industrialization fundamentally changed the culture-based connected-ness of human life and society. We have seen how the technical division of labour spread throughout a society as a prerequisite for mechaniza-tion and industrialization, and how this strengthened the technology-based connectedness at the expense of the culture-based connectedness. This is, as we will see, a defining characteristic of industrialization. People's brain-minds continue to symbolically map the structures of experience of their lives. Mental connections continue to be made, but their strength weakens as a symbolization of the parallel weakening of the social fabric and its connections to the biosphere. These and other developments lead to what will be examined as a process of desymbolization. The system of myths as a component of cultural unity remains as strong as ever, but the hierarchy of values loses its symbolic depth.

To understand the progressive weakening of the culture-based con-nectedness of human life, it is essential not to conceptualize society as a mechanistic whole. There is a fundamental distinction between mecha-nistic wholes and living wholes. This is particularly true when we examine the relationship between a whole and its constituents. All mechanistic wholes come into existence first as isolated parts. These are integrated into sub-assemblies, which are in turn interconnected to constitute the whole. The elements of the whole exist independently, each in its own region in space and time, and they interact by external contact.

In a living whole, on the contrary, there can never exist any indepen-dent 'parts.' Each 'part' has emerged within the whole by progressive cell differentiation made possible by the fact that each cell remains a 'local' expression of the whole by means of its DNA. The human body develops from an embryo by the progressive differentiation of cells, and the brain-mind evolves from the progressive differentiation of experiences. Growing up enfolds the cultural 'design' for making sense of and living in the world into the brain-mind, with the result that each individual is a unique manifestation of the social whole. Each cell is internally connected to all others by means of the DNA, and each experience is related to all others as an expression of the self via a culture. Each 'part' is internally connected to all others and the whole as well as externally connected to adjacent 'parts,' although no distinct boundaries exist. Think, for example, of the study of human life. No clear boundaries exist between the physiological and psychic components of a person's life since the brain and mind are enfolded into one another. Similarly, each member of a society is internally connected to everyone else he or she knows by the metaconscious images resulting from shared experiences. These people are not merely 'out there,' as something of their lives has become an integral part of the person's brain-mind. The person is also internally connected to society by means of his or her metaconscious images of its way of life, cultural unity, and so on. To a lesser extent, the person is internally connected to the local ecosystem. There is no separate society 'out there,' as each of us helps constitute the society for all others, as they do for us. Socialization transforms us from natural into cultural beings enfolded into a society. In the same vein, there is no separate environment 'out there,' as we help constitute the biosphere for other persons as well as other life forms, as they do for us. Hence, a society is an enfolded whole because of our symbolic links with reality, and each member is an individual manifestation of that whole.

Since culture is the symbolic basis on which a society mediates its relationships with reality, its science, technology, economy, social structure, political and legal organizations, morality, religion, and art are not interdependent components but rather *dimensions* of that mediation. These dimensions of mediation do not exist in isolation from one another but are enfolded together into the way of life that is the basis on which a culture mediates all its relationships with reality. When, for example, we speak of the technology of a society, we include a whole range of activities, institutions, and artefacts that have a specific focus assigned to them by the broader cultural context and that derive their meaning and purpose from the foundations of the culture into which they are rooted. Archaeology has long put this to good use.

In order to maintain their way of life, the members of a society must have a certain knowledge of their environment, derive from that environment whatever material things are necessary to maintain their way of life, distribute scarce resources, create a social diversity appropriate to its way of life from the natural diversity of each new generation, make political decisions, stabilize their way of life by providing its institutions with a legal basis, guide all actions by an appropriate morality, persuade the powers of their symbolic universe to cooperate with the community by creating a religion to communicate with the sacred, and symbolize the deeper meaning of what is happening in human life by means of aesthetic expression. As a result, the cultural mediation of the relationships between the members of a community and their world may be regarded as comprising several dimensions of mediation, including the scientific, technological, economic, social, political, legal, moral, religious, and aesthetic dimensions. Since no single activity is exclusively related to a particular dimension of mediation, the activities of the members of a society form a seamless fabric having a highly enfolded character. The activities we habitually call social, for example, contribute at the same time to other dimensions of mediation. As I have explained in chapter 1, the twin necessity of creating 'pure' work (separated as much as possible from everything else) and of using the Market to divide the world into internalities and externalities led to the progressive unfolding of the dimensions of mediation into relatively distinct spheres in a modern society.

Prior to industrialization, the way of life of a society evolved as a tradition based on the accumulated experiences of previous generations, which were carefully adapted to new circumstances. A culture implied a precautionary principle. Its way of life had proved its social viability and environmental sustainability; anything new should be treated with some caution to ensure that it would not undermine them. A tradition was enfolded into customary ways of carrying out particular activities. With industrialization, this tradition-based evolution of a way of life began to break down. There were no customary ways of building spinning machines or power looms, or of supplying them with power. This had never been done before. It must not be concluded that technological traditions had nothing to offer, but considerable creative leaps were required. These, in small or big ways, broke with customary ways of doing things. The coherence of a technological tradition was weakened, and when the same development occurred in other areas of life, a new coherence had to be established. As we have seen, it was not based on culture, with the result that the enfolded character of society was undermined. For example, technological traditions increasingly detached themselves from the culture-based connectedness, to reattach themselves to the technology-based connectedness, thereby weakening the former. Also, the wedge driven by the technical division of labour between the technology-based connectedness and the culture-based connectedness went deeper as a result of the necessary economic, social, political, and legal accommodations.

What I am suggesting is that the level of enfolding is itself a characteristic of any whole. Enfolding is non-existent in machines; it is high in living organisms; and it varies greatly in socio-cultural wholes. In the same vein, a relationship between socio-cultural wholes (such as a group) and the context constituted by the fabric of a society can vary according to the level of enfolding between the two. This is of particular importance for the present study. Generally speaking, the process of industrialization qualitatively changes the connectedness of individual and collective human life from a higher to a lower level of enfolding. This enfolding is the result of the technology-based connectedness weakening the culture-based connectedness as a consequence of developments such as a growing technical division of labour, the separation of work from the remainder of people's lives, and the dominance of the economic and political dimensions of mediation. Modern civilization is characterized by the highly unenfolded complexity of mass societies. The low level of enfoldedness does not make a society a mechanistically connected whole. Nevertheless, it makes it easier to legitimate the intellectual division of labour in the social sciences, which behave as if the economy, social structure, political institutions, religion, and other dimensions of a way of life are relatively distinct social components. Despite desymbolization, modern societies remain dialectically enfolded cultural wholes under great duress from technology-based connectedness.

2.6 New Cultural Moorings

We are now ready to examine how technology changing people involves changing a society's moorings in the sky as the abode of the gods. I have already examined how industrialization changed the moorings of a society in the biosphere by extending the ecological footprint through trade, colonization, and the accompanying military support for these activities. I will now show that industrialization involves an equally radical change in the cultural moorings of societies manifested by, among other things, the universalization of modern technology, the evaluation of technology on its own rather than in socio-cultural terms, the appearance of new secular religions, new forms of art and literature, and a desymbolization of language and culture. It will gradually become evident that the process of industrialization helped to place human life in an entirely new life-milieu, and that once again this resulted in the development of a new kind of awareness, integral to a culture, for making sense of and living in that life-milieu.

Having examined the role of culture in human life and society, we are now in a position to understand why industrialization is not merely the building up of an industrialized production system in a society. In the previous chapter, I noted that in England this chain-reaction-like process required changes not only to the technological dimensions of mediation but also to the economic, social, legal, and political ones. The new urban and industrial setting in which a growing portion of the English population lived as industrialization advanced was radically different from the rural context from which most people came. Along with this new setting came an entirely different lifestyle. Initially this amounted to a very hard life for many people. Consider the following extracts from reports on working and living conditions in France, where, like everywhere else in Europe, industrialization brought great suffering:

When you go through the workshops for dyeing, reeling ... weaving, etc., you realize that almost everywhere the children's schooling is neglected: there is not the slightest attempt to arrange the timetable to allow the child who has not received a primary education to go to school during the week; and if on Sundays he is allowed the opportunity to take an elementary education, no one bothers to make him do it. Almost everywhere the regulations concerning work hours, which should be suited to the age and strength of the child, are not observed. Often the apprentice is bound to an adult by a common task, in the shawl-making workshop for example, where a child throws the shuttle, or in the weaving workshop where a child, often less than 8 years old, puts on the stitches.

It is a matter of common knowledge that in many workshops the lodgings are unhealthy. The workers (the girls and women who reel the silk threads) are packed together in alcoves or recesses where the air does not circulate and is always stale; they sleep usually two to a bed, on a straw pallet; they receive nothing for personal cleanliness and have no possibility for even the most elementary measures of hygiene. Finally, although the room where they work is, because of the raw materials, kept relatively clean, negligence and dirt reign in the room where they sleep and take their meals.

The insufficiency of the food is borne out by the many complaints brought annually before the Conciliation Board. Sometimes destitution (and how can one avoid it when unemployment and rising food prices make the low salaries more and more inadequate?), sometimes the sordid greed of the workshop matron is the cause of this deplorable malnutrition. In many workshops they never eat meat; in others, animal foods appear twice a week at the noon meal; usually they do not drink wine ...

The apprentice cannot find a mother figure in her employer; maltreatments are in fact alleged every year before the Conciliation Board to obtain the termination of a certain number of contracts.

Poorly housed, malnourished, badly treated, our young girl finally wears out her constitution through excessive work. Whatever her age, she works regularly from 5 in the morning until 9 in the evening, and this 14-hour day, when the pressure is on, is prolonged until 11 or 12 at night. The matron's excuse is that repeated shutdowns of the Lyon factory oblige her to make the most of the times when it is busy. It is this same excuse, the irregularity of work, that is given to justify the disregard of a weekly rest: for many workers there is no Sunday. Need I add that concerning primary education the law is no better obeyed? One matron was not afraid to answer me that she demanded in her contracts a clause forbidding the apprentices to go to school: the children, according to her, are spoiled and come back more insubordinate.

According to medical statistics, the most common illnesses of our working class are consumption (tuberculosis) and stomach ailments, which begin at the period of apprenticeship. If we consider the Red Cross hospital, we find that in 5 years, out of 2,024 deaths, there were 771 due to consumption (26%) and, out of these 771 deaths, 408 belonged to the working class ... Those with consumption who survive the longest reach the age of 25 before dying.[28]

You enter these houses [in Rouen] only by low, narrow, dark alleys, where a man cannot walk upright. Down the centre runs a fetid stream of greasy water, and filth of all kinds rains down from the upper storeys; this water lies in stagnant, pestilential pools in small, badly paved courtyards. You climb spiral stairways without railings or lights, spiky with petrified garbage; and so you reach low, sinister hovels ... almost all devoid of furniture and household utensils.[29]

The community dwellings [in Lille] were scattered around these pestilential centres, out of which the locals tried to squeeze a small income. As you enter the enclosure of the small courtyards, a strange population of children – sickly, hunchbacked, deformed, pale and ashy-faced – crowds around the visitors and begs for alms. Most of these unfortunates are almost naked, and even the best outfitted are covered with rags.

But these at least breathe in the open air; and it is down in the cellars that one can judge the suffering of those who, because of their age or the harshness of the season, cannot go out. They usually lie on the bare earth, on a pile of colza straw, on dried potato skins, on sand or scraps laboriously gathered during the workday. The pit where they vegetate is completely devoid of furniture; and it is only the most fortunate who possess a small stove, a wooden chair and a few household utensils. 'I'm not rich,' an old woman told us, showing us her neighbour stretched out on the damp cellar floor, 'but I have my bundle of straw, thank God.'[30]

The conditions for the working people were such that their children had a life expectancy of 2 years, while for other children it was 29 years. All human dignity was eroded, and many turned to alcohol for escape. Dire poverty forced girls and women into prostitution and to abandon newborn babies: in Lyon from 1829 to 1835, 48% of all births at the charity hospital were illegitimate; from 1825 to 1829, out of 9,713 births, 5,270 babies were abandoned.[31]

The problems associated with these inhumane conditions required more government intervention. Protests and strikes were not infrequently dealt with by brute force, and of course the workers were blamed for being lazy, irresponsible, disrespectful of law and order, and so on. The industrial leadership who created these conditions sometimes justified their actions with the conviction that in the long run everyone would benefit. Groups like the Luddites argued that new technologies should be adopted only if they were beneficial to the community as a whole. Although such a position was eminently sensible (contrary to the modern usage of the term 'Luddite'), it was not heeded. The reasons are related to a profound cultural transformation that was gaining in scope. According to the values and myths inherited from the pre-industrial culture, the Luddites' position was moral and sensible. However, from the point of view of the new cultural unity that began to emerge, first in the industrial elite and only later in other sectors of society, their position was reactionary because it obstructed progress and happiness. The new cultural orientation frequently helped create euphoric extrapolations of the potential of new technologies that rarely materialized. The euphoria lasted long enough for these technologies to be implemented, however, with little attention given to the actual influence on human life, society, or the biosphere.

It must again be pointed out that the poverty of the industrial working class was not limited to a lack of money. Many peasants were also money poor, but conditions in industry were such that people could not live normally. Long working hours, fatigue, malnutrition, and slum conditions made it impossible for people to have a normal family life. They had no time, energy, or emotion for their spouses and children. Their poverty was first and foremost an existential one that included, but was not limited to, money poverty. Humanity has paid a heavy price for the brutalities associated with industrialization. International conflict, racial tensions, the destruction of non-Western cultures, the political split between left and right, the adversarial relationship between management and labour, as well as other problems, might have developed very differently had western European societies discerned the all-or-nothing character of industrialization. It is true that conditions in the industrial centres eventually improved, partly because things got so bad that governments had to deal with some of the worst problems. Child labour was outlawed (withdrawn from the regulation of the Market) and some social security provisions were made. Living standards, which had seriously deteriorated owing to industrialization, did eventually improve, and most importantly, the new cultural unity began to give new meaning, direction, and purpose to the whole society.

I will next examine this cultural unity in some detail. Imagine how the people who lived through the first phase of industrialization in England experienced their world. What features of that world must have appeared to overshadow everything else? Each year their world was populated with more machines; each year these machines were becoming faster and larger; each year there were more factories full of these machines; and each year the total output of goods was growing. Everything was directly and indirectly affected by these developments, from people's livelihoods to their communities and their physical surroundings. There was nothing on the horizon of their experience that might slow down these developments, let alone stop or reverse them. The metaconscious knowledge built up in the organization of their brain-minds as a result of interpolating and extrapolating these experiences suggested to many people that, despite some appearances to the contrary, the output of the economy would soon reach a point where everyone's material needs would be met. All the human creativity and ingenuity that had made this possible could then be refocused on the non-material aspects of human life and society, with the result that the material improvements would inevitably lead to social improvements, and these would very likely lead to spiritual improvements. It was difficult to imagine any limits. Nor did there appear to be any limit to the happiness people could achieve as a result of these improvements, as long as everyone worked hard to bring to it about. All this was presided over by capital, which was constantly translated into more machines and factories, which in turn produced more capital, and so on. With such an accumulation of capital, material progress was assured, and social progress would soon follow. As long as everyone worked diligently the day would soon come when everyone's material needs would be met, and social (and possibly even spiritual) progress would begin; and who would not be a happier person because of it? Nothing within the horizon of human experience could threaten these absolute certainties.

The emergence of this new metaconscious knowledge did not occur uniformly across the entire society. Those responsible for creating and evolving the new system, and who therefore had the most immediate experience of it, led the way. Those less directly involved but who still benefited from the above developments followed. Those who were victimized saw no way out. Initially many placed their hope in secular prophecies that the collapse of this system was imminent and that something better would inevitably follow. At first, some of the differences appeared irreconcilable. There was a conflict between at least two 'societies within a society.' Those who built the new way of life were also the most changed by it, while those who clung to what little was left of their world were least changed by it. These 'societies within a society' therefore differed between those who made sense of and lived in the world by means of a new emerging culture because they found the old one wanting, or those who to varying degrees clung to the disappearing way of life and its culture, so as to still keep something stable, meaningful, and moral in their lives. In other words, to the extent that social groups participated differently in 'people changing technology,' they also experienced 'technology changing people' differently. Nevertheless, these fundamental differences all disappeared as the new metaconscious knowledge gradually spread throughout society, eventually re-establishing a single cultural unity. The conflict between these groups did not end, but its focus shifted from the kind of society that was to emerge, to the distribution of the wealth produced by the new way of life. A cultural reunification based on a new cultural unity and way of life created an industrial society.

The development of the new metaconscious knowledge had two important implications. First, when the extrapolation and interpolation of people's lives in the world of that time metaconsciously identified what was most valuable and important in their experience, it became impossible to regulate by means of negative feedback what had been declared good, or even the ultimate good. This metaconscious declaration meant that any problems created by the good could not possibly be attributed to it. The source of these problems clearly had to be something other than what had been declared as good or even the ultimate good. It became next to impossible for people to recognize that, in addition to bestowing a great many benefits, the declared or ultimate good also produced many undesirable consequences that needed to be addressed. Nevertheless it remained a human creation: good for certain things, harmful for others, and irrelevant to still others. For the people of that time, the value of capital, material advances, and work were undeniable, but when (metaconsciously) nothing more important and good could exist beyond them, they were absolutized and thus turned into myths. These converted elements of human experience that were good for certain things, harmful to others, and neutral to still others into something that could do no wrong. It was next to impossible to conceive of any limits to the good such entities would deliver. To put it into traditional religious language, they were metaconsciously made omnipotent.

Again, all this had the practical consequence of making their regulation by negative feedback next to impossible. A simple analogy may illustrate the point. If a person was convinced that the heat delivered by his furnace could do no wrong, a sense of not feeling well could not be attributed to the house simply being too hot. He might then take an aspirin for his headache, drink juice for his dry throat, turn on a fan to stir the air – in sum, do anything except install a thermostat that by means of negative feedback 'criticizes' the furnace by comparing room temperature to the set point. When 'technology changing people' includes the transformation of the cultural unity of a society, it in effect changes the 'set point' constituted by its values and hence what the members regard as the legitimate exercise of power, the laws that should be obeyed, and the boundaries of what they will take to the street for. For the first industrializing societies, the new metaconscious knowledge meant that there was no decisive intervention in the new 'system.' Before examining some aspects of the rise of the new metaconscious knowledge and its consequences in further detail, it is worth pointing out that some of this reaches to the present. To varying degrees all governments act as if throwing money at some problem will help it go away, revealing a greatly weakened but intact belief in the omnipotence of capital. Particularly during the 1960s and 1970s, advertising echoed the cultural unity of the industrializing societies of the nineteenth century. Are you lonely? Buy this beer, toothpaste, or convertible and you will soon find yourself surrounded by beautiful people in luxurious surroundings who are all very happy. Material things are depicted as having the magical power of banishing loneliness forever. Tired of a drab existence with never enough money to spend? Buy lottery tickets and soon you may have the financial key that can unlock a new life and a happy future. Worried about risky technology? Don't be anxious, because those who scientifically study these risks have demonstrated that the public is simply irrational and does not (scientifically) know anything. The examples are easily multiplied, and we shall return to some of these themes in Part Two.

The emerging cultural unity became the basic symbolic blueprint for a system commonly known as capitalism, which was a historically unique way of life encompassing the kinds of changes associated with the process of industrialization. (State capitalism is the communist version of much the same thing.) For those who built and controlled the new system, capital became the lifeblood of their society: everything was organized for the accumulation and growth of capital. Everything happened by means of money, not only in the economy but also in the other spheres of life. The other sectors of society comprised those who (initially involuntarily) had to submit to the new capitalist order and sell their labour for money, with which they had to procure everything necessary for life. In the new economy everything had to be expressed in terms of money, which, as a result, became the measure of all things, the value of all values, and thus the highest value. However, the monetary value of everything did not place it in relation to everything else in individual and collective human life but instead in relation to everything else in the Market. As such, money first undermined all traditional meanings and values and then transformed them into economic equivalents. Life within the capitalist system was first and foremost economic, and, to a much lesser extent, cultural. The economic dimension of mediation was transformed into the base of society with the market as the new organizing principle. Human behaviour increasingly resembled that of *homo economicus,* maximizing the utility derived from all resources – especially money, since it could procure everything else. The Market became the organizing principle of society, replacing culture and its hierarchy of traditional values. For example, only paid work was regarded as 'real' work, and time became money. All societal activities, no matter how essential, were excluded from economic consideration in the national household if they could not be expressed in terms of money. Even governing the nation gradually became little more than management in keeping with the technological model. Human well-being also came to be expressed in terms of what the system produced. Thus the experiences of different sectors of society led to the sacralization of capital as the first secular sacred, although it happened in each sector via a different route because of different experiences. The more industrialization advanced, the more the deep metaconscious patterns of the members of society manifested a recognition best expressed in religious language: Who would we be, how would we live, what would our world be like without capital? Capital was the creator and provider, in which all things had their being and through which all things could be accomplished. This metaconscious recognition profoundly affected the religious institutions of the time.

The metaconscious knowledge implicit in the structures of experience of the members of society would therefore, in the course of generations, point to capital as that element through which almost all socio-cultural patterns were created. By the late nineteenth century, capital had become the first secular sacred. Along with it came a system of myths that extrapolated self-evident tendencies, resulting in an absolutization of reality as it was known and lived. At the centre, we find the myths of progress, happiness, and work. These myths interpolated and extrapolated those experiences so self-evident to everyone that no one at the time could imagine the present or the future to be any different. Of course, this was first evident to the industrial leadership most directly involved in establishing the new system, but it eventually became evident to everyone, including the industrial workers. Even one of the most famous critics of the new system, namely, Karl Marx, took progress to be entirely self-evident. Without it, his entire interpretation of history (including the 'fact' that socialism would inevitably succeed capitalism) collapses. Marx, like any other person of his time, place, and culture, could not subject progress to a critical scrutiny because it was as yet a true myth (in the sense of cultural anthropology) deeply buried in the metaconscious of the structures of experience of the people of that time. His successors followed this secular idolatry in the twentieth century, at the cost of millions of lives.

No decisive political intervention was forthcoming at that time. People with convictions on the left of the political spectrum believed that capitalism stood in the way: capitalists used the benefits of the new technologies for their own ends rather than for the advancement of the entire community. This situation, so they believed, would be temporary because capitalism would collapse under its own weight to be succeeded by socialism, in which the means of production would be owned by and used for the benefit of the whole community. It should be noted that Karl Marx was against labour unions because they would 'soften' the system and delay its collapse. For people on the right of the political spectrum, the problems of the time were simply the growing pains of the new system. Eventually the benefits of capitalism would trickle down to everyone. Social inequalities would soon reach an acceptable level no different from those of any other society. For them, the only threat to a bright future was socialism, seen as capable of destroying everything that had been gained. Whether one believed that a revolution was required or that a better society would be reached automatically, it mattered little in the long run. Progress was self-evident. It was therefore no longer necessary to ask: Progress in which area? Progress measured according to which standard? Progress became omnipotent as all limits disappeared; it became good in itself, and a value in its own right. This was closely associated with the sacralization of capital, which overvalued its usefulness to the absolute. Values ought to express the relative position of something in relation to everything else in the life of an individual and a community. When something becomes overvalued through sacralization, its relationships with all other values are ruptured or distorted. Myths must then reintegrate it into the symbolic universe by connecting what has been absolutized into an interrelated reality, bringing a 'secular eternity' into human life.

With hindsight, it might be argued that if the exploitation of people and nature was so extreme, how could the members of society individually and collectively maintain their self-respect and have a sense that they were engaged in something meaningful and valuable? Surely it must have been too painful for too many people to look reality in the face. Once again, religion and ideology came to the rescue. By anchoring that reality in a sacred and myths, they offered hope where there was little hope, consolation where there was little consolation, and a bright future where there was enormous suffering. It is supposedly the ruling elites that produce and sustain ideologies to advance their interests. However, it could be argued that both the left and the right were sustained by ideologies that, on the fundamental question of technology, differed little. The 'opiate of the people' appears to have worked for everyone. Given that this strikes close to the heart of our current vantage points and commitments, there will never be any consensus on this matter. Nevertheless, intellectual honesty requires that all positions think through the influence technology has on people, including the cultural ground on which they stand. As we have seen, the formation of the myth of progress in the metaconscious of the structures of experience of the people of that time, place, and culture first occurred in the industrial leadership. At that point, factory workers who were experiencing only misery clung to their traditional values and culture. A century and a half later, however, they too had internalized the new cultural unity. For many, this process was helped along by the secular gospel of communism, which promised salvation in the world beyond, where all alienation would disappear.

Alongside the myth of progress stood the myth of happiness. Happiness became the motivating central image. Earthly happiness could now be achieved by hard work rewarded with the ownership of things produced by the system. Being became having, so that one's identity was largely shaped by what was owned. In this way, the production of mere things was turned into the pursuit of human happiness. By progressively eliminating the burden of meeting material necessities, the pursuit of happiness would receive everyone's full attention. The myths of progress and happiness masked the fact that the people of that time could not face reality without destroying their sense of self-worth. It was not possible to accept that all the energies and resources being poured into production, including the sacrifice of many lives, would yield only things and nothing else. Yet only by means of a myth could something more be achieved. Prestige and social status were based on the things one owned, and everything could be bought with money. Through myths, being became having. The myth of happiness was supported by other myths. Comfort, for example, was transformed into its material dimension, and excluded all physical effort.[32] Comfort relaxed the hard-working person, who became happy by possessing material objects.

The world, it came to be believed, was made for humanity and for its happiness. But this could only happen by means of action and hard work. The limits traditional societies had placed on action fell by the wayside. This reinforced the myth of progress. No longer was progress limited in scale or to a particular sphere, and no longer did it have positive and negative aspects. Progress became good in itself, and this development could happen only by linking human experience to a sacred. History became the progress of humanity through hard work. Civilizations could be classified simply on the basis of their materials and tools. Progress was obvious everywhere. Abundance would replace poverty, science would replace religion, and democracy would replace oppressive regimes. All this could be achieved simply by means of hard work. Work made everything possible. The Judaic and Christian traditions had always taught that work was the consequence of sin, and their influence led the Christian Middle Ages to regard work as a curse. It now became the mother of all virtues, and laziness became the origin of all vices.[33] There could no longer be any explanation for poverty except laziness: poverty resulted from a failure to contribute to progress through work. Few of us today believe that work is capable of accomplishing all these things. After all, the material and monetary benefits of work depend not so much on what we do as on whether we happen to be in the right place at the right time. Recognizing that 'the time is ripe' and seizing the opportunity often amounts to taking a chance. We do not know what the outcome of our work will be, and a meritocracy is unachievable. Work rarely receives its proper reward and hence distorts the fabric of human relationships: those who work hard are not necessarily rewarded, and rarely do the poor get their chance. Worse, we do not know what others will do with our work, and it is not uncommon for this to be the opposite of what we intended. Early industrial societies not only overvalued work but made it into a myth. Everything was assimilated by the myths of progress, happiness, and work. *Homo economicus* was the mythical human nature that was lifting humanity out of its dark past.

As it turned out, the future was not a linear extrapolation of the present, and this gradually led to the weakening of some of these myths. These myths didn't take into account that the relationship between people and technology is a reciprocal one. As people changed their technology and, through it, their society, they themselves as members of that society were changed as well. Needs, expectations, and values began to change, imperceptibly bringing about a divergence between the myths and the linear extrapolation of experience on which they were based on the one hand, and the emerging reality on the other. The material needs of the members of society did not remain modest, in keeping with the standards of that time. Had this been the case, material poverty might actually have been eradicated. The influence technology had on people created a new 'human nature' whose material appetite gradually increased, eventually becoming almost insatiable. Comparing this phenomenon to any other time, place, and culture suggests that, from a historical perspective, this has almost never been the case: 'human nature' is apparently not fundamentally materialistic. However, I am again getting ahead of my story, and for now I will simply point to the fact that in the context of the new 'system' material goods gradually took on an entirely new meaning. This was inevitable to make the 'system' work. How could the enormous human effort and sacrifice be justified if the 'system' could do little else but produce industrial goods, and if these could only make a limited contribution to human life? They had to gradually be endowed with a greater value than was warranted. Otherwise, the entire 'system' could never have been made liveable. Overvaluing industrial goods necessarily meant greater material expectations. This was the threshold to a consumer society, which was crossed during the second phase of industrialization.

It is not until the myths of consumer societies take hold of their cultures that people are willing to work as hard as they do. Without the myths of progress, happiness, and work, the attitudes so essential for industrial production and consumption cannot emerge. It goes without saying that the failures of many technology transfer projects were partly blamed on the laziness, irresponsibility, and unreliability of local workers. What should have been recognized is that these people arranged their lives in time, space, and society via a cultural design that was very different from modern ones. In traditional cultures people worked to live, and when this was accomplished they would turn to more important things.

Whatever is sacralized is withdrawn from public criticism, political debate, and conflict. The sacred transforms whatever is believed necessary, realistic, or inevitable into 'the good.' As a result, genuine political imagination with regard to capital, science, and technology was pushed to the fringes for the past hundred years and banished from the central political arena. I am suggesting that the problem of idolatry took on a new reality in what people hoped would be a secular age. One might have expected that an age that developed control theory might also have learned something about the dangers of idolatry to a community.

What a culture declares as very good or as the ultimate good can do no wrong. This situation places a community in a potentially dangerous situation. It is highly likely that those creations which it declared to be good or the ultimate good will produce some desired consequences, but also some that undermine or even destroy life. The problem is that such consequences cannot be attributed to the creations that have been declared good or as the ultimate good. As a result, these creations can no longer be evolved on the basis of negative feedback, because this involves criticism of what has been declared good. For example, the thermostat on the wall of our living room could not regulate the temperature according to the desired set point if it were not allowed to 'criticize' the furnace by switching it on or off according to the actual temperature in the room. Over time, the destructive consequences of the creations declared as good or as the ultimate good by a community will grow, and along with them, possible compensatory measures. These will only burden a community without addressing the root problem. For example, it has trapped contemporary civilization in a labyrinth of technology, made up of ever-expanding compensatory mechanisms that cannot help us address our root problems.[34] It is for this reason that the creation of new secular religious attitudes present the same kinds of dangers to a community as did their traditional equivalents.

Although another sacred and another system of myths have sprung up since the nineteenth century, they still contain many of the same elements although they are integrated in a different way. We therefore begin to uncover some of the roots of our own existence, and this is threatening. Nobody likes being uprooted. Our reactions may well be: Have things not always been this way? Which civilization was not interested in happiness and progress? Yet this was not the case, at least not in the manner of the nineteenth century. The fact that we have difficulty in imagining a civilization uninterested in nineteenth-century happiness and progress reveals that we too are to some extent still held and guided by these myths. Myths transform something that is relative and hence temporary into a permanent aspect of human nature. By looking at the past via our own structures of experience, we tend to project our own myths on it.

Other civilizations have had quite different pursuits, and happiness a totally different content. In the Middle Ages, happiness consisted of a just relationship with God. In other civilizations with different religions, it may be a just relationship with the world or the universe. These same differences occur in Utopias. In the Christian new creation, happiness means being reconciled with God. Many other paradises have also been void of nineteenth-century happiness.

The same can be argued about the myth of progress. For example, it can hardly be argued that the Middle Ages were against economic development by means of technology. Yet revolutions at that time sought a return to a traditional way of life. Any new development was subjected to a set of values to see if it was just, and not automatically seen as good in itself. It is still an open question whether economic development is sustainable because of limited resources, the fragile ecology, and the growing problems of the developed and underdeveloped worlds. If a conflict erupts over access to scarce resources, for example, and if such a conflict leads to a total war, we may well destroy human civilization. Any development has positive and negative aspects, and there is no scientific way of determining the net benefit to humanity.

The cultural unities that emerged in the industrializing societies of western Europe during the nineteenth century paved the way for the technology-based connectedness of these societies to dominate their culture-based connectedness. The sacralization of capital ensured that its renewal and accumulation received the highest priority and that nothing could legitimately be put in its way. The myths of progress, happiness, and work made it self-evident that the strengthening of the technology-based connectedness was a prerequisite for a 'higher' culture-based connectedness. As a result, material things could accomplish much more than when they were symbolized through traditional cultures. They now acquired social and spiritual significance. The kinds of limits traditional societies had always placed on their technologies now appeared irrelevant and even superstitious. During the nineteenth century, technology was still regarded as essentially neutral and thus required the tutelage of either privately owned capital and the Market, or of publicly owned capital and central planning. Technology had not yet become a factor of production in its own right. This happened with developments soon to follow, as described in Part Two. Finally, the cultural unity of the industrializing societies in the nineteenth century also implied that progress and happiness could be achieved in a piecemeal fashion, since each and every effort of hard work was symbolized through these myths. It became unthinkable that technological and economic growth might encounter saturation levels beyond which their effects on human life and society might be minimal, non-existent, or mostly negative. The new cultural unities essentially implied that societies should first seek to advance their economies through technological development, and everything else would be granted to them. All this prepared a new secular religious terrain.

2.7 Religion, Morality, and Art

As noted previously, a society's religion is built on its cultural sacred. Since this sacred has its origin in the body of experience of a society and since nothing in human experience is absolute, reality keeps impinging on a society's sacred and hence on its religion. As a result, the evolution of a society or civilization is bound up with an endless process of sacralization and desacralization. I will briefly describe this process for Western civilization, because it will help us understand the kinds of religious transformations occurring in western Europe during the nineteenth century. When Christianity began to permeate the Roman Empire, it desacralized the symbolic and cultural basis for that world. In the course of that process, it helped to build a new sacred, namely, that of the Christian religion. Christendom was a confusion of the natural sacred and the religious constructions of the Roman Catholic Church. To openly live outside that sacred was to risk being burned at the stake. This sacred began to decline in the seventeenth century with the rise of science. At that point there was still no replacement, so that monarchs, even when they no longer believed in the Christian sacred, continued to lean on it. Even Napoleon, after the French Revolution, said that there could be no authority without the sacred, and that there was no other sacred than the Catholic religion.[35] The rise of science, commerce, and industry changed all that.

Science showed itself capable of explaining a great many things that traditionally had been explained by religion. There was nothing to indicate that this trend would not continue. New metaconscious patterns began to link up in people's minds, but since science did not yet affect their daily lives to any great extent it could not constitute a new sacred. Nevertheless, people began to place their hopes in what was implied in the new patterns of experience. The capabilities of science were extrapolated far beyond what experience could justify. People trusted science to lead them to the 'truth,' to freedom from religion and the power of the Catholic Church. They hoped they would be able to enter a rationally explicable universe free from all supernatural beings. This belief was intuited and accepted by many people because of the metaconscious patterns that had formed in their minds. As self-evident as the extrapolations of certain experiences were at the time, they became just as religious in character as those made by other societies. Science desacralized the explanatory role of the Christian sacred, and in doing so eventually contributed to a new sacred. No one today would identify science with truth, but we have firm experience where people of that epoch had only tendencies. Science, however, did not replace the Christian sacred since it could not permeate the socio-economic and political order of society, and hence could not be lived. For any phenomenon to be able to constitute the sacred of a way of life it must permeate the daily-life experiences of the members of society to become the source of all metaconscious patterns in their minds. Capital could do just that, and it therefore succeeded the Christian sacred. These transformations profoundly influenced the religious conceptions of both Christians and non-Christians in nineteenth-century Europe.

Hand in hand with the secularization of the sacred came a secularization of the Christian religion. More and more people intuited that something new had begun to constitute the very core of their existence and, consequently, began to treat it as something very special by exhibiting religious attitudes towards it. Feuerbach and other Christians tried to save the Christian religion by eliminating its dependence on the existence of God. After all, why should this aspect, so offensive to people who considered themselves rational and scientific, drag down with it the social contribution Christianity had made to society? Christianity had to be reduced to those elements that were socially useful. Hence all its 'horizontal' teachings (related to how people were to deal with others and themselves) were to be preserved, while the 'vertical' dimension of its teachings (dealing with the relationship between people and God) needed to be eliminated if Christianity was to have a future in a scientific rational age. Thus these Christians created liberal Christianity, which provoked a strong reaction from traditional Christians. The latter argued that the vertical dimension was primary and that without it the teachings about the horizontal relationships could have no meaning or purpose. They, also under the influence of the new cultural unity, often went to the other extreme of fundamentalism: emphasizing only one's personal salvation and an individualistic morality at the expense of community and the traditional Christian concerns for freedom, justice, and peace. The social groups that were building the new way of life typically exploited whatever belief in God still existed to advance their cause. Christianity became a morality, a ceremony, and a theology. At the end of this process a century later, God was proclaimed dead. The development of liberal and fundamentalist Christianity will not be dealt with here. I will simply state that the theological transformations that took place during the nineteenth century in a large measure reflected a change in the cultural unity of society; but this, as I have said earlier, neither confirms nor denies that Christianity is a religion like all other religions.

From the perspective of those who regard Christianity as revelation, the events of the past two hundred years had a different meaning. Christianity had indeed assumed the role of a religion during the Middle Ages. As a result of the emergence of a new cultural unity, Christianity lost its roots in the sacred. Feuerbach and other liberal Christians attempted to salvage portions of its religious functions. Conservative Christians, in emphasizing the vertical dimension, made it first and foremost a private affair and explained most social problems as a consequence of the abandonment of Christianity as the religion of the West. For those outside of this tradition, the abandonment of what by means of the myth of progress had become the 'highest' religion could mean nothing else but the dawn of a rational and scientific age. The result was almost complete confusion, rooted in a failure to acknowledge the possibility of the distinction between religion and revelation.[36]

The new cultural unity essentially divided society into one sphere dominated by the economic approach to life, roughly corresponding to the economy, and another dominated by the cultural approach, roughly corresponding to people's private lives. People's behaviour in the economy increasingly resembled that of *homo economicus*. This new human nature spilled over into private life, which continued to rely on the cultural approach, now dominated by a new cultural unity reflecting the influence of technology on people. Christianity continued to have an important influence, and theologians attempted to make sense of the situation in manifold ways. They did so in the only way they could, namely, by means of their cultural 'glasses' and religious traditions. In one way or another, God had to be related to wealth, progress, happiness, and work. This imperative produced theologies and moral understandings that survived well beyond the religious role Christianity continued to play at the time. They are alive and well in the secular economic theology that rules much of the world today.[37] The new morality that emerged embodied these developments.

At the deepest metaconscious levels of people's structures of experience, changes in the cultural unity began to affect their system of values by relating it to capital and the central myths. There was a second important reorientation of the value system, namely, the change from a group morality to an individualistic morality. Recall that in a pre-industrial society work was not separated from the extended family, which tended to be the basic building block of society. The uprooting of rural society with the migration to the urban centres, the separation of work from the family, and the social problems in the industrial centres prevented a vital new social structure from forming. It became more essential for individuals to make moral decisions on their own. Their responses in thought and deed, and reactions from others, were reflected in their metaconscious patterns. A group morality spells out what is right and wrong for the group as the primary unit within which the individual acts, while an individualistic morality guides the individual as the primary social unit. The transformation of the traditional group morality to an individualistic morality was associated with other developments, culminating in the appearance of a mass society towards the end of the second phase in the process of industrialization.[38]

In interpreting what was happening at the time, it must be remembered that the wedge industrialization had driven between the technology-based connectedness and the culture-based connectedness at the expense of the former was only gradually justified by the new cultural unity. There were plenty of moral and religious attempts to justify it in terms of the old crumbling cultural unity, but these were doomed in the long term. The old cultural unity implied that people should not live by bread alone, while the new emerging cultural unity implied the exact opposite because doing so was the only way to ensure that everything else could be accomplished as well. Despite its materialistic focus, Western civilization now regarded itself as being spiritually superior to all other civilizations. Regardless of whether or not one holds that Christianity is a religion like all others, it is evident that the behaviour of most Christians reflected a profound servitude to either vanishing or rising cultural absolutes as opposed to freedom.

The transformation of the cultural unity of society also had a profound influence on art and literature.[39] Artists are capable of intuiting something of the deep metaconscious patterns in their own minds. Since their art symbolized aspects of the changing human condition, its message was threatening to the new society. This led to a devaluation of art, literature, and drama by the emerging system. The wedge industrialization had driven between the technology-based connectedness and the culture-based connectedness legitimated the removal of art from the mainstream of society and its relegation to museums. Its message was devalued by a change in society's attitude to art, which was reflected in a shift in the meaning of the word 'culture,' for example. It now implied a distinction between the core of society, comprising technology, the economy, business, and the state on the one hand, and the 'icing on the social cake' made from 'culture' on the other. One could admire, enjoy, or be angered by culture, but then one had to return to the mainstream of life and get on with the 'real' things. Thus culture became a mere spectacle, detached from the 'real' society.

2.8 The First Generation of Industrial Societies

I have noted that the chain-reaction-like process of industrialization put enormous pressure on the tradition-based cultural approach to life because it could not readily respond to a flood of new and mostly unprecedented situations. People had to analyse and think through these situations by means of reason, thus gradually creating an alternative economic approach to life for dealing with the technology-based connectedness of society, now relatively distinct from the culture-based connectedness. Consequently, the technological transformations brought about by the industrial entrepreneurs required a societal transformation far beyond what these entrepreneurs had anticipated or foreseen. A knowledge of the enfolded and interconnected character of both technology and society was required to understand how the voluntary and involuntary consequences of the initiatives of these industrial entrepreneurs had a profound effect on technology and, through it, on society. Because of that interconnectedness it was not possible to stop the process in mid-course on the grounds that the objectives of these entrepreneurs had been achieved. Things quickly went far beyond their control as the consequences of their actions spread throughout technology and society. The grass-roots actions in England achieved what in France could not be accomplished from the top down. In France there was a grand vision of what the new society should be like, but the governmental initiatives were relatively ineffective.

In the present chapter, our attention has been focused on the reverse interaction between technology and society. As the entrepreneurs changed technology and society, a reverse interaction took place simultaneously. Through capitalism, technology and industry changed people's lives, now built up from very different experiences. When these became internalized in their minds, the culture on the basis of

which they interpreted their experiences and structured the relationships between one another and the world became affected. In time, the
very cultural and symbolic ground on which they stood shifted, giving
new meaning, purpose, and direction to their lives. When people change
technology, they themselves become affected as well. This reverse interaction led to a cultural reunification of society, thus producing the
first generation of industrial societies. Class struggle taken in the genuine sense of a struggle between sub-societies, each with their own
cultural unity, had come to an end.

The term 'industrial society' is somewhat erroneous because these
new societies were characterized more by the growing technology-
based connectedness and the tensions between it and culture-based
connectedness than they were by the development of industry itself.
The latter could not have taken place without a technical transformation of much of society. Yet these technical transformations were largely
controlled by the need to preserve and accumulate capital. The resulting capitalist order greatly constrained the social roles performed by all
parties, and this is still the case. This is obvious for the working people,
but is equally true for the owners of industry. To be a capitalist in the
new way of life is to exercise a relatively constrained social role. I am
not suggesting that this absolves these industrialists from all responsibility – on the contrary. What I am suggesting is that there are no simple
solutions. Certainly the solution of having the means of production
collectively owned, as was the case in communist industrial societies,
changed very little. The Soviet state was often so strapped for capital
that it could not pay its workers at all. Solzhenitsyn has described how
people were randomly arrested and shipped off to work camps to build
the necessary railroads and mines.[40] They were nothing but modern
slaves, regardless of the perverted ideology that claimed otherwise. In
many developing countries today, similar constraints lead to the exploitation of a significant portion of the population.[41] In sum, traditional morality provides little guidance in a society in which the
technology-based connectedness is temporarily regulated by the market as an organizing principle, thereby imposing itself on culture-based
connectedness; but this hardly means that human beings do not face
serious and profound ethical dilemmas.

Terms such as 'industrial society' or '(private/state) capitalist society' suggest that, whatever the initiative towards modernization may
be, it will trigger a diversity of interrelated changes that reduce historical and cultural differences. I have argued that the process alternatively

referred to as industrialization, rationalization, modernization, and secularization refers to the emergence of new societies whose similarities appear to outweigh their differences because of two constraints: a technology-based connectedness that anchors societies in the biosphere and a culture-based connectedness related to living individual and collective human life within a symbolic culture. There is no technological or industrial destiny, but these constraints influence the patterns of evolution of industrializing societies. As a result, societies belonging to this new category of industrial societies share the following characteristics:[42]

1. Production is carried out by privately or publicly owned firms that separate work from the family.
2. These firms introduce a new division of labour that is technical in character.
3. The renewal and accumulation of capital is essential. Each worker uses a lot of capital.
4. The renewal and accumulation of capital can only be assured by rational calculations. Cost prices must be calculated. Resource allocations must be based on calculations. Not all technical improvements can be adopted. They must be profitable to avoid the loss of capital. These calculations are not technical but economic in nature. Not all technical improvements can be adopted due to economic limitations. The way these calculations are made may vary, but all regimes must carry them out to avoid the loss of capital.
5. The population becomes increasingly concentrated in the industrial regions, beginning a process of urbanization.
6. The concentration of production requires a major expansion of the ecological footprint, as the services rendered by the biosphere as the ultimate source and sink of the networks of flows of matter and energy can no longer be supported by local ecosystems alone. The enlarged ecological footprint requires the expansion of trade, colonization, and military support. It also demands ongoing agricultural reforms to increase the productivity of the land, of which a growing proportion has become the hinterland of the industrialized urban centres. Colonization also permits the industrial societies to deal with their rapidly growing populations by means of emigration. The moorings of these societies in the biosphere increase in both diversity and intensity. This goes hand in hand with a different symbolization of the biosphere as a separate environment. The

dramatic expansion of the ecological footprints of industrializing societies is the beginning of what today is referred to as the process of globalization.

7. The cultural moorings of these societies are fundamentally different from their traditional precursors. A technology-based connectedness, temporarily shackled to the Market as an organizing principle, undermines the culture-based connectedness. This necessitates a cultural mutation anchored in a cultural unity with a secular sacred that gradually brings the 'gods' of religions from the sky to the earth. The weakening of the culture-based connectedness of human life and society begins a gradual process of desymbolization, reflecting a weakening fabric of relations. All this begins the evolution of an entirely new kind of society in which the culture-based connectedness has been weakened, to the point that it needs to be supplemented by technical means. A civilization that had once been taught not to 'live by bread alone' now enters into a new spiritual commitment to do precisely that, in the confidence that everything else will be granted to it in the process.

chapter 3

Living with New Moorings to the Earth and the Gods

3.1 Serving Technology

Some long-term implications of the process of industrialization become apparent when we investigate how the progressively strengthening technology-based connectedness gradually enveloped the correspondingly weakening culture-based connectedness of human life and society. It changed society's moorings to the earth and the traditional gods – a change of which a relatively distinct economy and new secular gods were the primary manifestations. It also set in motion a much more long-term process: technology gradually began to function as a relatively distinct system within human life and society. The technology-based connectedness linked together all the material elements of technology, and at the same time it distanced itself from the culture-based connectedness of society. In response, the culture-based connectedness had to be accommodated to the technology-based connectedness if society was not to break apart. It will gradually become apparent that the more the technology of society began to function as a system, the more society and the biosphere were turned into relatively separate environments whose raison d'être was to provide the system with the resources it required. Both society and the biosphere had to sustain this new system – tasks that would lead to unparalleled human, social, and environmental crises. Of the two components of industrialization, 'technology changing people' gradually overshadowed 'people changing technology,' as is evident from the fact that industrializing societies had to increasingly accommodate themselves to their technological creations. Counter-intuitive as this may appear, people ended up serving one of their own creations, which had been designed to serve them.

Eventually, this development led to a reversal of the relationship between people and their technology. It is this double transformation of the relationship between technology and society and the role of society in technological and economic growth, including its implications for human life, that is the focus of this chapter.

Much of this can be illuminated with the development of a schema for classifying technologies according to their relationships with their societies. The technology of a society is composed of a diversity of elements that may be thought of as forming a spectrum ranging from its 'hardware' (physical components including tools, instruments, machines, bridges, and railroads) to its 'software' (non-physical components comprising procedures, theories, models, journals, professional organizations, and institutions).[1] Each of these elements comes into being by means of the cultural cycle. The portion dealing with 'people changing technology' will be developed as the technological cycle. Although the terms 'life cycle' and 'life-cycle analysis' are common, they are once again based on a confusion between living beings and inanimate machines. Technologies do not have 'lives.' The phases of the technological cycle can be distinguished from one another on the basis of the role any element of the technology plays in society. During the first three phases, an element is born as an idea in someone's mind (invention), after which it must take on concrete form (innovation and development), before it can enter the mainstream of a society (application). During the last two phases, its use will first spread (diffusion), then decline until it stops altogether (displacement), thus ending active participation in the technology of a society. This most elementary form of the technological cycle may, of course, be terminated by the premature end of the element when, for instance, an idea cannot be made to work or proves to be economically unfeasible, socially unviable, or environmentally harmful. It can also be extended when, during the diffusion phase of an element, a new innovation leads to further application and renewed diffusion.

Networks of individual technologies, technological ensembles, and technological systems also find their way into a society and disappear from it by means of technological cycles.[2] Such entities may be 'invented' when it becomes evident that a number of individual technologies must function together because a technology-based connectedness imposes itself on their culture-based connectedness, which separates them from their surroundings. It may also become apparent that some essential links in these networks, ensembles, or systems are missing and must be 'invented' in order to have them perform satisfactorily. Such cumulative inventions that help to complete such technological entities or improve their functioning must be distinguished from non-cumulative ones that may lead to radical changes in these entities or their replacement and disbandment. The same comments may be made for the second phase of innovation and development. It is worth emphasizing that technological networks, ensembles, and systems include but are not limited to interconnected 'hardware,' and they simultaneously participate in society and nature in such a way that the technology-based connectedness dominates the culture-based connectedness. For example, modern factories and homes both involve a great many technologies, but factories are characterized by their technology-based connectedness overshadowing their culture-based connectedness, while this is still not quite the case for homes.[3]

Each of the constituent elements of a technology in any of the five phases of their technological cycles can affect and be affected by any sphere of human activities (dimension of mediation) that help make up the way of life of a society. These include, but are not limited to, the scientific, technological, economic, social, political, legal, moral, religious, and aesthetic (artistic and literary) spheres, which may be more or less distinct from or enfolded into each other. This simple schema identifies forty-five potential aspects of the relationship between any individual element of a technology and its socio-cultural context. Each of these interactions can occur in two directions, depending on whether the element affects or is affected by its context, thus creating ninety different kinds. A similar basic classification can be created from the categories of interactions between the constituent elements of a technology and the elements, cycles, and ecosystems of the biosphere.

Briefly consider a few examples of these interactions, beginning with invention. The creative process was extensively studied in the United States following the Russian success in putting the first satellite into space. Different conditions affecting the creative process were explored for the purpose of increasing the level of inventive activity. It was found that the creative process comprises several distinct phases. The first is that of intuition, when someone senses that something can be done better through an improvement or an alternative approach. Such an intuition must be explored and worked out in detail before its merit can be assessed, and this comprises the second phase. It is not unusual to encounter obstacles during this phase; and it was found that a third phase, incubation, is often very helpful, during which the problem is set aside and a person goes on with other activities. As the brain-mind continues to develop metaconscious knowledge from these and other experiences, another intuition may occur. Anecdotes of such intuitions, recounted by well-known scientists, mathematicians, and professionals, are numerous in the literature. They all show that, while a particular problem was incubating, a flash of illumination may suddenly spring up as a person gets off a bus or wakes up with an idea, for example. Once again, this intuition needs to be worked out. The experiences of the mathematician Poincaré are among the best-known examples.[4] It appears that the most creative people tend to use the process of incubation, going on to other work when a particular problem does not yield in order to return to it later.

The creative process is significantly affected by its socio-cultural contexts. A work environment in which creativity flourishes requires trust and cooperation. There must be a team spirit so that one can be confident that sharing one's ideas in order to stimulate others and to benefit from their contributions is in the interest of each member of the team. This is the opposite of a highly competitive environment in which people must protect their ideas or face the risk that someone else may take credit for them. Within a particular work environment, creativity may be stimulated by means of brainstorming or bringing in non-experts to ask so-called dumb questions.

It is clear that the creativity of a society varies greatly in the course of its history. It is an open question why some societies constitute the cultural centre of a civilization for a brief period and then decline, causing the centre to shift elsewhere. During the last millennium, almost every society in western Europe had its turn. Some economists have been very interested in these processes. By examining the number of patents applied for in a particular year as an indicator of societal creativity and correlating it with variations in economic, social, or other conditions, they have tried to create various kinds of models. Do economic booms stimulate inventive activity? Do recessions dampen it? What effect do wars have? Is necessity the mother of invention? The literature reveals some interesting correlations between inventive activities (measured by patent applications) and the conditions in a society. Insofar as these conditions are experienced by the members of a society and internalized into their minds, they constitute the most immediate context of creative behaviour.

The innovation and development of an invention during the second phase of the technological cycle may also be influenced by immediate or remote contexts. Suppose someone comes up with an idea for a new internal combustion engine. To fully explore this takes a great deal of time and knowledge, and frequently requires the building of a prototype. The inventor rarely has all the intellectual expertise and financial resources to bring this about. In such a case, she will have to approach others for support. They will ask questions and assess the viability of the idea in terms of its advantages over alternative designs, the related economic potential trends in car emissions regulations and their implications, as well as other conditions. In other words, the conditions in a society as seen by these people will act as a filter through which the idea is passed in order to determine whether it is worth supporting. Such people may include a spouse annoyed about spending too many evenings alone, a boss drawn in by a request for a leave of absence to work on the invention, a bank manager approached for a loan, the owner of a small machine shop in which the prototype could be built, and so on.

When the invention enters the third phase of the technological cycle, the resource implications become even greater. Suppose the inventor decides to set up her own business to produce and market the engine. Many different kinds of expertise as well as considerable financial resources will be required. Once again, the invention will be passed through many filters as a part of the decision-making process involved. What will the market for this engine be? Is it likely to grow rapidly? To what extent does the success of the engine depend on its own technological characteristics as compared with those of other engines? To what extent will it depend on other factors? Is the market for engines strong or in a slump, and how will broad economic changes affect these trends? What effect will interest rates have? To what extent are export markets limited by environmental regulations? These questions explore only some of the ways in which the conditions in a society and the biosphere can influence this phase of the technological cycle.

Many of the same conditions will also have an effect on the diffusion of the invention. I will highlight one example. For a long time, inventors quietly and secretively exploited their inventions, thus limiting their diffusion. This was neither in the interest of the inventor, who could make only limited use of the invention, nor in the interest of society, whose technological development was not fully exploited. As noted, patent legislation was an important step in solving this problem. The inventor's rights would be protected for a certain period of time, after which the invention would pass into the public domain. It was designed to benefit both parties, but the reality was often quite different.

Large corporations typically sought to prolong the life of their technologies, tying up entrepreneur inventors in the courts by challenging their patent applications. This frequently led to the complete or near bankruptcy of inventors, compelling them to sell their rights for a soft price. Another legal invention, namely, that of the corporation with limited liability, was essential to permit the ever more capital-intensive kind of technological development associated with industrialization.

A particular constituent may cease to be a part of the working technology of a society when it is displaced by new inventions that lead to superior alternatives. The latter can be of a technical nature or result from the societal conditions affecting its use. Displacement may also result when there is a significant change in the way of life of a society, which can make certain constituents of a technology irrelevant or useless. This does not mean that the particular constituent has to disappear entirely. It may continue to exist in museums, books, private collections, and may even be demonstrated and used in a marginal setting such as that of a pioneer village.

This brief discussion illustrates that the host of elements that make up the technology of a society are part of a complex network of relationships that binds these elements to one another and to their socio-cultural context. I will refer to an internal relationship as one between any two or more elements of a technology, and to an external relationship as one between any element of a technology and any element belonging to its socio-cultural context. It is now possible to distinguish between different kinds of technologies according to their structures and the way these are embedded in that of a society, to yield the following spectrum. Towards the one end we find technologies whose development depends more on the external relationships than on the internal ones. In other words, the elements of these technologies do not interact with each other in a significant way: they are each directly embedded into the socio-cultural context and interact with each other via this context. These technologies do not really constitute a whole, and to speak of them as technologies risks projecting something from the present on historical situations in which we find a diversity of weakly related individual technologies. The elements of these technologies tend to come into being largely as a response to the needs and conditions of society. Their evolution is the result of ongoing adaptations made for a variety of reasons (including considerations of a religious, moral, or magical nature) typically based on the experiences of their use. Considerations of their technical effectiveness (measured by performance values) do not dominate, with the result that their development tends to remain appropriate for the way of life of their society. The 'cultural DNA' is embedded in each 'technological cell' – a fact on which all archaeology is based. This does not mean that unexpected or undesirable effects cannot occur, because the overall impact on society of these individual technologies is a complex non-linear combination of their individual effects due to positive and negative synergies. Generally speaking, these technologies are an integral part of, and a response to, the development of a society as opposed to acting as something to which society adapts for better or for worse. Although the interaction between technology and society is reciprocal, the influence of society tends to be decisive. The culture provides the regulatory and control mechanisms, including the values on which they are based. Little technology assessment, planning, or forecasting is necessary. There is no need to distinguish between appropriate and non-appropriate technologies since the latter are virtually non-existent. Technology, or rather individual technologies, are culturally unique, which means that a society tends to have a greater influence on its technology than its technology has on that society. Nearly all the 'technologies' of pre-industrial societies have these characteristics. Today such 'technologies' can be found only in remote areas, almost always in a somewhat modified form.

The fact that the influence of external relations on the evolution of these technologies is much greater than that of internal relations also characterizes their dependence on the biosphere. The above argument may be expanded simply by including all the relationships between any constituent element of a technology, and the elements of a local ecosystem or the biosphere on which it depends for all flows of matter and energy, as external relations. Since the constituent technologies rarely depend on others, the chains of human activities via which matter and energy are borrowed from and returned to the biosphere minimally involve other technologies and tend to be as directly connected to the local ecosystem as possible. In other words, the technology-based connectedness is minimal, and where it occurs it is entirely integrated into the culture-based connectedness. Hence, the constituent technologies tend to be appropriate to society and sustainable by local ecosystems.[5]

Imagine an indigenous culture having a variety of technologies, including canoes for travelling and fishing, clay pots for the preparation and preservation of food, weapons for hunting and war, thatched huts for sheltering families, clothing for warmth, and jewellery for adorn-ment. An improvement in one technology that affects another may occur from time to time, but any such occurrence is likely to be infre-quent and highly limited. For example, an improvement in a tool used for the carving of canoes may also affect weapon-making, but it is highly unlikely that an improvement in a glaze for clay pots will affect anything else. As a result, individual technologies evolve on the basis of the experience of numerous generations, each one transmitting the accumulated experience to the next. For example, the shape of the canoes reflects, among other things, many attempts to make them manoeuvrable and stable under a wide range of conditions. Careful observations lead to changes, of which the successful ones are retained. For example, canoe travel in rough seas may compel the occupants to spend more time bailing than paddling, but they will surely reflect on this experience as a basis for making some changes. Such ongoing negative feedback between the technology and the uses to which it is put in a variety of contexts results in the course of many generations in the extraordinarily well-adapted indigenous technologies. Technologi-cal knowledge develops as people elaborate and refine portions of their structures of experience related to the building and use of a particular technology. Such knowledge is embedded in experience and culture, as I will examine in Part Two.

Towards the other end of the spectrum, we find technologies whose elements affect each other's evolution as much or more than they are conditioned by their societal context. They tend to have a strong inter-nal mechanism of development, which in the twentieth century was created by the flow of technical information. Each area of specialization acts as both a transmitter and a receiver of information. When an invention or innovation takes place, it is communicated in a variety of ways, including personal communications, conference presentations, and professional journal articles. When such information is received by other areas of specialization, it frequently triggers further develop-ments without any invention having to occur. Thus, in addition to the invention itself, there is a fallout of additional developments in the system. This is the case for the technologies of all the industrially advanced nations.[6] In these cases, the internal relations are as decisive as, and possibly more decisive than, the external ones in the evolution of these technologies. If, in addition, a society assigns a very high value to its technology through its culture, almost all technological advances will be accepted either as good or as inevitable because of the argument that if we do not pursue it, others will and we may fall behind in the global competitive race. In either case, the internal mechanism of technological development is stimulated as much as possible, and any objections to a particular innovation as potentially harmful or risky are dealt with as coming from a secular heretic to the faith and as undermining confidence in technology and progress.[7]

The presence of both a strong internal mechanism of technological development and a society attributing a very high value to its technology via a culture leads to technology taking on the properties of a system with respect to its societal context. Technological development now has two components. The first is the traditional one, resulting from inventions and innovations mostly in response to societal needs. The second derives from the flow of information within the system. As a result, the relationship between technological development and societal needs is much more at arm's length, requiring substantial societal adaptations. The technologies towards this end of the spectrum tend to take on a measure of autonomy with respect to society, thus requiring technical means for their regulation and control, means no longer spontaneously provided by a way of life and culture. For a while, it was thought that this could be dealt with by attempts to anticipate major technological developments in order to prepare society for the impacts they might have. However, technological forecasting proved difficult and unreliable. This state of affairs requires a very different societal context for technology. It may be created and imposed by a powerful state using technology to enhance and extend its control over society as part of a cause-and-effect relationship, which was the case in some ancient empires such as the irrigation civilizations. It may also emerge when a society spontaneously turns to technology for answers to its problems. This development is generally first promoted by a particular social group recognizing the technological potential, and it may become self-supporting when a new cultural orientation emerges. This happens when the members of a society become convinced that technological development will usher in a better future. They therefore spontaneously turn to technical solutions. Much of the resources and energy of society then goes into the development of technology. As a result, the value the culture attributes to technology continues to rise relative to everything else. The relationship between technology and much of what exists in its societal context is changed, producing a mutation in the culture's way of life. No technology can take on the properties of a system without these profound socio-cultural changes. The reciprocal interaction between technology and society is now dominated by the influence of the technological system, and society tends to serve the very thing it created to serve it. Technological performance values dominate context values within the cultures of these societies. The diversity of technologies, each a unique creation of a culture, now collapses. Technology becomes global and universal as an expression of its own technological values. From a historical perspective, this kind of technology did not emerge until the twentieth century.

The internal mechanism of development resulting from the flow of information within technology may be conceptualized as follows. Imagine a graduate student in electrical engineering finding a way to improve the frequency response of a particular solid-state device. This innovation may, in and of itself, be rather modest. However, when her supervisor encourages her to present her findings at a conference and also publish the results in a journal, the innovation may trigger many others. For example, someone working in hot-wire anemometry may hear of the paper and realize that this could lead to a modest improvement in measurements and consequently the understanding of various phenomena, which in turn may suggest ways of advancing various devices. The same pattern of events may be repeated in many other areas of application, with the result that even a modest innovation in one area can indirectly trigger many additional ones in other areas, thus contributing to technological growth much more indirectly than it would have directly. Developments like this jointly create a strong internal dynamic in an increasingly global technological system whose areas of specialization are connected by flows of information regarding millions of innovations that each may trigger others, and so on. The result is that even the best experts in a particular area of specialization will be unable to predict the major breakthroughs in five or ten years. Because of the endless combinations and interactions of countless innovations, technological development becomes extremely non-linear and unpredictable. Not even a group of specialists is capable of tracking the fallout within the global technological system of the millions of innovations that are made each year. Many of these find applications, but this does not mean that in any meaningful way they are a response to a human need or an unsatisfactory situation. Also, those who have invested in the research are not likely to be its primary beneficiaries. As a result, the development of the global technological system becomes relatively autonomous from the contexts of human life, society, and the biosphere. This autonomy is partly rooted in a new kind of technological knowledge that is no longer embedded in experience and culture. It is no longer acquired by learning to do things under the watchful eye of a master, but by learning in a classroom or by reading textbooks. It is what, in Part Two, will be examined as knowledge separated from experience and culture.

This technology is a result of the process of industrialization as constituent technologies are linked together in a technology-based connectedness that grows within and imposes itself on the culture-based connectedness of a society. As the throughput of matter and energy increases, so does the dependence on the biosphere. The ability of the local ecosystem to act as the ultimate source and sink of the required matter and energy is soon exceeded. Constituent technologies must then borrow matter and energy from non-local ecosystems via longer chains of activities, mostly involving other technologies. The concentration of production exceeds local needs, with the result that equally long chains of activities return much of the borrowed matter and energy to ecosystems other than the local one. The ecological footprint becomes geographically larger and also presses down more intensely as flow rates increase and the proportion of non-natural materials grows. Since the expansion of the technology-based connectedness reflects thermodynamic constraints instead of human values and aspirations, it increasingly imposes itself on the culture-based connectedness of a society. Contrary to the claims of industrial ecology, even if the flow of materials in modern economies could be made almost circular, the corresponding technology-based connectedness would not be comparable to, nor compatible with, those of the biosphere.[8] In sum, when the internal relations become more decisive than the external ones in the reorganization of a technology, it will impose its technology-based connectedness on society and nature, making it less and less appropriate to both.

In the middle of the spectrum of historical technologies, we find the ones in transition from the non-systematically structured technologies to those that behave as a system towards society. This portion of the spectrum of possible relations between a technology and society frequently involves an incompatibility between the structure of a society and that of its technology. It is the dilemma of the so-called underdeveloped and developing nations. Colonialism created underdevelopment by completely distorting the technology-based connectedness of these traditional societies by incorporating much of this into the ecological footprints of the industrializing nations. As a result, the technology-based connectedness of these colonies became separated from, and increasingly in tension with, their culture-based connectedness, throwing out of balance the dynamic equilibrium that had existed between their populations, productive capacities, local ecosystems, ways of life, and cultures. The result was an incoherent mixture of fragments that, compared to the situation before colonization, represented a process of development in reverse, thereby creating underdevelopment. After gaining independence, the process continued because trade and foreign aid involve technology transfer, in which technological constituents are taken out of one societal context and introduced into another with which they are not compatible. This is one of the primary reasons why so many technology-transfer projects fail to meet expectations.[9] In a significant number of cases, the recipient society may even be worse off than before. This marks the triumph of technology-based connectedness over culture-based connectedness.

When the constituent elements of the technology of a society begin to link together into a whole behaving with a measure of autonomy with respect to its societal context, the resulting situation is clearly undesirable from a human point of view. Technology is a human creation, designed to serve human purposes. The prerequisite to this situation is that the influence a society has on its technology is greater than the influence technology has on it. When the reverse is the case, the members of a society end up serving the very system they created to serve them. For this reason, it is important to distinguish between technologies that exist as a whole and those that exist as a set of relatively independent individual technologies, each directly embedded into the societal context and indirectly interacting with one another via that context. The former require an internal mechanism of technological development largely independent from society, and a culture in which performance values dominate context values. In terms of the principle of negative feedback for controlling a device, such a situation is analogous to one in which a furnace can influence the set point on the thermostat. It is only when the occupants of the house can really choose the desired temperature that the furnace will keep them comfortable. When technology influences the value system of a society, it becomes controlled on its own terms and not in terms of human values and aspirations. (See table 1.)

The end points of the above spectrum of possible relations between a society and its technology are ahistorical. Because of the reciprocal relationship that exists between people and their technology, it is im-

Table 1
A Spectrum of Possible Historical Relationships between Society and Technology

Most Traditional Societies	Industrial Societies
Non-systematic technologies	Technologies as systems within society
External relations most influential	Internal relations most influential
Culture-based connectedness regulates technology-based connectedness	Technology-based connectedness dominates culture-based connectedness
Cultural values dominate performance values	Performance values dominate cultural values
Highly appropriate with respect to a way of life	A way of life adapted to technology
Generally sustainable by local ecosystems	Unsustainable by local ecosystems and the biosphere
Technology as a resource for a society	Society and the biosphere become human and natural capital for technology
Technology socially determined and relatively neutral	Technology highly deterministic in its influence on culture
Technology regulated by culture (negative feedback)	Technological system based on positive feedback
Technology serves human ends	Technology shapes human ends (autonomy)
Technology as an element of culture	Technology as universal
People change technology	Technology changes people

possible for a technology to be entirely socially determined (neutral) or deterministic. These terms at best describe extreme tendencies in the spectrum of possible relationships between a society and its technology. Neutrality would refer to those situations where the influence society has on its technology is much greater than the influence its technology has on it. Determinism would describe the reverse situation. The position of any technology on the spectrum depends on the relative influence of technology-based connectedness and culture-based connectedness. On the micro-level, the relationship between people and the constituent technologies they encounter in their daily lives can be alienating if they are much more influenced by these technologies than they are able to influence them. For example, the work setting of a modern plant or office is alienating if its influence on the people working there is much greater than the influence they have on it.[10] A city is alienating when the demands it makes on the mental and physical resources of people by means of stressors including social overload, crowding, noise, and pollution exceeds these resources. The reactions range from resignation leading to a sense of helplessness, anxiety, and frustration to an attempt to as best as possible dominate this environment, greatly increasing stress levels that over time translate into severe psychological and physiological strain, producing a variety of illnesses.[11]

Because the relationship between people and their technology is necessarily a reciprocal one, it is essential to make a distinction between technological growth and technological development, between standard of living and quality of life, and between micro- and macro-level improvements. More is not necessarily better, because of the reciprocal interaction. We know this all too well from our relationship with food. On one end of the spectrum of possible relationships, the amount of available food is inadequate to sustain human life. The result is malnutrition and even death. Being hungry dominates everything else in life, and people lead a miserable existence. In the middle range where people have enough to eat, a small surplus may permit an occasional feast or celebration with good company, greatly enhancing the quality of life. Beyond this range, when the frequency of lavish meals becomes too great, food may once again have a negative influence on human life. By eating far more than their bodies require, people become overweight. This is frequently accompanied by a variety of health problems, thus negatively affecting the quality of their lives. This non-linear relationship between the quantity of food consumed and the quality of life may be illustrative of how more may not be better in the case of technology as well. Is it possible that in the modern world the density of technology-related experiences constitutes a harmful diet for the human mind and thus for culture? Since it is by means of culture that we make sense of and live in the world, can this negatively affect our quality of life? Many traditional cultures have answered such questions in the affirmative.

A human journey in time and reality presupposes a direction, if not a destination, aspired to. By valuing everything in relation to everything else, the cultural design of a community determines what is most important and valuable, thus creating the 'good' towards which the community can journey. Once again, because no higher good can lie beyond the horizon of the known, it becomes the absolute good of the universe and the moral capital that powers the community on its spiritual journey.

3.2 On Becoming Human Resources

Today it is common to speak of social and natural capital, and even of human resources, as if everything has become mere inputs for ongoing technological and economic growth. It is difficult to understand much of what is happening in our contemporary world without appreciating this necessity of turning everything into active resources, without which industrialization cannot succeed. By examining why industrialization began where and when it did, it will become apparent that the historical conditions of the time functioned as active resources for the new emerging 'system.'

To uncritically observe the beginning of industrialization from our present vantage point will get us into trouble. The notion of pre-industrial societies being first and foremost preoccupied with material advances is a distortion and oversimplification of the past as a consequence of projecting our own vantage point onto it. Traditional cultures did not put technology and the economy centre stage in their design for living in the world. Their materials (stone, bronze, or iron) cannot be used as benchmarks for assessing cultural evolution. Traditional cultures differed from their contemporary counterparts in that their cultural unities did not assign a very high value to technology. Technology was regarded as being like any other human creation: good for certain things, useless for others, and irrelevant to much of life. There was a strong sense of limits, somewhat analogous to the relationship health-conscious people have with their food. Such relationships occupy the centre of a spectrum in which at the one extreme people starve because they do not have enough to eat, while at the opposite extreme people's health suffers because they eat far too much, including a great deal of unhealthy food. In the same vein, traditional cultures have always limited the role of technology in human life, and it was not until the new cultural unities of the first generation of industrial societies began to take hold that these limits gradually disappeared.

David Landes sought to explain the origin of industrialization, using the following methodology.[12] First, we need to determine which conditions prevailed in the societies of western Europe, and particularly in England, on the eve of the Industrial Revolution. These conditions must then be compared to those existing elsewhere in the world at that time, and to the conditions that existed in earlier societies and civilizations. If it turns out that these conditions are unique, it is reasonable to hypothesize that, while they may not have caused the Industrial Revolution, they at least provided a fertile soil for it. In a second stage of the analysis, this hypothesis must be tested. It would be confirmed if industrialization began in those countries and regions where these conditions were most developed, and if differences in the rate of industrialization of countries and regions correlate with the degree to which these conditions were present. Such a procedure shows that the conditions in England on the eve of the Industrial Revolution were historically unique, and that they occurred to a lesser degree in other parts of Europe. England industrialized first, followed by other countries according to the degree to which these conditions prevailed. It will become evident that these conditions, though necessary, are not sufficient to get the process of industrialization under way. In other words, they can support a process of industrialization but do not cause it.

In reviewing the conditions that prevailed in England prior to the Industrial Revolution, there is no attempt to give an account that would satisfy historians. Our account simply aims to shed further light on why many nations today experience such great difficulties when they set out on the path of industrialization. The choice of England as our starting point is relatively arbitrary, and I could just as well have chosen another nation. Our story begins with the medieval way of life being lost in western Europe. This is a process in which a growing number of events and developments no longer appear to fit the patterns that characterize a way of life. This in turn affects the many individual lives that derive meaning, purpose, and direction from those patterns. The situation is not unlike that of a motorist who is finding her way to another city by means of a road map. A problem arises when she consults the map at an intersection in a small village and notices that it is indicated as being in the countryside. Since the homes are old, she decides she must be lost and retraces her steps. However, everything checks out and another explanation gradually begins to impose itself: the map must be wrong. She decides to buy another one at the earliest opportunity, but also to get to the bottom of the problem. As it turns out, the map is distorted as a result of a cartographic error. It cannot be corrected simply by shifting over the position of the village as marked. It is not a question of a small cumulative change, but of rearranging an entire portion of the map.

A similar situation can occur in the history of a society. I have already shown that one of the functions of the brain-mind may be compared to that of a map by means of which the members of a society orient themselves in the world. Although any culture has many defence mechanisms to deal with experiences that for one reason or another do not fit on the map, there are times when such defences are overwhelmed. Consequently, people may have a sense of a loss of meaning, direction, and purpose, which reflects a weakening of their symbolic roots in reality. The absolutization of reality as it is known into reality itself is undermined by anomalous experiences. The meaning, direction, and purpose of a variety of events and developments needs to be changed, and this in turn begins to affect behaviour. When these processes become self-reinforcing over the course of generations, new mental maps and a new way of life may begin to emerge, along with a new epoch in the history of a society. Failing this, stagnation, decline, and even the collapse of a culture could occur.

The transition from one historical epoch to another is never cumulative. Because reality as it is known is absolutized, any event or development that cannot be added onto that mapping of reality violates something of the absolute character of the original map. The very possibility of the event or development was ruled out by the map. This is not unlike something we know well in science. In a Newtonian mapping of physical reality, mass cannot change; and when it was discovered that it did, not only was the very concept of mass called into question but also the map it helped to construct. The absolutizing of reality as it is known into reality itself (minus the cumulative discoveries yet to take place) means that anything that suggests that the world may be different comes as a shock, from which we are protected most of the time by defence mechanisms. The situation is not unlike the one a student might encounter in a laboratory when, after plotting experimental results, he discovers a certain correlation except for one or two data points that appear to be way out of line. Most of us would simply disregard these points or write them off to some fly having sat on the instrument, and we may not even bother to report them. This would certainly be the case in an undergraduate student laboratory. If the student were pursuing doctoral research, he might check these measurements, and if they could readily be replicated, chances are that the world might turn out to be much more complex than he had assumed when he designed the experimental apparatus. After all, our mental maps imply images of nature as behaving in a relatively regular manner.

When we grow up in a particular culture, we learn to map the world through our experiences of it. Everything in that world has its place and value in relation to everything else. This is what makes our world coherent and liveable. If, however, a growing number of events violate what we explicitly or implicitly experience the world to be, then we have a feeling of being lost, which drives us to search for a new way that can make sense of our experiences. When many individuals are thus engaged it becomes a part of collective human life, and a transition is in the making.

As in any other historical epoch, people in the Middle Ages lived each and every event as a local manifestation in time of their (cultural) world. However, near the end of this era a growing number of events could no longer be lived in this way. For example, in the world of Christendom there was no room for a science that could, independently of theology and the Roman Catholic Church, develop reliable knowledge. There was no room for a king founding his own church, following a disagreement with Rome. There was no room for a people getting away with beheading its king, who was regarded as God's representative. There was no room for a declaration that the soul could be found nowhere. There was no room for a technology not subject to moral guidance. There was no room for economic activity outside of the religious order of things. Yet all these things and much more occurred, decisively weakening the enfolded character of the world of Christendom. This made it progressively easier to transform some elements of that world on terms other than its own, gradually leading to a new cultural approach and, eventually, a new world. In other words, beginning in the late Middle Ages we find western European cultures whose elements were more weakly enfolded into the cultural whole and were therefore more easily transformed and adapted to something else. Other elements had moved beyond this point to help create a 'cultural climate' for change, thereby sustaining that change. Still other elements became so 'loose' within the cultural ensemble that they were essentially neutral to change.

When a society loses its way of life, new and meaningful responses must be created to both old and new situations. There tends to be a greater reliance on human intellect and creativity. What was unique about the situation after the Middle Ages was that the role of reason became the focus of much attention and study. Some thinkers argued that reason would provide society with a reliable alternative to religion and culture. Initially, these ideas, first of the Renaissance and then of the Enlightenment, affected only a very small portion of society. However, an important seed had been planted in the cultural soil of post-medieval society. Soon the Industrial Revolution would create the conditions that necessitated an expanding role for reason.

Some societies applied reason to develop techniques of war, others, architectural techniques, and still others, legal techniques.[13] After the Middle Ages, reason was gradually applied to a growing number of spheres within the way of life of western European societies. A pragmatic and utilitarian spirit began to emerge. It was practical in its orientation, and materialistic. Western European cultures became characterized by what Ellul has called a clear technical intention.[14] No longer was the religious aspect of society's way of life the one that gave meaning and direction to all the others. Theology was no longer the queen of the sciences; the Church could no longer give order and direction to society. Gradually, people sought to fill the vacuum by relying on reason, which eventually led to the scientific, technical, and economic aspects becoming dominant and replacing religion. People had faith in reason to deliver them from the evils of a religious era. Rationalism and science would show the way to a better future. Eventually, rational and secular societies would replace the traditional religious ones. Thus, rationalism became a secular religious faith. Not until much later did it become apparent that this secular religious faith produced the same negative consequences that traditional religious ones had. Rationality overvalued reason by expecting from it much more than it could possibly deliver. It thus blinded people's eyes to the limitations of reason until enough time had lapsed for major problems to rudely awaken them. To see this, a distinction must be made between reason, rationality, irrationality, and non-rationality. As will become evident in Part Three of this work, traditional cultures operate in the domain of non-rationality, that is, they form another kind of rationality based on context. Reason, on the other hand, is a kind of rationality based on abstraction and separation from that context. The potential complementarity between culture and reason was destroyed by rationalism, which overvalues reason at the expense of culture, leading to a technical orientation of life.

The implication of all of this is that the concept of an Industrial Revolution overemphasizes the application of reason to machines and material production.[15] This was only a small part of the technical orientation that permeated the cultures of western European civilization. In the political domain, for example, the state emerged as the product of the French Revolution. In France, the state developed military technique by taking a rational approach to strategy, organization, logistics, and recruitment. The state also developed rational methods for administration, to harness the resources and energies of the nation for the transformation of society. The French state under Napoleon sought to take rational control of the laws governed by custom and tradition, and developed the well-known Napoleonic Code. These legal transformations were quickly adopted by other western European nations, except England. Weights and measures were standardized. A rational approach was used for the planning of roads. Science also became rationalized in the eighteenth century. The Industrial Revolution was, therefore, but one expression of the technical orientation that now permeated western European cultures, which relied on reason rather than on tradition.

Many people at that time understood that machines were only one aspect of the emerging new order, which aimed at making everything rational. The new orientation was manifested, for example, in the encyclopedia of Diderot. It showed how people sought to master things by means of reason: making things clear and precise, quantifying the qualitative, and bringing rational order to chaos and religious superstition. On the continent this new technical spirit was most strongly embodied in the state, which often lacked the means to execute its intentions. In England, on the contrary, it had a more popular base, particularly when it became apparent that one could greatly enhance one's social position and power through it.

The common failure to see the Industrial Revolution as only one aspect of a larger technical revolution can perhaps be explained by the fact that the transformations in the various areas of society by the rational approach did not come together into a relatively coherent historical pattern until the beginning of the mechanization of human work and the accompanying proliferation of machines. The chain-reaction-like process of industrialization demanded a rational transformation of one sphere of life after another.

The general reorientation of western European cultures created a fertile ground for industrialization. For England, I will briefly review some aspects that characterized the scientific, technical, economic, and social dimensions of mediation of its way of life, including the general technical orientation manifested in them.

First, consider the scientific dimension of mediation. The new science was characterized by three orientations: it was mechanistic, reductionistic, and mathematical/quantitative in character. In other words, the universe was now regarded as being fundamentally like a gigantic machine whose regularities could be expressed in terms of laws of nature. To explain any phenomenon, it was only necessary to decompose it into the most elementary building blocks of the universe and the relationships between them. Galileo said that nature was like a book written in the language of mathematics.[16] It was believed that everything should be quantified, and what could not be quantified was not worth bothering with. Among all the creatures, human beings were said to have large brains to enable them to understand nature. This new scientific orientation both reflected and reinforced the mechanistic world views of the cultures of the time, which held that everything could be conceptualized in mechanistic terms. Hence, it was favourable to technology, although the marriage between science and technology did not come about until the late nineteenth century.

Next, consider the technological dimension of mediation. Medieval societies gathered an enormous pool of technological knowledge and devices, which were put to extensive but not unlimited use.[17] For example, they were the first to systematically employ windmills and waterwheels, which had been known for a long time, and they also developed ways to make better use of animal traction. Yet the technology had to fit into the medieval religious conception of human life. This placed limits on its usage. When that way of life became extensively undermined, the pool of technological inventions and devices could be put to still greater use. It cannot be said, however, that it was the Christian religion that was opposed to technology, since typically the most advanced use of technology occurred also in the monasteries. The gradual emergence of a technical intention, and a growing awareness of the importance of the economy of a society, provided the motivation to put to greater use this enormous technological inheritance. As the cultural decline became widespread, the role of the monasteries in preserving and conserving what was seen as good and valuable from antiquity and the decaying way of life greatly contributed to the pool of technological devices and inventions available to future generations.

As far as the economic dimension of mediation is concerned, it should be remembered that an agricultural revolution had occurred some time earlier. This made it possible for a smaller portion of the population to grow enough food for the entire society, so that more people could be occupied in other activities. Second, through overseas trade, England had developed a mercantile capitalism, creating a pool of capital available for investment. Third, England was relatively wealthy. Its per capita income was significantly higher than that of many Third World countries today. This meant that a small portion of the population had a modest surplus, which it could invest. Fourth, the influx of precious metals, stolen from Central and South America, was put to use by the Portuguese and Spaniards to purchase goods from other European countries, thus stimulating these other economies. The economic aspect of life became increasingly important.

The social dimension of mediation continued to be affected by a demographic expansion that had begun prior to the Industrial Revolution. However, its magnitude was moderate compared to what we see in many Third World nations today. It gently stimulated the economy because there were more mouths to feed, more people to clothe, more houses to build, and so on. Traditional ways of doing things were no longer adequate to meet the needs. Hence, this demographic expansion reinforced the technical orientation in its culture, and the growing awareness that money could be made through the application of technology. Whenever there were more people than could be accommodated into the social roles of the emerging way of life, there was the opportunity to emigrate to the colonies. This acted as a safety valve whenever the population pressure on a way of life became too great. It should be noted that this demographic expansion, while substantial, was accompanied by a much greater expansion in the productive capacity of these societies, yielding a very different situation from the one found in many Third World nations today. The pressure of population growth on the economies of these societies is so large that many of them are unable to bear it.

Another important characteristic of the social dimension of England's way of life was the presence of a growing social plasticity: social institutions and structures became less rigid and, therefore, more open to change. In the Middle Ages, the social organization of society had been ascribed to God, who put everyone in their proper place at birth. It was widely accepted that God had delegated authority to the pope, and through him to the political rulers. This ultimate legitimation of the social order had gradually been eroded. This was largely the result of the undermining of social taboos by events such as the execution of Charles I by Cromwell. The king no longer represented divine authority, and he was no longer able to resist the nation. The notion of an unquestionable social hierarchy had disappeared, and society entered into a state of flux. It became unstable in all its structures. Religious taboos were also broken. Christianity was not the stabilizing influence it was on the continent. When Henry VIII could not obtain a divorce from the pope, he decided to make himself head of a new church to get his way. Such an act undermined the ultimate authority of the religious order, with the result that the Church of England never had the authority that the Roman Catholic Church enjoyed on the continent. Successful efforts were made to reduce the power and influence of religious orders. English Protestants were essentially divided into the Church of England and the Puritans (this was before Methodism). The Puritans, even after their political failure, were still the predominant influence. In keeping with the trend of the Reformation, they exploded all traditional religious taboos and developed a practical and utilitarian mentality. They emphasized the use and even exploitation of the good things of this earth, given by God to humanity. The relationship of this trend to capitalism is well known. The Church of England favoured tolerance from the end of the eighteenth century, and had social utility as its leading principle. Here too, there was a kind of secularization of religion. Religion was no longer the framework of society, and could no longer impose its taboos on it. Rather, it now sought to integrate itself into society, to adjust to it, so that the notion of social utility became its justification.

Another characteristic was an undermining of the basic groups of English society. This was not brought on by the state, as it was in France, but by the destruction of peasant society, which began in the early eighteenth century. The people who had become rich through trade bought out the old estates, and they behaved very differently towards the peasants than did the gentry. Everything by then was done for the sake of money. Technological development in England was rapid because the new rich discovered that with technology one could make a lot of money. In contrast, in France it was the state that sought to promote technology, but it did not have the means to carry out its plans. In addition, the enclosure movement in England began around 1730. It paved the way for new agricultural techniques that were very much superior. It also forced a lot of people off the land, which provided a potential labour pool for industry. With the disappearance of the commons, the pastures, and the forests, peasant society was destroyed. These uprooted peasants had to find a new way of existence: later on they could readily be absorbed into the industrial movement as wage-earners. There had also been movements against the guilds, against parental authority and the family, and other social institutions. All this helped weaken the social structure, putting it in a state of flux. This social plasticity was essential for sustaining the enormous transformations English society was about to undergo.

A further aspect not restricted to any particular dimension of mediation of England's way of life, which began to characterize the ways of life of all western European societies, was the appearance of a technical orientation in their cultures.[18] Drawing once again on the metaphor of the role of the brain-mind as a mental map, this orientation expressed the new cultural unities acting as 'cartographic principles of mapping reality' that were metaconsciously developed by unfolding the new emerging ways of life. For example, in France technological development was encouraged by the state. It took a scientific form and was promoted by the scientific academies. Under the directives of the state, the nobles applied these techniques. However, they did so indifferently. In England, on the other hand, the technical movement came from the grass roots of society. Profit was, from the beginning, its main motive. Commercial interests in agriculture and industry found techniques to be advantageous. It should be noted that the meaning of advantageous became increasingly defined in terms of profitability, making it much less contextual in orientation. With hindsight, it is clear that the appearance of a technical orientation in the cultures of industrializing societies marked the beginning of a fundamental change in the role of culture for making sense of and living in the world, which began to permeate all dimensions of mediation.

This analysis could be continued for other dimensions of the way of life of English society. However, we have identified the most important conditions, which together created a fertile climate for industrialization. Their combination was not found in earlier societies. They were (1) a technological pool of inventions and devices; (2) a suitable economic climate; (3) a demographic expansion; (4) social plasticity; and (5) a technical orientation based on a mechanistic world view and a keen economic awareness. These conditions were most highly developed in England, which was also the first society to industrialize. This confirms Landes's hypothesis that the conditions in England on the eve of the Industrial Revolution were closely associated with the frontiers of industrialization.

The above conditions created a situation in which individual economic behaviour and the development of economic institutions could proceed with decreasing reference to, and concern for, the context in which they took place. In other words, the value of the economic dimension of mediation of England's way of life gradually increased with respect to all others, to the point that, for some nineteenth-century observers, it was entirely self-evident that the economy constituted not only the base for western European societies of that time but for all societies throughout history.

It should also be emphasized that it was the conjunction of these five conditions that is historically unique. For example, a Third World nation today could photocopy the relevant contents of all our libraries and collect all our technical information, but that would not enable it to industrialize. In fact, the difficulties encountered by many Third World nations wishing to industrialize can often be associated with the absence of one or more of the above conditions. The structure of a society cannot accommodate just any kind of technology, and vice versa.

The above historically unique conditions created a fertile soil for industrialization. As noted, in England the chain-reaction-like process began as a consequence of attempts to overcome the instabilities in the putting-out system of textile-making. It did not respond well to sudden increases in the demand for textiles when, for example, overseas markets opened up again after a war. Entrepreneurs sought to increase their control over the work; the successful response to this problem was to mechanize textile-making and to house the machines in factories.[19]

In North America, the process of industrialization occurred in societal conditions very different from those in Europe. We encounter a frontier society pushing back native societies and the wilderness. Although the predominantly western European settlers brought with them their own languages and cultures, they found themselves in entirely new circumstances for which there were no ready-made answers. They had to adapt their lifestyles and cultures. There was no real tradition for doing so, and they had to invent a new society. Social plasticity was therefore very great, as were the economic opportunities.

A second important difference relates to the ecosystems in which industrialization took place. North America had an abundance of resources, particularly wood and, later on, hydroelectric power. What was in short supply was the highly developed base of skilled labour prevalent in western Europe. This situation characterized industrialization in the United States. Sawmills provide an obvious example of the difference this made to the path of technological development. In Europe with its severe scarcity of wood, but abundance of labour, saws were designed to waste as little wood as possible. In North America, on the contrary, saws were primarily designed to economize labour.[20] On the frontier, machines were often difficult to repair if highly skilled fitters were not available. The response was the invention of inter-changeable parts that made machines easier to build and repair.[21] Simi-larly, Taylorism and Fordism, which made it possible to produce complex products with as low-skilled a workforce as possible, were North American inventions of necessity.[22] This led to a period in which American industry developed a model of industrial production copied the world over.

Conditions in the so-called Third World are very different again. The classic distinction Galbraith made several decades ago between differ-ent kinds of underdevelopment in essence identifies the kinds of re-sources required for industrialization that a particular group of societies lacked.[23] On this basis, he distinguished between Latin American, Afri-can, and Asian underdevelopment. This classification confirms what we have learned about the process of industrialization and the kinds of conditions that supported it in western Europe.

The conditions that prevailed in England on the eve of the Industrial Revolution cannot be interpreted as prerequisites for industrialization. However, the complete restructuring of both technology and society requires enormous resources that must be drawn from society and nature. Hence, the conditions under which it takes place either sustain or impede it. No society can hope to complete such a massive transfor-mation without turning itself and its ecosystems into the necessary human and natural 'capital' from which the transformation process must draw. If a part of this 'capital' is absent at the beginning of industrialization, it must either be immediately generated in the pro-cess or obtained from external sources. Doing so was much more diffi-cult in the twentieth century than it was in the nineteenth century. We have noted that a portion of the English population that could no longer work the land because of the enclosure movement and that could not be absorbed into the factories had the possibility of emigrating to the colonies, thereby taking a great deal of pressure off the system.

Third World societies have no such safety valve today. For this rea-son, industrialization in the twentieth century frequently involved to-talitarian regimes that had no scruples about essentially enslaving a portion of their people as free labour and in propagandizing the popu-lation to induce the necessary psychological and social conditions and corresponding behaviour patterns.[24] Even in democratic societies, we have become so accustomed to being the human and social 'capital' required for ongoing technological and economic growth that we find terms such as 'human resources' entirely normal. The need to turn ourselves into 'resources' once more reflects the fact that, as people change their technology, it simultaneously changes them. In the course of generations, being restructured in the image of the machine becomes more and more self-evident and normal. It is true that, in early prehistory and at the dawn of history, human life was profoundly influenced by the life-milieu; but today the fact that many of our aspirations result from the influence of a life-milieu of our own making instead of being an expression of human freedom is hardly recognized. Human sustainability requires that we symbolically distance ourselves from our life-milieu, as humanity has done twice before; but with growing desymbolization our ability to do so has become significantly impaired. However, I am again getting ahead of my story.

3.3 Technology and the Human Journey

Much has been made of images of 'spaceship Earth' as seen from outer space, and their influence on human consciousness. On the positive side, this view marks an impressive technological triumph and a growing awareness that humanity shares one home and one common future. At the same time, it makes us uneasy because of what we are doing to our home and that of our children and their children. This view also reminds us that the journeys of different societies and civilization have become interdependent and intertwined on a level unprecedented in human history. While differences remain, the universal use of modern science and technology has produced similar ecologies of relations everywhere, with comparable consequences for human life, society, and the biosphere. It has become possible to use terms such as 'globalization,' 'modernization,' and 'international economic order.' The world has become divided into the so-called underdeveloped, developing, and developed societies, implying that they all share one common human journey, so that their only distinguishing mark is how far they have progressed down that road. History has become the story of this journey of humanity, which has been projected back into the past as a single journey towards our common present.

I will briefly examine some aspects of the technologies of totemic societies at the beginning of human history. Their cultures interrelated science, technology, magic, and religion. Their conception of nature is the key to understanding much of their symbolic universe. These people were competent observers of nature. Their strong scientific attitude was focused in a way that was quite different from that of modern science, but its results were equally impressive. Without them civilizations could not have emerged in so many places in so short a time. Agriculture, animal husbandry, pottery, weaving, the conservation of food, and the discovery of many useful natural materials were all prerequisites. Imagine, for example, the systematic research that went into the discovery of the possible uses of the many varieties of plants, which was so extensive that we have added little since then.[25] These cultures knew from experience the extent to which the natural world is interrelated.

The situation of prehistoric people at one time may have resembled the one we would be in if we had to survive in the jungle after a plane crash. Assuming that this would be an entirely unfamiliar environment for must of us, we would face the challenge that roots, plants, or fruits do not come with labels such as 'edible' or 'poisonous.' For our group to survive, someone would have to be selected to eat unknown berries and the result carefully observed by everyone. This is a microcosm of the challenge undertaken by prehistoric people.

The lives of the members of prehistoric groups were totally embedded in the natural life-milieu, and nearly every experience was permeated by it. Everything these societies needed for their sustenance, as well as everything that endangered their existence, came from that life-milieu. The structures of their minds built up from experience implied the basis of a meta-language concept of a family of life. When a member of such a society intuited this, a name might be given to it. This concept (quite different from that of 'nature') found its way into the language as other members recognized its validity and significance, because it corresponded to something deep in their mind's structures. Once such a concept entered into the language, the society began to elaborate its meaning, which helped shape its entire way of life.

Ever since Leonardo da Vinci's painting of the *Mona Lisa* symbolically portrayed the separation between humanity and nature, it has become very difficult for us to imagine living in a totemic symbolic universe. Such a universe was symbolized as one family of life within which human beings were the 'talking animals.' At the same time, the group's experience showed that this family was a life-giving and life-sustaining force providing them with everything they needed but also constituting everything that could endanger their lives. It traced the boundaries of what was possible and necessary, of what was good and bad, thus dominating people's entire existential condition. In fact, the natural life-milieu so permeated all human experience that it became a means of organizing the experience of totemic societies. These people took it for granted that what we call nature was one family of life, so that everything was populated with spirits and thus had the ability to respond to human communication. In terms of our own understanding, these people regarded the cultural sphere as integral to the natural sphere and the human group as part of the society of nature.

Totemic societies lived as if the natural and cultural orders reflected one solidarity of life. The zoological and biological classifications could thus be used to render the human order intelligible. Totemism was, in fact, a system that classified human experience by incorporating it into the natural diversity of animals and plants. Cultural relations were regarded as a projection or reflection of nature. The structure of such a world has been described by Lévi-Strauss.[26] Totemism was of immense practical value. When at the dawn of history societies began to be constituted, new possibilities opened up. By pooling resources and knowledge a society could do things that were beyond the capabilities of small prehistoric groups. But almost everything that could be undertaken depended on an adequate knowledge of the natural milieu, be it in order to find a better protection from dangers or to make better use of the natural inventory. This led to the massive scientific effort mentioned above. Under these circumstances totemism was an effective way of organizing society to mobilize its resources for the acquisition, preservation, and transmission of knowledge of the natural life-milieu. Society was organized by drawing together clans, each having a totem. By associating, say, an animal and plant with a totem and giving the clan a special association with them, each clan became a group of specialists in some of the constituents of nature. A division of scientific labour thus became possible. In their early phases of evolution, totemic societies appear to have acted as large scientific communities drawing everyone into this scientific undertaking. Their science, unlike ours, preserved context in its intellectual division of labour. It may be that these scientific benefits were an accidental by-product of very different social intentions, but the results were the same.

Consider the implications for science of a totemic symbolic universe. As noted, for totemic societies there was no nature as we know it. Nature was populated with spirits, powers, and deities who had jurisdiction over specific constituents or phenomena. Typically, this jurisdiction was spatially delimited. There was no purely material world, and this made science as we know it impossible. After all, what guarantee would one have of reproducing an experiment if the powers that controlled the phenomenon in question could change it at will, as they might when angered by someone's behaviour? Clearly, the regularity of nature expressed in natural laws based on a mechanistic conception of nature would be unthinkable. Science was therefore inseparable from magic and religion. The regularity observed in nature could be counted on only if the powers that be willed it so. Imagine the power religious leaders derived from this situation when, as in the Aztec culture, for example, they mediated between an agricultural community and the spirit of the sun, who could decide to take a rest and not return the next morning.

The possibilities of technology were equally limited in a totemic symbolic universe. Much more could be achieved by manipulating the powers of nature by means of magic than by technology, although the former certainly did not exclude the latter. In totemic symbolic universes magic was the technology par excellence. Consider a simple example: the hunting of large animals. By means of magic a contract could be made with the spirit having power over the kind of animal to be hunted. In a ritual the animals were symbolically given over into the hands of the hunters. The outcome of the hunt was assured, and this gave the hunters the courage to tackle animals that were much stronger than they were. This clearly did not eliminate the need for good hunting techniques and weapons, but their potential was limited. An unsuccessful hunt could be blamed on the counter-magic of a hostile tribe.

This brief sketch of the world of totemic people is of course highly rationalized; consequently, it can be fundamentally misleading. We may ask, for example, how it is possible that these so-called primitive people, being capable of systematically analysing nature, did not know the futility of magic. In posing such questions we touch upon the difficulty of abstracting and analysing an aspect of another culture by means of our own. The coherence of a symbolic universe does not depend on the factuality of any aspect. So-called primitive people knew very well that magical techniques often failed, but accounted for this failure as being the result of the counter-magic of hostile tribes, errors made in the ritual contract, or the spirits being angered by the village. To understand a symbolic universe we must, as it were, attempt to live inside it and try to understand it in terms of itself. For example, as children we knew that plants could not respond to us since it was implicit in everything the people around us said and did. Much of what we learned was based on these kinds of beliefs. Even what we learned in science class was ultimately accepted at face value. We believed our teachers, and these beliefs are rooted in the myths and sacred of our culture. Any symbolic universe based on a system of myths becomes incoherent, primitive, and irrational when regarded by a mind embedded in another culture based on a different cultural unity.

To sum up, the technologies typical of the transition from prehistory to history appear to differ from modern ones by the way the cultures of the time symbolized reality. There existed no separate 'nature' or 'environment' as distinct from humanity. Reality was essentially seen as one gigantic family of life, thus necessitating the unique relationship between religion, magic, science, and technology. Both the role of technology and the constraints placed on it derived from the way people made sense of and lived in the world by means of their culture.

Traditional technologies also behaved very differently from modern technology. They were not easily diffused beyond cultural boundaries, and when they were it was not without some significant adaptations to themselves or to the recipient cultures. The diffusion of a particular constellation of technologies depended on the diffusion of the culture (including its religion) in which it was embedded. Within a culture there is a certain unity, but between cultures there is little or no diffusion of technology except when accompanied by cultural or technological changes. The result was an enormous diversity of technologies associated with a corresponding cultural diversity. Technology tended to be so deeply enfolded into a culture that the two were almost inseparable.

Until relatively recently in Western civilization, it apparently did not occur to anyone to compare the different technologies to determine which ones were more efficient. This would have limited diversity. For example, many cultures made swords. Their members knew that neighbouring cultures made their swords differently, but they did not compare them to see which ones were the most efficient (as measured in terms of performance values). Swords continued to be made according to certain traditions, and this is why we can speak of traditional technologies.[27] Of course, people were concerned about the effectiveness of their swords in battle, but this is an entirely different matter than our modern preoccupation with efficiency.

Whenever an empire politically dominated a diversity of cultures, as did the Roman Empire, the cultural ensemble became more unified by developments such as the construction of roads and the growth of commerce. The result was a melting-pot phenomenon that caused the erosion, transformation, or collapse of many cultures. In the case of the

Roman Empire, Greek culture did not survive, while Jewish culture did. At this point a certain technological diversity began to appear on the local scene. Technological development now depended less on a specific culture, and artisans and craftsmen became more experimental. They tried different things as they were exposed to various influences, which broke down or at least reduced the weight and authority of tradition by relativizing it. The large cities within the empires created a demand for a diversity of things that were either made in the vicinity or imported. In other words, technological development in some cases began to respond to commercial demands that were largely independent of the local culture.

3.4 No Detached Observers

I have sought to convince my readers that what is usually referred to as the Industrial Revolution is only one prominent aspect of a mutation in Western civilization. In many respects, this mutation began well before the Industrial Revolution and continued long after, producing a fundamental change in the way human beings are born, grow up, work, play, and die. On the level of culture and society, little remains untouched. Fundamental changes in technology continue to be accompanied by significant changes in the economy, the social structure, political institutions, the legal framework, morality, religion, art, and literature. In the English language, for example, many words underwent a fundamental change in their meaning during this period. It represents the birth of the world as we know it today.

In a sense, it may therefore be said that this mutation gives rise to who we are individually and collectively. We can hardly pretend to be neutral and detached observers of our own cultural birth and growing up. We are in a sense looking at our family album as we begin to explore a new leg in the journey of Western civilization, which is ongoing. More and more societies within Western civilization embarked on the same journey, and the consequences were increasingly felt all around the world. Today, all of humanity appears to be embarked on this journey, referred to by a variety of concepts such as industrialization, modernization, and development. Within the so-called industrially advanced nations, there is an ongoing debate around sustainable development. Can we continue the present journey and, if so, for how long? Does a finite planet impose limits on this journey? What kind of world is this journey helping to create? Is it the world we intended, or are we expecting another for ourselves and our children?

In any case, the bottom line is clear. The vantage point from which we will continue to examine the transformation of individual and collective human life that has taken place during the last two centuries is itself the outcome of the journey thus far. It is not a question of putting ourselves into the picture; we *are* in the picture. It is a question of critically examining how we are a part of this picture, and how this affects our perceptions and our work. Not to do so is to surrender to the myth of science. It holds that science is carried out by detached observers, which no human being can ever be. No myth could be more dangerous for our future.

Each one of us will have to put ourselves into the picture that will emerge as a consequence of the present analysis. It is important to destroy any pretence of being a neutral and detached observer. A critical examination of our vantage points will reveal those areas in which we are involved and how our own presuppositions and convictions play a role. By making this role as small as possible in our dialogue and in discussions with others, we can become more critically aware and thus as objective as possible in our investigation. I am not calling for an abandonment of science just because it is human in order to plunge into a complete relativism. On the contrary. I am calling for a restoration of science and our participation in it to the status of a unique human activity. None of us lives within reality but only within reality as we have learned to symbolize it by means of our culture, formal education, and experience. Objectivity may be seen as a commitment of the members of a free and open society, or of a scientific or professional community, to journey together as far as they can while being critically aware of their differences and how these unavoidably affect their work.

Although the reverse interaction between technology and society ('technology changing people') has not been given much weight in modern scholarship, it is implicit in the way we experience and talk about our modern world, as will be explored through a few examples. When we speak about industrial societies or advanced industrial societies, are we not implying that they have more in common than what divides them? For example, does this term imply that democratic, socialist, and communist industrial societies are essentially identical? Are the present developments in eastern Europe and in the former Soviet Union a manifestation that Marxist ideology has held back necessary adjustments in society, to the point that tensions between it and the industrial base have become so great that rapid major adjustments are now inevitable? If these adjustments are inevitable, can we speak of freedom or must we admit that these societies are acting out of necessity? What about the Japanese 'miracle'? Is Japan's traditional social structure, with its emphasis on the group rather than the individual and its respect of authority, a major factor in its success? If so, will the continued erosion of these traditional elements help reduce the differences between it and other industrialized societies? Is the American way of life primarily the result of 'people changing technology' or of 'technology changing people'? Will the rhetoric about free trade and globalization contribute to legitimating growing global competition, and will this force all industrialized societies to constantly restructure in order to make themselves into efficient nation state 'corporations' regardless of their way of life, culture, and history? What does all this imply for future generations and the biosphere?

Think about another terminology in common usage, namely, that of under-developed, developing, and developed societies. Does this terminology not imply that all the nations of the world today can be classified according to the success they have achieved in going down the same road, referred to as industrialization, modernization, rationalization, secularization, development, and progress? Are all nations more or less trying to do the same thing, with unequal success? If so, we are in a historically unprecedented situation. Some two thousand years ago it would have been impossible to compare the Greek, Chinese, and Aztec civilizations. They had totally different goals, expressed in radically different ways of life and cultures. They were incomparable, going in very different directions. Does the above terminology imply that things have fundamentally changed in the twentieth century? Yet it is not uncommon at international conferences to hear delegates from the Third World make a plea for the transfer of science and technology, while at the same time arguing that they do not want Western culture, which they regard as dehumanizing and destructive. Is it possible, however, to use identical means (modern science and technology) to achieve different cultural ends?

Answers to the above questions depend on the relative strengths of the two interactions involved in the process of industrialization. If, in the course of this process, the effect a society has on its technology is much more profound and far-reaching than the reverse effect of the newly created technologies on society, then it is possible for the same means to be used to achieve different ends. If, however, the latter interaction is much more profound and far-reaching than the former, then a society may find itself in a position where the overall consequences of the process of industrialization may be very different from the ones intended. In this case, the means may dominate the ends.

In the case of technology transfer from the industrially advanced world to the developing countries, there is a great deal of evidence to suggest that not every kind of technology is compatible with every kind of society. Much well-intentioned foreign aid has been a complete failure because of this.[28] It has been noted, for example, that before a traditional society can take advantage of modern technology, it must undergo fundamental changes in its culture and way of life.[29] Some authors have argued that intermediate or appropriate technologies must be designed that require fewer and less substantial changes in the recipient societies.[30] In other words, it appears that although the relationship between the transferred technology and developing societies is not deterministic, at least some kind of compatibility must exist.

It is thus essential to regard the influence society has on technology as inseparable from the influence that technology has on society. If the latter influence is at all significant, then a certain convergence may take place between all societies turning to modern science and technology to achieve their goals. Furthermore, since not any kind of technology can exist in any kind of society, the structure of the one constrains the possible structures the other can take on, and vice versa. Hence regional differences in Canada and the United States, for example, may in part be understood in terms of differences in the relationship between technology and society that correlate with different levels of industrialization. Similarly, differences between the northern and southern hemispheres can in part be understood in these terms. To put it somewhat simplistically, it is as though different levels of industrialization coexist in our modern world. Notice that we have not claimed that a given stage of technological development determines the kind of society of which it is a part, but that the influence that technological development has on the society may be significant enough to constrain the differences between societies that have achieved similar levels of technological development. When industrialization is seen as a reciprocal interaction between technology and society during which each transforms the other, then the necessities to which the members of a society must respond are in part created by that process. However, this in no way determines their response to them.

Finally, the interpretation of the behaviour of individuals or social groups during the process of industrialization (or at any other time, for that matter) should take into account their societal context. Individual freedom and choice are inevitably constrained by the necessities of the 'system.' To emphasize one at the expense of the other could cause a distorted view of the situation. By way of illustration, consider the parable of the good industrialist. After a long struggle, Mr Jones could no longer watch his employees live and work under the kinds of conditions all too typical of English industrial towns in the early part of the nineteenth century. He had undergone a conversion, in the etymological sense of that word, of seeing that reality must be symbolized in a different way. Putting his newly found beliefs into practice, he decided to pay his employees higher wages to enable them to improve their living conditions and to make his factory a better and safer place to work. With the constraints of the 'system' of the time, he had essentially two ways of accomplishing this goal. The first was to recover the costs by charging higher prices for his products. They would become much less competitive, sales would drop, and profits would turn into losses. The likelihood of the business going bankrupt would be considerable; and when that happened, Mr Jones would cease to be an employer. The second option was to use his personal wealth to cover the extra costs. However, all personal fortunes being finite, it would again be just a question of time before Mr Jones ceased to be an employer. To make a distinction, therefore, between 'good' and 'bad' employers makes little sense if the 'system' so constrains the social role of employers as to make it virtually meaningless.

Social roles increasingly expressed the influence of a technology-based connectedness on a culture-based connectedness. Although the new cultural unity made this somewhat more bearable, a heavy dose of ideology was required to make it liveable because the ability of a society to express its humanity through its culture became constrained. This did not mean the end of morality or the end of history. We should see the situation for what it was. The 'system' did not impose itself overnight, and during the early phases of its formation, human freedom could have been exercised in the face of necessity to ensure a more humane outcome. As time went on, this became more difficult. The first act of human freedom is to recognize a situation for what it is. There is no point in underestimating the necessities and constraints of a particular historical situation. The next step in exercising human freedom is to attempt to find some play in the 'system,' and when this becomes a grass-roots movement, a mutation may occur. The worst slavery to the 'system' comes from those who mislead themselves and others by claiming freedom and ignoring necessity and constraint. It leads to false hope and ignores the seriousness of the situation, thereby limiting the creative energy to impose human values and aspirations.

PART TWO

Disconnecting from and Reconnecting to Experience and Culture

PART TWO

Disconnecting from and Reconnecting to
Experience and Culture

People Changing Technology: Severing the Cultural Moorings of Traditional Technological Knowing and Doing

4.1 Transcending the Limits of Technological Traditions

The transformation of traditional societies into the first generation of industrial societies, as described in Part One, gradually put enormous pressures on their technological traditions. This was to be expected since these societies found it next to impossible to continue their tradition-based ways of life. Too many of the required adjustments were without any precedent, and their linking together into an expanding technology-based connectedness greatly weakened the culture-based connectedness. Since technological traditions were integral to these societies, they were weakened both from within and from without. They experienced an additional limitation resulting from their need that anything of technological importance must manifest itself (directly or indirectly) to the senses in order to make it accessible to their experience and culture. Such was the case for the leading industries in the first phase of their development. For example, it is entirely apparent when a spinning machine breaks threads or a particular part overheats. Most construction materials provide warnings of one kind or another of extreme stress before they fail. And metallurgical processes tend to have sufficient observable manifestations that correlate reasonably well with what is happening. In contrast, much less can be observed about chemical processes, and, in many cases, what can be observed does not provide much indication of what is really happening. This is even more so for electrical phenomena, particularly those occurring in circuits.

Such limitations are symptomatic of a deeper structural problem, which was gradually resolved by traditional technological knowledge being replaced by something new: technical knowledge. In Part One, I showed that the process of industrialization strengthened the technology-based connectedness of a society at the expense of its culture-based connectedness. Traditional technological knowledge had been developed to deal with situations where the technology-based connectedness largely overlapped with, and was subservient to, the culture-based connectedness of a society. With the birth of the technical division of labour, mechanization, and industrialization, technologies increasingly had to be symbolized in relation to the socio-technical entities within which they functioned, and through these to human life and society. Thus the traditional culture-based symbolization that sought to understand everything in relation to everything else in human life and society was not very effective. What was really required was a way of symbolizing a technology in relation to the local technology-based connectedness of a society, and through it to the culture-based connectedness. The new cultural unities had prepared for this possibility. A new intellectual division of labour in science and technology had to be invented to help bring this about. How this emerged and how it further strengthened the technology-based connectedness of a society at the expense of its culture-based connectedness is the subject of this second part of our narrative concerning where we are going with science and technology. Traditional technological knowing and doing based on experience and culture were replaced by non-cultural alternatives. Their emergence required further changes in the role of culture in individual and collective human life, which we will examine in Part Three.

The limitations of traditional technological knowledge were transcended by a transformation that I will refer to as the separation of knowledge from experience and culture, or simply the separation of knowledge from experience. It in turn led to technological doing that was separate from experience. Such a transformation is somewhat different from the conventional view of a strong interaction between science and technology in the latter part of the nineteenth century. The separation of knowledge from experience has even less in common with the perspective that holds that technology became applied science. These and related subjects will be revisited in the chapters that follow. Our focus will be on how the chain-reaction-like process of industrialization, as described in Part One, encountered a major bottleneck in the latter half of the nineteenth century, which is directly related to the limitations of traditional technological knowing and doing. It was joined by a number of economic limitations roughly during the same period.

Some observers, including Karl Marx, were convinced that, based on the economic trends of the time, capitalism would be unable to overcome its limitations and would consequently weaken and collapse from its own internal contradictions. Such a collapse did not materialize because the separation of knowledge from experience and culture transformed capitalism and transcended many of its limitations.

Although England was the leading industrial nation at the time, it did not pioneer the transformation of technological knowing and doing. This was Germany's accomplishment, and it succeeded in this by forging a new relationship between the university and industry – a relationship which proved so effective that Germany quickly became the leading industrial nation and memorably flexed its industrial-military muscle during the First World War, which was also the first technological war.

In order to examine the separation of technological knowing and doing from experience and culture, our focus will shift to the 'software' of technology. For a nation to take full advantage of the new technological knowing and doing, another transformation of both technology and society was required, once again taking the form of a chain-reaction-like process of interdependent adjustments. This led to a universal technology and a new kind of industrial society, the nature of which sparked considerable controversy at the time. The complexity of this double transformation was much greater than that of the transformation described in Part One. Part Two therefore focuses on 'people changing technology,' primarily in relation to industry. The present chapter examines the phenomenon of knowledge separating from experience and culture, beginning with culture-based acquisition of skills. The next chapter will examine scientific disciplines and technical specialties as transcultural elements. The following chapter will describe the adjustments a society must make to its technological, economic, social, and political dimensions of mediation in order to make use of these new transcultural elements. The analysis of 'technology changing people' will be taken up in Part Three. In developing the generic model of this transformation of technology and society, I will shift the historical and geographical focus from England in the latter part of the eighteenth and nineteenth centuries to the United States, primarily following the Second World War. It was there that the Fordist-Taylorist system of production and the requisite changes to society were most pronounced, making the United States the pre-eminent manufacturing nation in the world for a number of decades. Once again, my references to the United

States may not satisfy sociologists and historians because these simply serve as a point of departure for developing a generic model of this new stage in the process of industrialization. These changes in the 'software' of technology, to be described in Part Two, were superimposed on the ongoing changes in the technological 'hardware.'

4.2 The Destruction of Technological Traditions

Generally speaking, until the end of the nineteenth century, technological knowing and doing represented an elaboration of a society's culture-based knowing and living in the world. Within traditional ways of life, technology had its own traditions made up of customary ways of doing things. Technological knowledge was transmitted from one generation to the next by apprentices who learned by doing under the watchful eye of someone who had mastered a particular craft. School-based learning complemented learning by doing: it provided rules of thumb for the calculation of particular details, for example. The mastering of a technological skill through an apprenticeship program was, in essence, no different from the processes by which children acquire daily-life skills for making sense of and living in the world on the basis of a culture. Each daily-life skill is learned as an integral part of learning to live in the world since it is symbolized by an ever more differentiated structure of experience. The metaconscious knowledge associated with a particular skill is integral to the metaconscious knowledge associated with making sense of and living in the world. The strongest evidence that this is the case comes from traditional technologies. Their cultural appropriateness and environmental sustainability exhibit their participation in symbolically knowing and doing everything in the context of everything else. I will seek to show that the acquisition of an individual skill is integral to the acquisition of a culture.[1] Stuart Dreyfus distinguishes five stages in the acquisition of skills: novice, advanced beginner, competent, proficient, and expert.[2] This theory focuses primarily on adult learning, which, for the first few stages, can be guided by rules. By reinterpreting my theory of how babies and children learn a culture as a dual process of acquiring individual skills and the skill of living in the world, it is possible to combine the two theories.

During the first stage, apprentices are taught to look out for certain features of situations that are relevant to the particular skill and to respond to them by means of simple rules. Initially, this activity fully engages their attention; anything else that may be happening is a distraction. For this reason, the rules must be context-free and apply to a whole range of situations. Initially, their experiences of situations are based on a foreground-background distinction in which the features relevant to the rules constitute the foreground. As these experiences are internalized, symbolized, and incorporated into the structure of experience by means of the processes of differentiation and integration, apprentices become more aware of their similarities and differences without being able to articulate what these are. They are refining the relevant portions of their mental maps, gradually permitting the consideration of more features in each case. As more of the context is drawn into the foreground of their experiences, the simple rules with which they started become increasingly inadequate. More complex rules that pay attention to more of the context of each situation are now required. Everything else remains a distraction with respect to the practice of a skill. The process of differentiation is now based on a more extensive range of prior experiences, permitting a greater diversity of situations to be easily recognized as essentially similar to previous ones. Other situations may be seen as somewhat dissimilar, and although the rules are still applied, it is now with some sense of caution, apprehension, or reluctance. Apprentices begin to realize that, in some situations, the literal application of a rule does not make sense. Little, if any, metaconscious knowledge has as yet been accumulated in that portion of the structure of experience of an apprentice that corresponds to his or her learning the skill. These developments characterize the second stage of skill acquisition, that of advanced beginner.

When people become competent in carrying out a particular skill, they no longer follow the rules automatically. They begin to recognize a range of situations where this makes no sense, leaving them with no other option but to analyse such situations in order to determine an appropriate response. They begin to develop plans for dealing with situations in which the routine application of rules is not satisfactory. In other words, the processes of differentiation and integration have expanded the structure of experience to the point where there is a growing number of situations to which routine rule-based responses are not appropriate. However, the experiences of such situations have not yet been sufficiently structured by the processes of differentiation and integration for prior experiences to function as paradigms. There are now two kinds of behaviour patterns: more or less routine responses to some situations, and responses that involve analysis and plans for dealing with unusual situations. Apprentices now become much more involved in what they are doing and begin to have a greater sense of responsibility if anything goes wrong. Routine responses also become affected. It is now much less a question of the apprentices mechanically applying rules that they did not invent to situations over which they have little control. There is a beginning of a greater emotional and intellectual involvement in applying the skill. This development is essential for learning because a successful response to a somewhat unusual situation brings satisfaction, and failed responses are less easily forgotten. In other words, the structure of experience is beginning to imply metaconscious knowledge and values capable of guiding the development of the skill. Processes of differentiation and integration enfold these values into similar past and present developments in the entire structure of experience of a person's life. As Stuart Dreyfus notes, during this stage people tend to behave like stereotypical puzzle-solvers, but this does not mean that human intelligence is first and foremost problem-solving that can be captured by heuristics, algorithms, scripts, frames, or other rule-based entities. That this is the case becomes evident during the next two stages of skill acquisition.

The transition to the fourth stage, that of proficiency, is marked by a qualitative change. It is as if the apprentice has an almost immediate intuitive grasp of the situation, but he or she still has to think about a response. The number of experiences that the apprentice has accumulated in a particular skill domain has grown to the point that the processes of differentiation and integration have been able to create an extensive symbolic structure of these experiences that is now beginning to act as a repertoire of models that are intuitively (metaconsciously) used to grasp new situations. It is only when a situation is sensed to be different for some reason or another that conscious analysis comes into play. There is a kind of holistic recognition of a variety of situations. Stuart Dreyfus calls it 'knowing how,' or intuition, as distinguished from the 'knowing that' based on rules and sets of isolated features deemed relevant for the application of a skill. The relevant portion of the structure of experience of the apprentice now functions as a well-developed mental map permitting most situations to be metaconsciously recognized as being similar to previous ones.

In the fifth stage, that of expert, there is also a growing incidence of responses to situations that reflect a fluidity, rapidity, and level of involvement not seen before. In other words, the metaconscious knowledge built up in the portion of the structure of experience related to the skill domain is now so extensive that fluid and intuitive responses to many situations can be made with the apparent ease that characterizes expert behaviour. The skill has now become such an integral part of the former apprentice's structure of experience that only very rarely do conscious choices and decisions need to be made. Under normal conditions, experts do not have to solve problems or make decisions, much as I have shown in the acquisition of culture-based daily-life skills. The development of the structure of experience corresponding to the skill domain is now so extensive as to be able to cover almost any situation. Insofar as analysis does occur, it is no longer addressed to the situation directly but to the intuitive sense the expert has of it. The processes of differentiation and integration have fully realized their potential for creating metaconscious knowledge. In other words, human intelligence is ultimately based on making sense of and living in reality by giving everything a place in our life in relation to everything else. It is this relational character of our being in reality rather than conscious problem-solving analysis that is the basis for human intelligence.

In a later paper,[3] Hubert and Stuart Dreyfus apply the five-stage model to the skill of morally coping with the world. From the above description, it follows that learning to do so is an integral part of acquiring skills and living a human life. My theory of the acquisition of culture and daily-life skills suggests that learning a skill ultimately consists in giving it a place in our life and our being in a way that is individually unique and culturally typical. Hence this learning contributes to the development of metaconscious values and thus to our morally coping with the world. In a subsequent work,[4] Hubert Dreyfus elaborates this by adding two more stages to the skill-acquisition model, namely, mastery and practical wisdom. As we further develop the theory of knowledge separating from experience, it will become apparent that this latter stage in particular becomes unattainable in cases where expertise has separated from experience and culture. This is because the metaconcious knowledge generated is also separated from experience and thus from the values of the culture.

Although a neural net is a fascinating analogy for understanding the development of a symbolic structure of experience in the brain, there is no separate self to which the outputs of the net can be presented. There is also a vast difference in complexity. Hubert and Stuart Dreyfus estimate that a chess master can recognize roughly fifty thousand different positions and that expert automobile drivers can probably intuitively respond to a similar number of situations.[5] When this is multiplied by all the daily-life skills that human beings acquire, it is obvious that our language does not have a sufficiently large vocabulary to make these distinctions explicit. Hence, much of human intelligence cannot be expressed in words – a problem that has plagued knowledge engineering from the beginning.

In order to understand how the first phase in the process of industrialization put pressure on tradition-based technological knowing and doing, we need to examine how the above process of skill acquisition was affected by the rapidly changing context in which it took place. First, consider the apprenticeship processes in those industries that led industrialization, namely, textiles, machine-building, and metallurgy. Initially, the features of particular situations relevant to a skill correlated very well with what was actually happening. For example, the inadequate strength of a machine component might be evident from excessive bending, deformation, and, in the extreme, failure. A visual inspection readily suggested what needed to be done. Similarly, the adequacy or inadequacy of most other design features of a machine could generally be correlated with observable symptoms, including excessive vibration of a workpiece or tool because of inadequate support; changes in colour and odour or frequent seizing up related to the level of lubrication; and the constant breaking of threads in a spinning machine. Similarly, in metallurgy the colour of the flames, ores, and metals correlated rather well with what was happening.

Now contrast this with the acquisition of the technical skills required by the chemical and electrical industries that emerged a little later. For example, brown smoke given off by a chemical reaction can mean a number of different things. The colour of a chemical does not correlate with its basic properties. Very little can be observed from the workings of electrical circuits. In other words, where the industrial processes occur out of sight and below the surface, far fewer features can be correlated with what is actually happening. Hence, carrying out a skill on the basis of negative feedback from what can be observed by the senses becomes much more difficult. Even where gauges are used, a theory is required to correlate their readings with what is actually happening.

The higher the level of technological sophistication in the textile, machine-building, and metallurgical industries, the more subtle were the observable features that could be correlated with what was happening, eventually leading to a situation that was somewhat analogous to that in the chemical and electrical industries. What could be observed on the basis of experience was increasingly inadequate to guide further technological development. Drawing on natural science and mathematics did help in coping with the limitations of technological *knowing how* acquired from the experiences of learning a skill under the watchful eye of a master by contributing to *knowing that*. It was not long, however, before this too became inadequate. Skill acquisition based on the world of experience and culture had to be replaced by an abstract world that lay 'underneath' experience and which could be accessed only in equally abstract ways.

The second pressure exerted on tradition-based technological knowing and doing resulted from the necessity of technologies having to be linked into networks and ensembles, for reasons I examined in Part One. A culture-based process of skill acquisition symbolizes the meaning and value of anything in relation to everything else in a person's life and thus in the context of the way of life of his or her culture. This is equally true for the process of skill acquisition, which relies on the same processes of differentiation and integration to create a symbolic structure of experience. I have suggested that, as apprentices advance through the five stages of skill acquisition, they are able to take into account the ever-broader context of a situation. However, the more a technology involved in the acquisition of a skill becomes part of a network or ensemble, the more difficult it becomes to metaconsciously symbolize its meaning and value in relation to everything else in individual and collective human life. In part, this is because the technology and associated skill now interact with human life indirectly via the functioning of the network or ensemble. It is also because, from the perspective of the apprentices and masters, what matters most is what might be considered as the *technological* meaning and value of the technology and associated skill, that is, their place relative to everything else in the network or ensemble. In other words, the technology-based connectedness has to take precedence over the culture-based connectedness, but this runs contrary to the process of skill acquisition. The conflict grows as more and more of the 'hardware' of technology is linked into networks and ensembles that, in turn, become linked together into one technology-based connectedness as the first phase in the process of industrialization is completed. To resolve this conflict requires a different kind of technological knowledge based on different meanings and values, the features of which I have described elsewhere.[6] This eventually leads to performance ratios replacing human values in technological and economic growth. These ratios measure how well a particular technology and the associated activities strengthen the local technology-based connectedness at one of its nodes by the success of producing as much desired output as possible from the requisite inputs. This implies a further potential for strengthening the technology-based connectedness of the socio-technical network or ensemble and, through it, the overall technology-based connectedness of a society. Performance ratios reflect the 'meaning' and 'value' of a technology within this technology-based connectedness in terms of the performance a technology contributes. In other words, the 'meaning' and 'value' of a technology in this context become closely related to its power (defined as the rate at which desired outputs can be obtained from requisite inputs). The influence of technology-based connectedness on culture-based connectedness essentially converts these performance ratios into values, thereby reflecting the dominance of technology-based connectedness over the culture-based connectedness in these societies. However, I am getting ahead of my story, and will return to this subject in detail in Part Three.

The third source of pressure on tradition-based technological knowing and doing was rooted in the dependence of the master-apprentice relationship on a technological tradition. These technological traditions were capable of extraordinary feats even by modern standards, as long as they were not overwhelmed with internal and external changes that were either too numerous and/or too different to permit the orderly adaptation of the traditions without losing their internal coherence. Under such conditions, it was very difficult to correlate the features of situations with what was actually happening because there was too much going on at the same time. Too many situations coming fast and furious overwhelmed the stock of precedents tradition had to draw on for its evolution and adaptation to new circumstances. From our analysis in Part One, it is clear that these conditions for an orderly adaptation of technological traditions were progressively eroded as industrialization advanced. As the limits of the adaptive capabilities of technological traditions were reached, it became clear once more that a different kind of technological knowledge had to be found. Technological traditions were an integral part of the entire tradition of a society, and the pressures on the former contributed to the growing ineffectiveness of the latter. The tradition-based character of the cultures of industrializing societies was coming to an end. The technological and societal contexts simply became too turbulent for the tradition-based mode of advancing technological knowing and doing.

Finally, a fourth pressure on tradition-based technological knowing and doing exerted by industrialization came from competition. This becomes evident when, within this complex turbulent environment, we seek to understand the evolution of the entrepreneurial firm, which was the basic component of industry. If we plotted a measure such as efficiency, productivity, or profitability over time, a learning curve would emerge in which three developmental stages could be distinguished. During the first, the number of factory-based enterprises in a particular sector of the economy was very small, so they competed against the putting-out system of textile production. It should be remembered that the mechanization of textile-making in England allowed businessmen to have a greater control over the process, which was important during times of rapidly changing demand. As this advantage became obvious to other entrepreneurs more textile factories emerged, which led to a second stage during which factory-based enterprises increasingly competed against one another.

At first the competitive pressures were met by making further technological advances. This was not very difficult. Intensive and prolonged experience with machines made their weaknesses and limitations obvious. With a little ingenuity, improvements could be made by many people. As the machines became more sophisticated and complex, greater human talents, more time, and thus a greater investment were required to make further improvements. Typically, such improvements produced smaller and smaller gains. Businesses thus entered the third stage of the learning curve, characterized by the fact that ever more talented people had to spend ever more time to realize ever-smaller gains. In economic terms, this meant that those businesses that could write off this investment over many machines could make it profitable, while smaller businesses could not benefit from these economies of scale.

At this point, competitive pressures forced entrepreneurs to review other aspects of their business operations. Ever-faster, more complex, and costlier machines necessitated changes in plant layout. It was not uncommon for a traditional factory to have machines of one kind grouped together in one area or room. Imagine that a worker had to use four kinds of machines for a particular production stage. As the machines became increasingly rapid, the portion of the cycle time spent operating machines in this production stage diminished. To keep the machines running required more people marching back and forth from one area or room to another. It was not long before the buying of more efficient machines became uneconomical unless they were rearranged so that the output of one machine went directly into another with minimal delay. Modifications to many other aspects in the operation of entrepreneurial firms, such as purchasing and marketing, also became necessary. What, for example, was the point of buying more efficient and costlier machines when purchasing procedures were relatively inefficient, resulting in the factory having to be shut down several times a year because of a lack of raw materials? Traditional marketing approaches also became inadequate. Shutdowns, because the warehouse and factory aisles were full of unsold goods, had to be avoided at all costs if faster and more expensive machinery was to be profitable. In other words, within the firm a kind of chain-reaction-like effect could be observed as well. The result was that a greatly expanded technology-based connectedness imposed itself on the culture-based connectedness of the traditional organization of the firm. Barring major breakthroughs, most aspects of the business soon ran into a situation in which ever-larger efforts representing an ever-increasing investment of human talent and financial resources were producing diminishing returns. What was true for particular businesses sooner or later characterized an entire industrial sector, and eventually all of industry.

During the last part of the nineteenth century, the development of the textile, machine-building, and metallurgical industries in England began to level off without competitive pressures being reduced in any way. It helped to drive the English economy into a recession, the severity of which was somewhat moderated by the rise of the chemical and electrical industries. Based on tendencies of that time some observers, including Marx,[7] expected that the plateaus reached by certain industries, coupled with ongoing competition, would soon cause capitalism to collapse under its own contradictions. There is no doubt that severe economic difficulties contributed to the outbreak of the First World War, the Great Depression, and the Second World War. However, the breakup of capitalism never occurred, in part because of an entirely new development that no observer in the nineteenth century could have anticipated. It was the transformation of the 'software' of technology – a result of the separation of knowledge from experience and culture – that superimposed itself on the ongoing transformation of the 'hardware' of technology. This led to the removal of the apparent plateau that had been reached by technological and economic growth resulting from the limitations of technological knowledge almost entirely based on experience and culture. This kind of technological knowledge was not very effective in dealing with the growing technology-based connectedness.

4.3 Parallel Modes of Knowing

What was the new kind of technological knowledge that began to play an increasingly significant role in the leading industrializing nations of western Europe during the late-nineteenth and early-twentieth centuries? The usual answers include technology becoming applied science; technology acquiring a scientific basis; and the interaction between science and technology becoming so important that each transformed the other without losing its distinct identity. When we address the question of the relationship between science and technology, the first thing to realize is that there is no science as such but simply a host of weakly interacting scientific disciplines that draw separate pictures of aspects of the world from different vantage points, which cannot be easily integrated.[8] The structure of science is therefore very different from that of modern technology, which, as I am in the process of showing, has a systems-like structure that integrates its many branches on several levels. It is these structural differences that make it very difficult to satisfactorily characterize the relationship between science and technology in one of the above three ways. I will seek to steer a new course in examining science and technology as human creations that play a role in a way of life. In the past, both science and technology were particular manifestations of their host culture. Today, science and technology have become universal, which is a result of their distance from culture-based ways of life. In a sense, they have become separated from experience and culture.

That this is the case is evident from our daily lives. Today, many of us know certain things in two parallel ways: one related to daily-life experience, symbolized by a culture; and the other from school-based learning. As a first example, consider the knowledge we have of the grammar of our mother tongue. Elsewhere,[9] I have shown how babies and children acquire language skills without being able to articulate them. A five-year-old has acquired many language skills without any explicit knowledge of grammar, which she will learn in school. The two ways of knowing grammar remain relatively distinct. We have been unable, for example, to have computers mimic daily-life language skills from rules and algorithms. Philosophy has never been able to find the rules that underlie our daily-life language skills, leading some philosophers to the conclusion that such rules do not exist. We recognize this ourselves when, for example, we help a friend who is a recent immigrant with a letter she is writing. At some point, we may have to say: I cannot tell you what is wrong with this sentence grammatically, but we just don't say it that way. Learning the grammar of our mother tongue clearly does not constitute the foundation for our language skills. These remain metaconscious, enfolded into the structure of experience. This metalanguage is the basis for communication between the members of a culture and helps to maintain a dialectical tension between individual differences and a cultural unity. However, this metalanguage is not a set of rules, algorithms, scripts, frames, or other rule-based entities. Hence, the foundation for our daily-life language skills is qualitatively different from the grammar rules we learn in school. I recognize that this argument would have to be considerably nuanced to satisfy linguists, but I believe this can be done and have outlined it in an earlier work.[10]

A similar case can be made for our school-based scientific knowledge of the world and our daily-life knowledge symbolized by our culture. Among the daily-life skills learned by babies and children is an ability to arrange events in time and space and to relate things causally. Once again, it can be shown that the processes of differentiation and integration operating within the structure of experience develop metaconscious knowledge of time, space, and causality. These are the metalanguage prerequisites for children to make sense of the explicit concepts. In the same vein, the daily-life experiences of manipulating objects lead to a metaconscious knowledge that some studies have referred to as 'intuitive physics.'[11] According to these studies, the intuitive physics of students in our culture resemble theories that were widely held before Newton. This is not surprising because these theories sought to elaborate and systematize daily-life experiences and thus made intuitive physics explicit. One such study showed that, even after students had been introduced to Newtonian mechanics, many of them continued to respond to problems couched in daily-life terms on the basis of their metaconscious knowledge of moving objects. The studies also showed that many students find Newtonian mechanics counterintuitive and that they get themselves into trouble because they continue to attempt to understand it in terms of their intuitive physics. The two modes of knowing continue to coexist. These kinds of studies point to important pedagogical difficulties in the teaching of mathematics and science that have not been overcome. The problem does not go away; even physicists have to rely on their intuitive physics when biking.

The findings ought not to be surprising. When students are introduced to Newtonian physics, they do not encounter the world of daily-life experience. Instead they encounter what can be referred to as a 'mathematical abstraction' of frictionless planes, point masses, weightless strings, frictionless pulleys, zero air resistance, and a great deal more. This abstraction occurs nowhere in the real world and cannot be directly experienced. In fact, it is rather difficult to simulate such a world, and many high-school experiments do so rather poorly. Is it any wonder, therefore, that students encounter this mathematical abstraction as counter-intuitive? Is it surprising that students who prefer experiential learning have above-average difficulty learning mathematics and physics? It is clear that Newtonian mechanics is not an extension of the metaconscious knowledge students have acquired through their daily lives. We ask these students to temporarily suspend their intuitive physics until they are much more advanced in the subject so that they can gradually develop the mathematical abstraction to approximate the real world a little more by compensating for the fact that planes do have friction, that pulleys have inertia, that strings have weight, and that moving objects encounter air resistance. However, this is not possible until they are well advanced in the subject. In any case, most students do not get that far.

The difference between the intuitive physics of the students and the Newtonian equivalent stems from several factors. First, the former builds on daily-life experiences that are metaconsciously incorporated into the organization of the brain-mind by means of neural connections, thereby creating a great deal of metaconscious knowledge. In other words, intuitive physics builds on and extends daily-life experience symbolized by means of a culture. Pre-Newtonian physics generally resulted from equivalent but conscious and explicit attempts to do the same. In this respect, Newtonian physics was counter-intuitive and its mathematical abstractions broke with experience and culture. Second, the frame of reference of daily-life experience is that of our physical embodiment in the world. What we experience is how the behaviour of physical objects is symbolized by means of our culture. When students begin to learn Newtonian physics, they cannot enter the world of the mathematical abstraction in the way they participate in the daily-life world. They must learn to act as detached, and thus disembodied, observers – something that some students find very difficult. It also implies a shift in the frame of reference that no longer maps meanings and values but concepts that can (at least initially) have meaning and value only in the mathematical abstraction of the world of Newtonian physics. Third, the shift from intuitive physics to Newtonian physics is made very confusing to the students because of the claims of science voiced by their teachers. Students are not encouraged to use their imaginations to enter this abstract world of Newtonian mechanics that clearly exists nowhere. On the contrary, they are told that science is objective, that they are studying the real world, and that their experiences are merely subjective, with a very limited reliability. Yet it is clear that when they leave their physics class to go to the gym to play volleyball, they had better not switch off their intuitive physics and switch on their newly acquired objective and reliable knowledge of mechanics. What the explicit and hidden curricula tell the students about science in general and what is happening in their physics classes in particular is not very helpful in understanding the meaning and value of the physics classes for their lives. We do not teach Olympic athletes the latest physics in the expectation that it will constitute a more reliable and objective foundation for their skills. Nobody doubts that this would be disastrous if they are to cope well with their daily-life sports world.

In technology, we also encounter two parallel and relatively distinct forms of technological knowledge. Consider the following example. While erecting the steel structure for a high-rise office building, an experienced worker observes a crane bringing up a beam that does not look adequate for the span it must bridge. Before riveting it into place, he calls his supervisor, who will do one of two things. She may remind him that he gets paid to erect steel and to leave the thinking to the engineers. Alternatively, knowing that this worker has a great deal of experience, she may quickly run to the field office to check the drawing and possibly call the design engineer.

The worker has acquired a metaconscious knowledge of the strength of materials that permits him to gauge a particular situation, but all he will be able to say is that it does not look right or that things appear to be fine. From countless experiences, his brain-mind has built up a great deal of metaconscious knowledge of the strength of beams and columns. However, this metaconscious knowledge cannot cumulate into the equivalent knowledge engineers have. This is acquired in the classroom or from the reading of books that introduce engineers not to the real world of structures with beams and columns, but to the abstract world of a continuum, which, unlike a real material, has its properties uniformly distributed throughout space. Students learn to draw finite elements as small portions of such a continuum with normal and sheer stresses, with little or no reference to their physical basis. They are then taught to write equations symbolizing what goes on in the continuum by balancing forces and moments, corresponding to the finite element being in equilibrium. The situation is very similar to that of learning physics in high school. The technological knowledge of materials can thus be built up either from experience and culture or from the study of a mathematical abstraction that, as the students advance in their subject, increasingly approximates but is never identical to the real world. The knowledge engineers have of the strength of materials is not an extension of the knowledge that experienced workers acquire by erecting structures. The knowledge of engineers is contextualized within the abstract world of a discipline or specialty and is thus separated from experience and culture. In an extreme case, someone could learn stress analysis and design bridges but never set foot on a construction site. He or she would be unable to eyeball a situation to see if it looked right. Such a person's experiences of bridges would be the culturally mediated ones related to the daily-life use of bridges. These are very different from the 'experiences' of bridges via drawings, design calculations, and technical specifications.

On the shop floor we find the same coexistence of technological knowledge embedded in experience and culture of machinery and processes, and of corresponding knowledge separated from experience and culture. Engineers who design machinery and processes do so in a highly idealized and abstract fashion. For example, they assume that the properties of a material are uniformly distributed throughout space, although no such materials exist in the real world. A batch of identical parts heat-treated in a furnace will be affected differently depending on their exposure to the heat. Sometimes these differences have no practical importance, but at other times machining such parts could break tool bits, requiring adjustments that experienced tool-and-die makers would, in many cases, make without thinking. They are constantly bridging the gap between the idealized and abstract world of engineering drawings and specifications on the one hand and the world of their shop-floor experiences on the other. In other words, tool-and-die makers, machine operators, and shop-floor workers know the technology in the shop differently from the engineers. Both kinds of knowledge have their limitations and could play a unique role in conjunction, but unfortunately this is rarely the case.

By way of illustration, consider the case of a manufacturer of photocopying machines.[12] It became apparent that the machines assembled during one of the shifts had a well-below-average incidence of customer complaints and service requirements. An investigation led to the finding that the foreperson on that shift would not release any machine until it sounded right when tested. Unfortunately, engineers tend to dismiss this out of hand, a reaction that shows a lack of awareness of the limitations of their knowledge. The engineering drawings and specifications of the photocopier depict an ideal machine with ideal dimensions, with the only concession to reality being the tolerances within which the real dimensions should lie. This means that when a large sample of machines is examined in terms of the distribution of their tolerances, some machines will have most of their tolerances below the specified dimensions and others mostly above. Most machines will have a fairly random distribution. It is entirely possible, therefore, that in the above case machines whose tolerances were predominantly over or under did not sound right because they were either a little too tight or too loose. Apparently, the shift supervisor had noticed this and had learned to correlate it with certain problems that could be reduced by making some adjustments, with beneficial results for the company and its customers. The shift supervisor had a knowledge of these photocopiers that the engineers did not have, and vice versa.

Shoshana Zuboff interviewed workers to gain a better understanding of how they perceived the effects of automation.[13] Pulp and paper mill workers reported that after computerization they had little physical contact with the process, whereas, in the past, they could handle the pulp to see, for example, if everything was working properly. They were now obliged to watch numbers on a computer screen. The new control technology thus significantly curtailed the development of their technological knowledge of the process, based on experience and culture, which could negatively affect their performance as operators. The contribution that technological knowledge embedded in experience and culture makes to industry has been little studied, and few engineers and managers are willing to acknowledge its existence, let alone permit it to play a useful role.[14]

To sum up, I am further developing a previously made distinction between knowledge embedded in experience and culture and knowledge separated from experience and culture.[15] With respect to technology, these parallel forms of knowledge apply to different worlds, use different frames of reference, and involve people differently. It should already be apparent that the development of knowledge embedded in experience enriched with more and more science cannot gradually become knowledge separated from experience. There is a discontinuity, because the latter begins in an abstract world mapped with a different frame of reference by a disembodied observer who can only be 'present' in his or her technological imagination.

4.4 The Technological Knowledge of a Society

Thus far, I have elaborated the distinction between knowledge embedded in experience and knowledge separated from experience by means of examples of technological knowledge acquired and held by individual persons. I will now turn to the technological knowledge that a society possesses by developing one of the above examples in relation to the knowledge of the strength of materials.

Consider the knowledge of the strength of beams western European societies had during the first phase of industrialization. For example, naval architecture, mine construction, architecture, and machine-building involved considerable technological knowledge of the strength of materials. A master shipbuilder might take a newly constructed sailing vessel into a storm to observe how it fares. Based on a long tradition, he would know where potential trouble spots might occur and direct the attention of his apprentices to a variety of things that could be observed such as the bending of the masts, the sounds made by joints that were too heavily loaded, the deflections of various structural members, cracks that might appear as a result of fatigue, and some characteristic sounds of different kinds of wood under severe stress. He would be able to tell them whether a particular situation looked right or whether something should be kept under careful surveillance. He might point out that, on a similar ship built previously, there had been some problems in certain parts of the hull and point to how he had tried to deal with them. The master builder might explain what, in these cases, one should look out for and what one should do if a retrofit was required, and what could be learned for the next design.

Now imagine that the master shipbuilder visited a friend who was supervising the construction of a mine. Once again, this involved a great deal of knowledge of the strength of materials, but it was different in several important respects. The kinds of construction materials used were different, as, for example, more softwood was used in mine construction than in shipbuilding. The loads that had to be sustained were also different. The way a mast transmitted the forces exerted on the sails to the hull to produce forward motion was very different from the static, continuously distributed load sustained by the walls and ceilings of a mine tunnel or the load carts laden with coal put on a railroad track, supported at regular intervals by ties, to facilitate moving mined coal to the shaft for raising it to the surface. The geometry of the structural members also varied a great deal: cylindrical tapered masts secured to the hull at one end were very different from the beams and columns in mines that had rectangular cross sections of constant size. It would be impossible to transfer the knowledge of the strength of beams acquired in shipbuilding to what was required in mine construction. From the perspective of experience and culture, these situations were too dissimilar for one to provide paradigms for the other. Each master builder possessed technological knowledge that could not be transferred from one area to another. It was an integral part of a technological tradition that he had mastered.

The same is true when we compare the construction of mines with that of buildings. Load configurations, methods of support, geometries, and materials generally differ a great deal. In buildings a great many beams can be supported with masonry walls, and stone arches may take the place of beams. The same kinds of differences are encountered when we compare machine-building to the other three areas of technological construction.

As long as the technological knowledge of the strength of materials remains embedded in experience and culture, it is impossible to gather together into one body of knowledge what master builders in different areas know. From the perspective of experience, things appear fundamentally different in each area of application. The metaconscious technological know-how is built up within each tradition associated with a particular branch of technology. This technological knowledge could be made explicit by means of extensive experimentation. For example, beams with rectangular cross sections could be supported at each end and a load applied midway between. The deflection can then be plotted as a function of load. This load could be increased until the beam fails. The experiment could be repeated by varying other parameters, including the cross-sectional area, beam span, load configuration, and the kind of material from which the beam is made. The relationships between the different parameters could then be plotted to yield a series of design charts. A part of the metaconscious knowledge of the strength of materials would now have been made explicit and systematized. Before these can constitute an empirical design manual, however, the experiments would have to be repeated for all possible configurations encountered in each and every branch of the technology of a society.

This would have little value since the design charts for shipbuilding would generally not be applicable to the kind of situations encountered in mining, architecture, or machine design, and vice versa. Hence, little is gained by making metaconscious know-how explicit. This body of technological knowledge embedded in experience cannot be systematized any further except through mathematics, but this would involve a discontinuous separation from experience and culture. As it stands, the design manual would be exceedingly voluminous, enormously expensive, and inapplicable to any significantly new situations. Such a manual would do absolutely nothing that the technological traditions in shipbuilding, mining, architecture, and machine construction could not accomplish better with far fewer resources.

From the perspective of stress analysis, the situation is diametrically opposite. The many different beams now appear to be fundamentally alike, with only superficial differences, because each one is now seen in terms of the same continuum with slightly different distributions of normal and sheer stresses resulting from different load and support configurations.[16] That this is so is evident from the free-body diagrams engineers draw of these different situations. All context is stripped away and replaced by the forces exerted on the continuum, now represented by a line. Whether these forces are caused by the sheets of a sail, the guy wires of a mast, the mast support in the hull, a bookcase sitting on a floor, the soil above the roof of a mine tunnel, the pressure exerted by a machine tool, or anything else makes no difference. Similarly, whether the continuum symbolizes steel, stone, or wood shaped into different geometries matters little. Such details are symbolized by other mathematical abstractions such as the moment of inertia of the cross section and the modulus of elasticity of the material. By making all the situations encountered by the different branches of the technology of a society related to the strength of materials fundamentally similar, modern stress analysis is able to systematize this body of technological knowledge. It is now separated from experience and culture. In some sense, stress analysis looks at situations from the inside out, whereas technological know-how deals with them from the outside in. However, this is only partially correct since stress analysis substitutes a continuum for the real materials. What happens within that continuum (as represented by differential equations) does not correspond to what happens in real materials. These do not have flat interior surfaces on which normal and sheer stresses are exerted. There is no correspondence between the mathematical finite elements and the 'building blocks' of real materials. Nor is there uniformly distributed elastic matter. It is not a question of reducing phenomena that may be observed by the senses to underlying constituent phenomena until the most fundamental ones are reached. It is a question of mathematically symbolizing what is happening as opposed to symbolizing on the basis of experience and culture. The symbolization based on mathematics requires an embodied detached observer 'experiencing' the continuum by means of a technological imagination.

The relationship between the two bodies of technological knowledge related to the strength of materials (one embedded in and the other separated from experience and culture) cannot be grasped in terms of one being obsolete and inferior and the other modern and superior. Each one has its own peculiar strengths and limitations. It is obvious that technological traditions on the basis of knowledge embedded in experience and culture have accomplished spectacular feats in human history. Examples abound. The bases of the pyramids were levelled to a degree of accuracy that would be difficult to match, and their mode of construction remains a mystery. Many great buildings of the past have endured much longer than any modern structure will. For the music of the time, the acoustics of the cathedrals rivalled that of the best concert halls today. The sound quality of a Stradivarius has never been equalled. These technological traditions generally produced technologies that were socio-culturally appropriate and environmentally sustainable to levels we would be hard put to match. On the other hand, such traditions quickly reach their limitations when what can be perceived by the senses does not correlate very well with what is actually happening, when undesired consequences build up gradually almost unnoticed until it is too late, or when a tradition is overwhelmed by a highly turbulent context. An obvious example is the gradual salinization of the soil as a result of irrigation by the early Middle Eastern civilizations. Modern technological knowledge suffers much less from these limitations by substantially limiting its consideration of context and being primarily guided by performance ratios that have no context significance. The result is spectacular successes in the domain of performance and equally spectacular failures in the domain of context compatibility, making new technologies frequently inappropriate and generally unsustainable.[17]

There is another important difference between technological knowledge embedded in experience and technological knowledge separated from experience. It may again be illustrated with respect to the strength of materials. A master builder might look at a situation to see whether it was 'right,' which included but was not limited to the ability of the materials to sustain the loads imposed. It simultaneously had to be 'right' functionally and aesthetically, among other considerations. In other words, the know-how of the strength of materials was integral to technological know-how. This is not the case for stress analysis. It is impossible for engineers to solve the differential equations of stress analysis and come up with a design for a technological component to perform a particular function. All engineers can do is to design such a component first and then analyse it after-the-fact to determine whether it will be strong enough. Design and analysis are distinct but interdependent activities that each have their own paradigms. Here we begin to unravel the source of so much confusion in the literature dealing with the relationship between science and technology. I have argued that the grammar of children's mother tongue, which they learn in school, does not become the foundation for their language skills. Neither do physicists use their science when they ride a bike. Because of its inherent limitations, the knowledge engineers have of a factory will never become the foundation for the knowledge shop-floor workers have. In the same vein, strength of materials, together with all the other engineering sciences, can never become the foundation for engineering design. The former can only analyse certain details of something that has already been designed, and this order cannot be reversed.

4.5 A Discontinuous Change in Technological Knowing and Doing

I have suggested that in the course of the first phase of industrialization, technological knowledge embedded in experience ran into limitations as pressures on its ongoing development continued to increase along with competition. But this did not happen uniformly throughout all branches of industry. When people who had mastered a particular technological tradition encountered limitations, they had little choice but to turn to science to see what it had to offer to help them. For example, science might provide them with a better understanding of the fundamental phenomena involved in a manufacturing process. Failing this, they could adopt the scientific approach of controlled experimentation in an attempt to gain some of the knowledge they needed. The use of science could also permit quantification or the development of rules of thumb. That this process was omnipresent in

England during its first phase of industrialization is quite apparent.[18] However, to jump to the conclusion that this was the beginning of technology becoming applied science, the marriage of science and technology, the scientification of technology, or the beginning of the engineering sciences is about as plausible as the idea of a physicist using her science to ride a bike. The situation may change somewhat if, after having bought a new bike, she constantly falls and resorts to her science to find out whether or not there is a flaw in the design of the bike. However, this will never result in a situation where physics becomes the basis for her bike-riding skills. I believe this is exactly the kind of thing that was happening in England during the first phase of industrialization. People at the frontiers of the technological knowledge embedded in experience turned to science in an attempt to overcome some of its limitations. As such, science undoubtedly enriched certain technological traditions, but it can hardly be said that it provided a foundation for them. Had this been the case, then the development of technological knowledge separated from experience would have occurred first in England. That this was not the case suggests a discontinuity in the development of technological knowledge. Landes argues that the new technological knowledge base first began to develop in Germany.[19] So great was the ability of technological knowledge separated from experience to overcome the limitations of technological know-how embedded in experience that Germany rapidly took over the industrial leadership from England. This historical discontinuity corresponded to a discontinuity in the development of technological knowing and doing.

In order to understand this discontinuity, it is essential to briefly return to basics, namely, our symbolic links with reality based on a culture. From this perspective, let us consider the following question. Can science create a foundation for technology that is independent from culture? To accomplish this, it would first have to create such a foundation for itself, and many have believed in its ability to do so. It is not so long ago that science was seen as an activity carried out by detached objective observers who had broken with daily-life experience and culture and no longer observed the world from a vantage point reflecting their physical and cultural embodiment in the world. A detached objective observer is clearly a mythical entity, since it is impossible to view the world without the vantage point of the human body and the cultural 'lens' of the mind. It matters little whether this is explained away by asserting that empirical investigations aided by a neutral observation language can bypass a culture, or that rationality is built into our brains and is thus autonomous from experience and culture, or that the world is ultimately mathematical in nature and is thus not accessible by experience and culture. It is also not very long ago that the positivist account of science was widely accepted. It suggested that science was a cumulative progression towards Truth. Scientists were closest to this Truth and could therefore act the role of secular high priests, guiding society to a bright future. It is still very difficult for many of us to come up with a list of human concerns to which science cannot make a contribution. We continue to behave as if science were omnipotent in the domain of knowledge, and no religious tradition has escaped its judgment. As we shall see in Part Three, the myth of science is alive and well in modern cultures.

Thomas Kuhn may be interpreted as restoring science to its status of a special human activity carried out by communities of scientists.[20] He rejected the possibility of a concept of scientific progress having a sociological or historical validity. Thus science was not about absolute knowledge or truth. Kuhn's work remains a watershed in the thinking about science, precisely because it restored science to the status of a fully human activity. What remains controversial is what Kuhn claimed was the foundation of scientific practice, namely, the paradigm, or what he later called the disciplinary matrix. This concept can be clarified by relating it to that of a culture in the recognition that the experiences of a scientific education are grafted onto the structures of experience of students, thereby converting their mind's 'lens,' acquired from primary socialization, into a bifocal one – the bifocal part being the paradigm or disciplinary matrix. As we will see, in this way science is separated from experience and culture without being able to replace them. There cannot be a scientific foundation for individual and collective human life. It is simply not possible for a scientist to skip primary socialization and directly acquire the 'culture' of a scientific discipline. Scientific education remains a secondary socialization that converts the structures of experience of scientists into a bifocal lens through which they make sense of and live in the world as human beings and professionals in ways that are different yet not independent. The metaphor of a bifocal lens finds support in the fact that modern science is no longer tied to any particular culture, having become global and universal. Once again, this points to its separation from experience and culture.

What this separation from experience and culture involves may be illuminated by highlighting some of the changes in the relationship between science and culture that have occurred in Western civilization since the Middle Ages. A full account clearly goes beyond the scope of this work. All I can undertake is to briefly sketch a few central developments that had their roots in the Middle Ages.

In most traditional cultures, time, space, nature, and money were essentially defined by cycles, including those of daily-life activities, religious holidays, and the seasons. The experiences of daily life, when internalized, formed a metaconscious concept of time that was unique to each culture. The invention and growing use of the clock gradually reversed this relationship. Today, all events can be precisely situated in a linear time that lies behind or underneath these events, constituting an independent frame of reference. This modern concept of time is independent from modern cultures, having become standard time. Time is now an independent variable against which everything can be measured. Hence, it may be said that time gradually separated itself from experience and culture, beginning in the Middle Ages with a papal decree that monastic prayers be held seven times a day at regular intervals.[21] The full impact of this separation of time from experience and culture came during industrialization when, for the first time, it became necessary to synchronize the lives of factory workers with the rhythms of machines and factories. That this is very difficult for human beings is evident from the endless and severe disciplinary measures factory owners had to take against those who did not start or stop on time or wasted this new time. Many of these measures appear draconian to us, but they indicate how difficult it was for human beings to become accustomed to a concept of time that had no meaning or value in their experience and culture, because it was completely different from their metaconscious concept of time.[22]

The same observation can be made for the relationship between daily-life experience and space. For example, in many medieval paintings the positions in space of the things that were painted did not determine their relative proportions to each other. People or things might be drawn larger than others if their meaning and value were considered greater. In some paintings depth was limited to almost a single plane, while in others it appears to be restricted to two or three, one situated behind the other. In other words, things were painted according to their meaning and value in human life, and space was filled accordingly. It was not yet possible to think of the painter as an observer, relatively detached from the world 'out there.' People still saw themselves as integral to creation and as belonging to the land. It was not until the late Middle Ages that this creation broke up into humanity and the perfect clockwork of a separate nature (what later became the environment).[23]

It has been argued that the *Mona Lisa* was the first painting to depict the split between an inner and outer world.[24] It was this split that resulted in landscape paintings and the possibility of tourism, as people went to see the Alps, for example. Space now separated the observer from what was observed. As was the case for time, space gradually became an independent frame of reference, taking the form of a three-dimensional Cartesian space within which everything in the daily-life world could be situated.

Once again, industrialization greatly accelerated and expanded these developments as machines took centre stage, compelling a rearrangement of human life around them. The contributions of particular machines have been examined in great detail. For example, the railroad played an important role in transforming the traditional time-space-society continuum.[25] Travellers were now embodied via the train into the landscapes through which they journeyed. This required a different kind of perception, referred to as 'panoramic,' comparable to what Simmel called 'urban perception.'[26] Gradually a new culture of time, space, and the social was created.[27] In the course of centuries, time and space separated themselves from experience and culture. For some time, culture continued to arrange individual and collective human life within that time and space. This also began to change with industrialization. As noted in Part One, the shift to a market economy required that almost everything take on a monetary value. It situated everything in relation to everything else within the market and the economy (the 'world' of capital was at one time based on the gold standard), undermining the way culture values everything in relation to everything else in individual and collective human life. To a considerable degree, the capitalist order separated itself from that of experience and culture. The chain-reaction-like process of industrialization put so much pressure on the cultural resources of society that it became difficult for a society to evolve on the basis of its traditions. For Adam Smith and other economists, this was a positive development since the common good would best be served by people acting rationally in their own self-interest. The other side of the coin was an undermining of the role of culture by which activities are interpreted and motivated by their meaning and value in relation to everything else in collective human life.[28] Human activities now first and foremost required time, and time was money. Activities that did not contribute to wealth creation were a waste of time. In this way, the old cultural order was replaced by another in which activities were embedded in time, time in money, and money in the sacred order of capitalism. To sum up, time, space, and the social gradually became separated from experience and culture. It was as if the symbolic universe of a society had been relocated in an independent time and space and the technology-based connectedness of human life had become the raison d'être of the culture-based connectedness. Within the cultural cycle, 'people changing technology' and 'technology changing people' played an increasingly important role in cultural change. For example, by the middle of the nineteenth century, English workers had internalized a great deal of this change, thereby appearing to be more reliable and effective in this new order than workers who had not because of their shorter exposure to industrialization.

The above developments gradually paved the way for an alternative way of making sense of and living in the world. Doing so by means of experience and culture proceeds from the 'outside in,' but a new approach gradually emerged that did so from the 'inside out.' As noted previously, the cultures of early Christendom involved a self-awareness and an awareness of the world that were enfolded into one another. A process of unfolding gradually caused this self-awareness and awareness of the world to create human observers detached from an 'autonomous nature machine.' Human life was thus placed on the stage of matter, space, and time. For some it gradually became apparent that this separate and autonomous world had to be approached on its own terms and hence from the inside out. Along with this new awareness of the world there necessarily emerged a new self-awareness as a detached and objective observer. This development first occurred in science, much later in technology, and still later in daily life as a consequence of 'technology changing people.' In a process that took centuries, space, time, and matter came to be symbolized as being 'out there' to be observed, quantified and later manipulated.

I will briefly examine some aspects of these changes, deferring a more complete analysis to later chapters. At the dawn of Western civilization, the concept of a law of nature was unthinkable, and Plato and Aristotle never used it.[29] Under the influence of Christianity, such a concept became thinkable. The role of God as creator could just as well be conceptualized as that of lawgiver. Such a conception was not without its theological difficulties: it undermined the concept of God's sovereignty since He would now be bound by His own laws. One way around this problem was to regard these laws as evidence of God's ongoing involvement in His creation, which He willed to continue. In any case, somewhere between the Greeks and Newton the concept of laws of nature began to make sense theologically and became widely accepted. Eventually, the metaphysical basis disappeared, but the concept of laws of nature now appeared self-evident. For example, Engels, speaking at Marx's funeral, noted as one of his accomplishments the discovery of the law of the development of human history with its five stages, each built on a unique base composed of the forces and relations of production.[30] Similarly, Darwin's work could be interpreted as the discovery of the law of the development of organic nature.

The medieval metaphysical basis for science paved the way for its mathematization.[31] During the late Middle Ages there was a growing conviction that the key to God's design of nature was its mathematical laws. Accordingly, Galileo argued that science was not to seek physical explanations but mathematical descriptions of experimental data. Newton pushed this to unprecedented levels. For example, his mathematical law of gravitation had no physical explanation at that time. Why two planets might attract one another was simply not understood. For some, this made his law of gravitation questionable, but others felt that the physical basis could and would be discovered later. For them, by describing experimental data in mathematical terms, scientists were simply discovering the mathematical regularities that underlay all experience and culture. Time and space separated from experience and culture became important aspects of this deeper underlying mathematical reality. In this sense, the cultural changes described above created a fertile soil for certain scientific developments, and these in turn reinforced certain new cultural directions during the late Middle Ages.

The importance of mathematics for science has continued to grow.[32] I do not want to defend the position that mathematics is the queen of the sciences, but its importance has certainly not received the attention it merits, particularly in the literature dealing with the relationship between science and technology. Mathematics continues to be the basis for organizing and interpreting experimental evidence, which, in a sense, may be considered as validating it. Of course, a great deal of modern mathematics has as yet no application, but this certainly does not impede its development.[33] Since the nineteenth century, mathematics has included a growing number of concepts and theories that contradict our intuitive experience of nature. Non-Euclidean geometry, complex numbers, and quantum mechanics are but a few obvious examples. Such concepts and theories can be defended on the ground that they 'work' empirically, but things have gone much further than that. In modern physics, for example, mathematics has gone well beyond replacing matter with theories of fields and is now drawing mathematical pictures of the universe. The gap between such pictures and our intuitive physics based on experience and culture has grown so large that it is increasingly difficult to make any cultural sense of these pictures. Will the end of science be a collection of mathematical formulae?[34] There are many fascinating metaphysical speculations as to what all this means, but I will limit myself to the observation that mathematics is playing a fundamental role in bringing about the separation of scientific knowledge from experience and culture. I am not denying that mathematics continues to organize what is empirically (i.e., physically) observed, but it interpolates and extrapolates such findings to a point where it yields a great deal of scientific knowledge that is not empirically accessible. Scientists, more than ever before, have to suspend their intuitive knowledge of the physical world when they look through the 'bifocal' part of the 'lens' of their minds. In terms of the foundation of mathematics and thus the foundation of science, I will simply point out that mathematics 'works.' However, this does not justify the extreme position that objective reality is purely a social construction shared by scientists. The subject invites fascinating debates, but I should get back to my story.

The essential role mathematics has played in causing scientific knowledge to separate from experience and culture is one and the same as that which transformed technological knowledge. Many interactions continue to occur between science and technology, but the influence of mathematics is probably more decisive. I have already implied the reason. When a traditional master builder looked at a situation and concluded that it was 'right,' he was simultaneously assessing a number of things that today would have to be dealt with separately by engineering design and engineering analysis. It is impossible to design something by solving the differential equations of stress analysis or those of any other branch of engineering, for that matter. Nor is it possible to design something by simultaneously solving all the relevant differential equations. It is only after a technological object has been designed that it can be analysed from the perspective of an engineering science to determine, for example, whether it will be strong enough or whether it will transfer the required amount of heat. In other words, the purpose of analysing this or that aspect of a design from the perspective of a particular engineering science is twofold: to determine whether a product, process, or system will work adequately, and if its performance is optimal. Approaches based on mathematics are more likely to get the job done than those of science, for a variety of reasons. Although in many cases the basic phenomena studied by scientists and used by engineers are the same, their occurrence in nature or in human-made things makes a decisive difference.

The task of engineering analysis and optimization usually takes the form of something being studied for the purpose of making it 'better.' The findings of the study are compiled into some kind of model, which is examined to determine under what conditions optimal performance can be achieved. The results are then used to make improvements. Within this framework, mathematics is extremely useful. In many instances, engineers know the boundary conditions within which the parameters of interest vary according to some well-established principles, such as that the sum of the forces or moments exerted on a finite element in equilibrium are zero, or that the matter and energy within an element are always conserved. Mathematical equations can then be written that govern the distribution of certain parameters within the boundaries of the analysis. Sometimes these equations can be solved if the boundary conditions are known. If this is impossible, mathematics can be used to interpolate or extrapolate experimental data to yield a mathematical description of the variation of certain parameters. It is then possible to examine how these distributions change if some of the design parameters are modified or if the boundary conditions are altered. It is clear that this is, first and foremost, a mathematics-based approach that may or may not have a physical explanation. My previous example of stress analysis is prototypical: there is no physical basis for a continuum within which finite elements are imagined with smooth sides on which normal and sheer stresses operate. Nevertheless, it is a useful model that 'works' and permits engineers to get the job done. However, the approach is much more than a kind of pragmatism.

The full implications of the above developments can best be grasped when we situate them in relation to the role of culture in human life. I have argued that the living of individual and collective human life fundamentally depends on metaconscious processes of differentiation and integration that symbolically interpolate and extrapolate individual experiences into larger structures whose functions include that of a metalanguage, a mental map, and a lens for making sense of and coping with the world. On the conscious level, mathematics interpolates and extrapolates the 'experiences' of the world mediated by experimental apparatus into rational concepts and theories. As such, mathematics helps to create a scientific and technological approach to the world that separates itself from experience, thereby making it possible for modern science and technology to become universal. It is difficult for us to appreciate the significance of this fact because we are so accustomed to interpreting alternative forms of science and technology embedded in experience and culture as fundamentally non-rational (i.e., cultural) and thus non-scientific and non-technological in the modern sense.

Simondon discussed the inevitable differentiation of mediating approaches to the world and their accompanying knowledge – what I have previously referred to as dimensions of mediation.[35] Gusdorf shows that this differentiation leads to ongoing attempts to reunite this knowledge.[36] The library of Alexandria, Diderot's encyclopedia, and the founding of the University of Berlin are some of the best-known attempts. In contrast, Napoleon separated the faculty of science from that of letters, resulting in a split between 'two cultures.' Further subdivisions followed, resulting in the present almost pathological fragmentation. From the perspective of culture, the knowledge fragments have become decontextualized, so separated from one another as to approach the asymptote of information, by becoming symbolized as isolated bytes in the memories of computers. From the perspective of mathematics, the situation looks quite different. It explicitly interpolates and extrapolates the experiment-mediated data of the world into internally consistent rational descriptions that may or may not have a physical basis. Hence, the external consistency of these descriptions can no longer be found in the real world but in the mathematical foundation they share which, in turn, may or may not have a physical basis. This development greatly contributed to *homo informaticus* being the human processor of disembodied information. Piaget, for example, emphasized the logical aspect of the development of children, but avoided the adolescent stage with its upsurge of non-rationality, out of which the adult personality emerges.[37] Modern art performs a requiem for the death of meaning and thus the human subject.[38] As I will show in Part Three, this decontextualization of knowledge removes a barrier to a divide-and-conquer knowledge strategy first developed in the university and then supported and utilized by the state and industry. In it, human beings play the role of logical, mathematical automata, which artificial intelligence failed to model. If it had succeeded, a totalitarian knowledge would have been the result. As it stands, scientific and technological knowledge have separated themselves from experience and culture without the ability to create an alternative universal culture.

To sum up, I am moving towards reinterpreting science and technology from a cultural perspective. The fact that we are cultural beings symbolically linked to reality is not restricted to immediate experience. Metaconsciously, we live each moment in the context of all others, which thereby makes the living of a life possible. This metaconscious interpolation and extrapolation of immediate experience points to higher levels of meaning of a world above and below, ultimately expressed in cultural unity. In prehistory, the world beyond was one of spirits that dwelled in everything. Judaism and Christianity swept the earth and the sky clear of spirits in response to the First Commandment, only to constantly recreate themselves as religions. In the late Middle Ages, the search for explanations of the world gradually shifted from God's presence 'above' to God's presence 'underneath' in the order of creation. For a time there was a struggle as to whether this order should first and foremost be accessed through theology or whether this could be done independently by empirical investigation. As the metaphysical justification for any underlying order weakened, a secular scientific theology stepped into the void. It affirmed that the fundamental order was in essence like a gigantic clockwork and thus of a mechanistic nature. It was a non-living, non-spiritual, and objective order coded in the language of mathematics. In the fifteenth and sixteenth centuries, there was a mathematical revolution, possibly more fundamental than the scientific revolution, and one without which the profound transformation of technological knowledge that ushered in the twentieth century would be unthinkable.[39] Experience and culture had little access to the new world 'underneath,' made up of fundamental particles and their motions. Another approach was required that could interpolate and extrapolate experimental findings in a search for new meaning, which was now essentially mathematical. Auguste Comte unintentionally highlighted its secular religious significance by pointing out that science was gradually impinging on the role traditional religions had always played as the arbiters of ultimate reality.[40] Industrialization greatly accelerated the imposition of the objective order on that of experience and culture. A new and objective time, space, and society were required as the 'cultural software' for the machines. Daily events had to be synchronized with this objective time. Work had to become as 'pure' as possible, by being dissociated from human life and converted into the abstraction of making money. The rhythms of work reflected those of the machines. All other daily-life activities must not waste time because time was money. All this reflected the fact that the ever-growing usage of machines consumed ever-larger amounts of capital, which became the lifeblood of society – a society that nurtured every cell in the body social as long as it did not waste time.

A.M. Smith refers to the above transformation as a shift to knowing things 'inside out.'[41] He quotes E.A. Burtt as follows:

> The world that people had thought themselves living in – a world rich with colour and sound, redolent with fragrance, filled with gladness, love and beauty, speaking everywhere of purposive harmony and creative ideals – was crowded now into minute corners of the brains of scattered organic beings. The really important world outside was a world hard, cold, colourless, silent and dead; a world of quantity, a world of mathematically computable motions in mechanical regularity.[42]

The transformation of knowing the world 'outside in' by experience and culture to 'inside out' by science situated human beings in a sense-less reality – senseless with the double meaning of excluding the senses and therefore excluding the possibility of being symbolized in the context of human life. Each human being became 'an unimportant spectator and semi-real effect of the great mathematical drama outside.'[43] For a time, much of this thinking occurred on the periphery of society, away from the mainstream. Nevertheless, it reflected something of what was happening deep down in the cultural unity of some western European societies. The principal impact on these cultures largely came as a consequence of 'technology changing people.' It is inappropriate to credit the great thinkers of the time as intellectual prime movers of cultural change. With hindsight, they were people ahead of their time who sensed something of the profound changes in their symbolic links with reality, resulting from the cultural evolution which they in turn had helped to articulate and advance.

Knowing things 'outside in' by means of experience and culture creates a life-milieu – a world of meaning and value in relation to which individual and human life evolves. Knowing the world 'inside out' makes it impossible for the integrity of individual and collective human life to be the cause and effect of such a liveable world. It should come as no surprise, therefore, that the transformation of technological knowl-edge during the latter half of the nineteenth century did not come easily. Many engineers resisted it after they saw no practical use for the theoretical knowledge, which went against their 'intuitive technology.'[44] Sometimes this resistance was overcome by failures. For example, boiler explosions forced a systematic investigation of the strength of materials.[45] Depending on the country and its conditions, technological knowledge embedded in experience usually dominated technological knowledge separated from experience, but there are notable exceptions, as illustrated by a comparative study of bridge construction in nineteenth-century France and America.[46] During the second half of the nineteenth century, the interactions between technological knowledge embedded in experience and culture on the one hand, and technological knowledge separated from experience and culture on the other, as well as the interactions between science and technology, became increasingly complex, but the end result was similar everywhere.[47]

chapter 5

Scientific and Technological Knowledge in Human Life

5.1 Scientific Education and Culture

Having introduced the concept of scientific and technological knowledge separated from experience and culture, I will now turn my attention to examining how these non-cultural elements are introduced into human life and how they coexist with the role of culture, particularly in the lives of scientific and technical specialists. The learning of science involves a process of secondary socialization into a non-cultural way of symbolizing and acting on the world. The fundamental premise is that when teenagers and young adults are introduced to science or technology in high schools, colleges, and universities, we can expect a continuity in the way they symbolize their experiences. Although we know very little about the higher symbolic brain-mind functions, it is not reasonable to suppose that the processes of symbolizing experience through neural connections suddenly change in a discontinuous manner just because people are taking courses in science or technology.

The further development of the organization of the brain-mind resulting from secondary socialization into a scientific discipline is not cumulative with respect to the structure of experience. We have shown that taking a physics course does not extend a person's intuitive physics, nor does taking a stress analysis course extend whatever intuitive knowledge of the strengths of materials a person may have acquired. The acquisition of a new skill is the consequence of a progressive differentiation of the structure of experience. Assuming no discontinuity in these processes during secondary socialization, we need to examine how a scientific discipline is internalized. What we will seek to demonstrate is that the experiences of secondary socialization are fundamentally different from daily-life experiences symbolized by means of a culture. Hence, the former are differentiated from the latter to constitute a substructure within the structure of experience. The meta-conscious knowledge built up while learning a discipline corresponds to Polanyi's concept of tacit knowledge.[1] A scientific discipline (as a unique way of making sense of and dealing with the world) is to these substructures of experience what a culture is to the structures of experience.[2] A substructure is the result of first symbolizing the experiences of secondary socialization, and later, those of professional practice.

What are the differences between the kinds of experiences being symbolized? Recall the studies of intuitive physics through which high-school and college courses in physics create what, to some students, appears to be a counter-intuitive alternative. Students do not enter the 'world' of physics the way they entered the daily-life world during primary socialization. During secondary socialization, there is no world. There is only a mathematical abstraction made up of symbolic entities such as infinite frictionless planes, forces, point masses, air without resistance, pulleys without inertia, and so on. They have to imagine this world, which the teacher implies is the real world 'underneath.' It can only be accessed from the 'inside out.' Nevertheless, simulating such a world is extremely difficult and requires a very complex apparatus.

The students' experiences of this mathematical abstraction differ from any daily-life experiences in three important ways. First, the foreground (the focus of their attention) symbolizes a mathematical abstraction that can be represented only symbolically and must be imagined as a 'world.' Their senses must symbolize the already symbolic constituents of this mathematical abstraction by means of diagrams, variables that may or may not correspond to physical entities, symbolic generalizations, and equations. The background (everything receiving only peripheral attention) is that of the daily-life world, such as the classroom and the blackboard on which the teacher develops the subject, or the library or bedroom setting where the students study and do their homework. This background is not continuous with the foreground, as is the case for many daily-life experiences where the foreground and background belong to the same world. These experiences cannot be lived the way others are. It may be supposed, therefore, that the processes of differentiation and integration operating in the brain-mind learn to distinguish between daily-life and science-related experiences, and this is indeed evident from the coexistence of intuitive physics and school physics. It is reasonable to suppose, therefore, that the experiences of secondary socialization into a scientific discipline constitute a substructure of experience, which can be metaphorically compared to three things: a mental bifocal lens, an insert into the mental map corresponding to the abstract 'world' of a scientific discipline, and a distinct metalanguage within the metalanguage of daily-life experience. That this is almost certainly the case follows from situations such as that of an experienced welder concluding that the beam just brought up by a crane does not look right. No matter how many work-related experiences this welder accumulates in his structure of experience, they will never lead to a knowledge of stress analysis. Stress analysis, unlike technological knowledge embedded in experience, does not represent a cumulative refinement of the organization of the brain-mind. It begins in the abstract world of a continuum. Once again, the separation of scientific knowledge from experience and culture must be symbolized by a substructure of experience within the larger structure.

The experiences derived from secondary socialization and those related to primary socialization and daily life are different in a second important respect. A culture creates a life-milieu from daily-life encounters of reality, but a scientific discipline does no such thing. The latter creates a 'world' of symbolic entities that are theoretically (and preferably mathematically) linked, creating a logical internal consistency that has no equivalent in individual and collective human life. Despite the importance of technology-based connectedness in modern societies, human life remains non-logical (that is, outside of the domain of logic).[3]

There is a third qualitative difference between the experiences derived from secondary socialization and the experiences related to primary socialization and daily life. Although metaconscious knowledge is built up by both, the one points to a deeper underlying unity in the reality a community has come to know by means of a culture, while the other points to a theoretical and possibly mathematical consistency behind a set of phenomena taken to be similar enough to be investigated by means of the same methods and approaches. Although the theoretical and mathematical consistency is shared by all scientific disciplines, the content of that consistency cannot be integrated (by means of a science of the sciences) into a scientific view of the world. Observing the world 'inside out' does not reconstruct that world. For example, when the members of cognitively distant disciplines communicate with each other, they may have no choice but to 'translate' what they say into the culture they share in order to establish some common meanings and values, even though these may not be scientific in nature. This communication is thus based on an appeal to common sense, by which I mean the body of experiences symbolized by means of their shared culture that can provide an intuitive explanation.

This situation is inevitable, given the very different ways culture and science deal with context. All science does is gather one particular kind of knowledge. Its point of departure is the recognition that the complexity and interrelated character of reality makes it unmanageable for empirical investigation. All that can be done is to abstract sets of phenomena that appear to be amenable to similar methods and approaches for study. Such sets are examined in the highly restricted theoretical context of a discipline or sub-specialty and in the equally highly-restricted physical context of a laboratory, each created through the experiences of secondary socialization and practice. The unmanageable complexity of reality is thus reduced to a manageable complexity in which a limited set of variables can be examined in a controlled fashion, preferably one at a time. A scientific discipline gathers a great deal of knowledge about sets of phenomena out of their context and in a context of its own making. We are unable to determine the loss of understanding that is bound to occur as a result of this process of abstraction.

Since the experiences associated with secondary socialization into a scientific discipline differ in at least three important ways from daily-life experiences, it may be expected that the former are differentiated from the latter to create a substructure within a person's structure of experience. As a result the metaconscious knowledge built up in the one is not coextensive with that built up in the other. Daily-life experiences are lived in the context of all other such experiences and only indirectly in the context of the experiences derived from learning and practising a scientific discipline. The reverse would also appear to be the case. This is manifested, for example, by the fact that scientific values are very different from the values of any culture. As a result, the implications for human life of scientific knowledge separating from experience and culture are far-reaching. Once more, I will take physics as an example. How can physicists live their lives when school physics coexists along with intuitive physics as symbolized by the presence of a substructure of experience within the larger structure? Why are their lives not split into two, with one part related to daily life based on a culture and the other related to doing physics based on a scientific discipline? How can they deal with the world 'inside out' in one sphere and with the world 'outside in' in another? From the behaviour of physicists, it is obvious that the relationship between their substructures of experience and their structures of experience has no common denominator with schizophrenia or multiple personality disorders.

Another way of posing this problem begins with the observation that physicists experience no 'edges' separating two symbolized spheres of their lives. We know that the two spheres affect each other in the following indirect manner. Particularly in the twentieth century, science had a significant influence on the life-milieu of industrialized societies. Physicists, like everyone else, experienced this life-milieu in their daily lives, which affected their awareness and the way they participated in their cultures, as evidenced by the well-known correlation in time between major changes in art forms and breakthroughs in physics.[4] I am not suggesting that Picasso studied Einstein or that Einstein drew inspiration from Picasso's paintings. The parallel changes in physics and art were more likely both cause and effect of deep and profound changes in culture and its role as a symbolic commons.

Understanding how scientists *live* the separation of scientific knowledge from experience and culture is essential for understanding modern societies. These societies have rejoined these two 'worlds' by means of beliefs, values, and myths. This is why the substructures of experience and the structures of experience of scientists can be regarded as being joined together as a bifocal 'mental lens' through which they experience the world. The 'bifocal' part is for experiencing it 'inside out,' and the remainder for experiencing it 'outside in.' The issue cannot be approached directly, given how little we know about the higher symbolic functions of the brain-mind. However, there is a great deal of indirect evidence that we can puzzle together to obtain a coherent picture.

The substructure of experience grows out of the structure of experience and develops in much the same manner. The model of skill acquisition developed in the previous chapter applies to secondary socialization, as long as we recognize that each new skill emerges by differentiating the substructure derived from learning a discipline. Each new skill helps to build up metaconscious knowledge, much of which is not exclusive to one skill. This is particularly true for scientific beliefs and values, which emerge in all scientific disciplines.[5]

The fact that intuitive physics remains distinct from school physics further illustrates that the substructure of experience and its metaconscious knowledge on the one hand, and the structure of experience and its metaconscious knowledge on the other, remain distinct but connected through beliefs, values, and myths that assign to each their appropriate role in the individual and collective life of scientific practitioners. Again, this is open to sociological analysis.

Because of the extreme intellectual division of labour in modern science, secondary socialization is always limited to a very small portion of science. Primary socialization, on the contrary, deals with the entire culture of a society. How the two are lived will therefore give us clues as to how the relationship between science and culture is symbolized in terms of the connection between the substructures of experience of scientists and their structures of experience. Once again, consider the fact that no 'edges' or limits are experienced around either science or culture. In both 'worlds,' scientists behave as if reality as they have come to know it is reality itself, minus the many details that will still be discovered. When they are, they supposedly will further elaborate the patterns of our knowledge without disturbing their form or structure. This provides an important clue as to how a society integrates science into its culture. In incorporating science into its way of life, the cultural unity must include the myths that make this possible. First and foremost, this inclusion must be realized in the lives of the members of society. Since scientists have a detailed knowledge of only a small portion of science, these myths do not need to permit scientists to live a relationship between science as a whole and culture, but only between a small part of science and culture. The absence of a science of the sciences and a scientific world view is something about which we can have fascinating discussions, but it is not something scientists need to *live*. Given their limited knowledge of science, scientists will extrapolate their experiences of a particular scientific discipline and sub-specialty to create a metaconscious image of science, even though in reality it does not constitute a single entity. Since, in their experiences, they do not encounter any limitations in gathering knowledge about the world, this extrapolation makes the metaconscious images of science omnipotent in the domain of knowledge. Our difficulty in finding examples of what science cannot do in the domain of knowledge is the cultural equivalent of the myth of science.

The myth of science represents an overvaluing of science,[6] as is evident from the fact that modern societies have permitted science to sit in the judgment seat to pronounce its verdict on everything cultural, including morality and religion. With hindsight, this is quite silly, since science is exceedingly effective at gathering an entirely different kind of knowledge, namely, that of things abstracted out of their context. Science is therefore limited by the loss of understanding that occurs as a result of this abstraction, which precludes a scientific view of the world. Science does expose the limitations of culture, which seeks to understand everything in its fullest possible context and thus has to deal with the unknown. It is now becoming evident that in revealing the dependence of cultures on myths, science has not eliminated them from human life, but instead has contributed to the creation of new secular ones. The myth of science helps us to recognize the limitations of culture-based knowing but obliterates from human consciousness and cultures the awareness that science, like any other human activity, has its own limitations.[7] This is why I am referring to a myth of science: it transforms the meaning and value of science as a human activity into something no human activity can ever be. The myth of science establishes a hierarchy between the two kinds of knowledge. Knowledge separated from experience is perceived as objective and reliable (although we no longer accept the connection with Truth), and knowledge embedded in experience is thought of as merely subjective opinions.

Examples of this abound. It is no longer possible to legitimately appeal to common sense as the body of mostly metaconscious knowledge we share, since it can be challenged by any scientist. This drama plays itself out most spectacularly in advertising and the courts. In advertising, it is the scientist in a white coat who must tell us something about a daily-life activity (like doing the wash) that everyone already knows but that the 'expert' supposedly knows better. In the courts, specialists with more or less the same background are frequently found as witnesses for both the prosecution and the defence because their knowledge, void of any daily-life context, can be integrated into opposing arguments. The common sense of citizens no longer has any status. In one extreme case known to the author, a schoolteacher was permitted to testify in a trial of someone accused of robbery. The only evidence held by the prosecution was a signed confession which this teacher, as a volunteer working with the accused almost from the beginning of his incarceration, knew he could not possibly have understood given his exceedingly limited knowledge of the English language. It was a simple matter of common sense to anyone who had known the accused, but that did not mean it had any status in a court of law. The precautionary principle seeks to re-introduce a measure of common sense into the courtroom by preventing polluters from claiming that the absence of a scientifically established link between a certain pollutant and health effects means no link at all, so the case should be dismissed. As our analysis proceeds to consider other forms of knowledge separated from experience and culture, further details of our new secular myths will become apparent. For now, it is sufficient to know that such secular myths integrate the substructures of experience into the structures of experience of scientists in a manner previously likened to that of a bifocal mental lens or an insert into the mental map.

In the present work, I am primarily interested in the many members of modern societies who have received a higher education in science but who will never become part of the very small number of people who make up the informal membership of the invisible colleges,[8] which are the communities that advance the frontier of a particular scientific sub-discipline. Instead, these people, after secondary socialization, will begin to participate in the organizations essential for maintaining and evolving modern ways of life that are increasingly dependent on a technology-based connectedness. These organizations apply a diversity of knowledge separated from experience to issues related to their goals. It may well be that, in the course of doing so, this knowledge is also advanced by applying it to new situations, but this is not the primary goal of these organizations. Invisible colleges, on the other hand, seek to advance the knowledge of a particular sub-specialty by informal collaboration between a relatively small group of people from across the world who carry out related research, are interested in reading one another's papers (frequently before official publication), attend the same international conferences, and have a stake in maintaining communication. The application of knowledge is not the primary purpose of such invisible colleges and is generally carried out by the much larger organizations referred to above, which will be analysed in the next chapter.

Consequently, I will limit my examination of secondary socialization to the acquisition of a first or second academic degree in a scientific specialty. The students who continue their education on a doctoral or postdoctoral level may eventually become members of invisible colleges. I am interested in secondary socialization as an entrypoint into participation in organizations that apply knowledge separated from experience. I will restrict my attention accordingly, and avoid the related controversies in the sociology of science. It must not be inferred from this that those who participate in modern ways of life on the basis of a first or second academic degree in science do not reach the highest levels of the five-stage skill-acquisition model presented in the previous chapter. Instead, their secondary socialization is the prerequisite for acquiring skills in manipulating and evolving the technology-based connectedness of a modern society. Secondary socialization prepares people to understand and deal with the world in the context created by a particular specialty. Especially in the applied sciences, this context has been shaped by the requirements of advancing the technology-based connectedness.

There are several ways in which a substructure of experience acquired from learning and practising a scientific discipline manifests itself in the practices of specialists. For example, a scientist may take a look at a problem and, like a chess master, know immediately how to deal with it. This ability corresponds to what Thomas Kuhn called exemplars.[9] These may be regarded as the skills acquired by the specialist. Symbolic generalizations are usually theoretical expressions such as Newton's laws of motion, or principles such as the statement that matter and energy can neither be created nor destroyed. They help to specify the variables and relationships that characterize the 'world' of a discipline, which is inseparable from their 'meaning' and 'value' in it. These symbolizations give shape to the theoretical and mathematical consistency of that world. Another manifestation of the substructures of experience is a commitment to a model, ranging in kind from onto-logical to heuristic. One model that is commonly presented in high school has gas molecules behaving like ping-pong balls that bounce off one another and the container walls in a random motion. This model, like many others, provides a permissible analogy to the world of daily-life experience. As such, accepted analogies and metaphors have an explanatory power in comparing certain situations to others that we consider as having been satisfactorily explained. Much of this is first contained in the build-up of metaconscious knowledge that constitutes the metalanguage for what may later become explicit. The buildup of metaconscious knowledge also provides the basis of what is explicitly known as values. These are widely shared between scientific disciplines and contribute more than anything else to scientists having a sense of community. These values underlie a great deal of scientific behaviour. There are values related to how good predictions, experimental techniques, theories, models, or beliefs are considered to be. Values related to the internal and external consistency of a field are particularly important for assessing whether something constitutes an anomaly, or for deciding between competing theories. Some values have an aesthetic character, such as a preference for the simplicity or elegance of a mathematical solution to a problem.

What the members of a scientific discipline may share as scientists is different, and possibly even very different, from what they share as members of the same or a different culture. Our beliefs, values, and myths hold that what is shared as scientists is objective, and what is shared as members of a culture is subjective. I have argued that this is far too simplistic. Objectivity has to do with the theoretical and mathematical consistency of a scientific discipline, and subjectivity with the cultural consistency of the way of life of a society. Modern cultures value the former above the latter, with the result that the bifocal part of the mental lens of scientists is regarded as affording a better look at the world. These same cultures imply that those who have not acquired a bifocal mental lens cannot expect to know the world as reliably as scientists do. Because of their primary socialization scientists know this, and this binds together the members of scientific disciplines who may come from all over the globe and who, in many cases, do not share the same culture. Their arm's-length relationship to their cultures provides a sense of community among scientists.

I will briefly comment on three additional aspects of scientific disciplines, namely, their evolution in time, their relationship to reality, and their relationship to reality as it is known by their culture. I have noted that a scientific discipline is to the substructures of experience of its practitioners what their cultures are to their structures of experience. Hence, we can expect both similarities and differences between the historical evolution of scientific disciplines and that of cultures. In an earlier chapter, I suggested that during any epoch in the history of a society, its culture is elaborated cumulatively to cope with internal and external changes. It is not a question of new discoveries and changes being added as new parts, but of further differentiating the culture so that the additions are made from within, thus reflecting the cultural unity as a unique local manifestation of it. The best analogy I can think of is that of a developing human embryo in which the DNA is enfolded into each new cell, tissue, and organ. In the same vein, the 'cultural DNA' is enfolded into each new cultural element by means of differentiation. In pre-industrial societies, this was accomplished by adapting a tradition, but this is much more complex in mass societies. Keeping in mind the fundamental differences between cultures and scientific disciplines, the latter may be regarded as having a scientific tradition elaborated and refined in the course of what Kuhn called normal science.[10]

Recall that a society undergoes a long transition period as it moves from one historical epoch to another. The process begins with a diversity of changes that each on their own can be fitted into the dialectical consistency of a culture. However, together they begin to subtly shift patterns of metaconscious knowledge, and when this development continues, it undermines the cultural unity of the society. Artists, poets, novelists, and scientists may begin to express something of this in oblique ways. As the process continues to gain momentum, other people may have a feeling of anomie.[11] Eventually, things come to the surface, and what Toynbee has called creative minorities may begin rearranging one or more spheres of a way of life in ways that make more sense and that usually are to their advantage.[12] When this is successful, the new metaconscious patterns grow in scope and lead to an embryonic new cultural unity that gradually displaces the old one.

In science, something similar can happen. At the same time, there is a significant difference because the theoretical and mathematical consistency of a scientific discipline is qualitatively different from the dialectical consistency of a culture. Suppose that a member of an invisible college has carried out an experiment, and the results turn out to be different from what was expected. An effort gets under way to explain the difference. A variety of hypotheses are checked, and when nothing obvious is found, word may spread and other members of the invisible college may begin to replicate the experiment. If their findings are similar, the experiment is likely to attract growing attention. Some scientists may feel that something is being overlooked and that, in due course, the difficulty will be resolved. Others may not be so sure, and see the result as anomalous with respect to the 'map' drawn by the scientific discipline. Because of the qualitative difference between the consistency of the map and that of a culture, anomalies do not really occur in a culture since it takes account of the fact that human life is full of contradictory elements. In other words, scientists have a different metaconscious knowledge of the consistency of their discipline's 'world' than they do of their daily-life world. Individual differences in the way the members of the invisible college have built up their substructures of experience within the unity of a scientific discipline add to the ensuing debate over how to interpret events, in which the metaconscious knowledge of the consistency of a discipline's 'world' plays an important role. Such intuitions guide scientists in their behaviour, and after a long and difficult transition, a new scientific mode of practice usually emerges. There is now a consensus that this new mode of practice has the potential to explain more phenomena with greater precision. Junior and senior members of invisible colleges do not make the transition with the same ease. It is impossible for anyone to directly change their substructure of experience. All that can be done is to add new experiences, many of which presumably have to do with rethinking old ones. The result is that non-cumulative changes or scientific revolutions are more easily initiated and coped with by those who have a less well developed substructure of experience.

One important difference between the social and the natural sciences is the constantly changing world of the former. The relationship between theory and reality becomes even more complex because of the impossibility of conducting the kind of experimentation that is done by the natural sciences. This also creates a problem for the intellectual division of labour. The communities of social scientists (or any other scientists, for that matter) rarely discuss the limits of their knowledge and where it depends on that of other disciplines. In this way, the knowledge of other disciplines is converted into knowledge externalities.

We have become painfully aware of the consequences of the externalities in modern economics and the heavy price we continue to pay for this 'economic fundamentalism.'[13] It was undoubtedly reasonable for the founding 'intellectual fathers' of this discipline to take for granted the ability of communities to limit self-interested behaviour that established marginal utility and profitability by *homo economicus*, but this is no longer the case in mass societies. They could take for granted the ability of local ecosystems to function as the ultimate source and sink of most flows of matter and energy, given the relatively small scale of the economy. They could equally take for granted that the effects of market transactions on third parties would be relatively modest. Similarly, free trade made sense: the mobility of international capital was low so that investment capital would simply flow from one part of an economy to another, thereby avoiding unemployment or underemployment. Capital was indeed becoming the prime factor of production, with technology as yet in the background. Producers could not yet influence consumers in the absence of advertising and a mass society. Had the development of economics taken into account the findings of other disciplines to critically assess what it took for granted, we might now have a better understanding of our economic affairs.[14]

A similar situation occurs in engineering. Despite the environmental crisis and social problems, engineering methods and approaches have remained fundamentally the same. Elsewhere, I have argued that modern civilization is in the grip of an intellectual and professional fundamentalism which claims that the theories and practices of our intellectual and professional fathers and mothers are still good enough today.[15] Knowledge externalities of all kinds are now so pervasive that the value of science for helping us better understand our situation and for overcoming the many difficulties we face is severely weakened. One consequence of this situation is that many students find it very difficul to find some meaning in what they are learning, particularly when they try to put it in the context of experience and culture. These ex-periences play an equivalent role to that of intuitive physics in the learning of school physics. Other students may have no difficulty in suspending their need for meaning in order to manipulate what with growing specialization is more and more like decontextualized information. By this, I mean that specialties increasingly deal with knowledge that has so little context that its meaning and value are difficult, if not impossible, to determine. What is happening in society is a filtering process that favours *homo informaticus*, essential in a world where the technology-based connectedness of individual and collective human life dominates the culture-based connectedness. In this sense, secondary socialization transforms human beings into *homo informaticus*.

To understand the role science plays in human life, it is essential to pay attention to the way scientific disciplines act as knowledge filters. For example, Newton's law of motion, expressed as force being equal to the product of mass and acceleration, implies a number of things about the world 'inside out.' It suggests something about the important constituents and phenomena that inhabit this 'world,' and by implication, that other constituents and phenomena do not matter. Chemical, biological, psychological, social, economic, and other forces are filtered out or banished to other disciplines, which is why the 'worlds' of scientific disciplines have no common denominator and preclude a scientific interpretation of reality. As is the case with primary socialization, what is *not* said in scientific practice is as important, and sometimes possibly more important, than what *is* said. The metaconscious processes of differentiation and integration cumulate these omissions into beliefs, values, and, ultimately, myths. The reason for this is that those entities and phenomena that are explicitly dealt with are turned into *all there is* – at least from the perspective of a particular scientific discipline. The world 'inside out' is thus converted into sets of manageable problems that have solutions. The fact that in daily life we so easily talk about our problems suggests the influence of science on modern culture. In real life, there are no such problems and certainly no such solutions. This highlights once again the unique nature of modern (universal) science in relation to all other (local) knowledge-related activities and to human history. It is the first form of human knowledge that has become separated from experience and culture.

A scientific discipline helps to constitute meanings of sorts for the constituents and phenomena in its 'world.' These will generally coexist with their intuitive counterparts, as we have seen in physics. The popular literature about science and its findings introduces some of these 'meanings' and 'values' into the world of experience and culture as being more accurate and reliable, thereby confirming the myth of science that the living of human life is merely subjective, and thereby making science a powerful ally of the forces of alienation in our modern world.

5.2 Contemporary Technological Doing Embedded in Culture

Under the influence of Taylorism and industrial engineering, the technological knowledge of so-called blue-collar, unskilled, and semi-skilled workers and white-collar clerical staff is barely acknowledged. There are few studies of this knowledge, but the findings of one of the more extensive ones converge with my own.[16] I will synthesize them in the following pages.

For reasons that will become evident in the next chapter, modern corporations are divided into two sectors: one characterized by knowledge separated from experience dominating knowledge embedded in experience, and the other characterized by knowledge embedded in experience dominating knowledge separated from experience. The interface between these two sectors is where two 'worlds' come into contact to reveal their differences. For example, the work of engineers based on the mathematical abstraction of a continuum encounters real materials, and their abstract descriptions of machines in the form of drawings and specifications become real machines. The existence of a possible gap is not acknowledged; nevertheless, it must be bridged, and it is the so-called unskilled and semi-skilled workers that learn to do so. Such learning by doing involves many experiences forming a work-related cluster within their structures of experience, which help build up a great deal of metaconscious knowledge or know-how of their jobs. The technological knowledge embedded in the experience of blue-collar workers includes, but is not limited to, the following kinds of metaconscious knowledge:

1. Machines perform operations on real materials, which in the work of engineers are symbolized by the mathematical abstraction of a continuum whose idealized properties are assumed to be uniformly

distributed throughout space. In some cases, such approximations are adequate, while in others the gap between the mathematical abstraction and the real materials causes problems that workers must learn to deal with. By observing how variations in raw materials or workpieces affect the operations of the machines they run, so-called unskilled or semi-skilled workers develop a feel for this and learn to compensate accordingly.

2. According to engineering drawings and specifications, all machines are created equal, and any differences that occur are assumed to be trivial. It is also taken for granted that these cause only non-essential differences between the ways machines are supposed to operate according to their design and specifications and how they actually function. In addition, there may be a significant gap between the original design and the way the machines actually operate, because the design was based on certain assumptions or calculated estimates of the magnitude of certain phenomena such as dust levels and how they affect lubrication, levels of vibration and the way these affect the ability of machines to 'hold' adjustments specified by maintenance requirements, seasonal or other variations in temperature or humidity, differences in the ways mechanics compensate for these or other problems, and so on. On the basis of their experiences, workers on the shop floor learn to compensate for these differences between the way the machines are supposed to work based on technological knowledge separated from experience, and the way they actually work.

3. Variations in the distribution of tolerances or other differences may lead to some machines exhibiting unique idiosyncracies, either from the beginning or in the course of gradual wear. The situation is analogous to someone who has learned to cope with the idiosyncracies of an old car and knows exactly what to do under certain circumstances. If someone else borrows such a vehicle, they may not know how to cope, with the result that, on a cold morning, they may be unable to start it.

4. On occasion, the gap between the way an engineering design intended a machine to operate and the way it actually functions is dealt with by operators learning to make subtle changes to the machine. Examples include clamping something over a nozzle in order to reduce the rate of suction or blowing by partly blocking its opening, installing a partial hood over an area prone to dust accumulation, devising a gadget to facilitate the regular checking of

something that is prone to misalignment, or using tape to raise a surface. These kinds of adjustments may have to be negotiated with maintenance mechanics to convince them that they will not produce headaches for them. For example, a mechanic may feel that the installation of a hood simply reflects the laziness of an operator who does not want to clean some parts of a machine as often as he or she should. A lot will depend on the reputation and skill of the operator. If he or she is one of the best, no such objections will be raised by the maintenance mechanic; while if the operator is regarded as lazy and a poor team player, the mechanic may insist that the hood be removed.

5. During start-ups, when equilibrium conditions do not yet exist, machines may present operators with other difficulties, encouraging them to reduce shutdowns for cleaning, adjustment, and other maintenance to a minimum. With a great deal of experience, they may develop a feel for the tradeoff between the benefits and the problems caused by such shutdowns and consequently take shortcuts by, for example, cleaning some parts of the machine while it is in operation, thereby exposing themselves to a much greater risk of accidents. Operators may take risks due to a certain pride in being able to 'beat' the machine, or have little choice if they are to meet their production targets.

6. With a great deal of experience, operators learn to spot defects. They develop a feel for something that does not look right. They may also become very good at inspecting large numbers of whatever they produce, and spot a problem almost immediately.

7. Another important gap that workers learn to bridge is the one between how things are supposed to be done and how they actually get done. The former is largely based on knowledge separated from experience as symbolized in the organization chart, job descriptions, and reporting lines. These together symbolize the optimal way in which work is supposed to be done and how each worker fits into it. The real situation is usually somewhat different for reasons that include an inexperienced co-worker who needs constant help, an opinionated maintenance worker who cannot hear what others have to say, an amiable supervisor whose lack of diagnostic skills and inability to resolve problems is such that others need to jump in, and so on. After some time on the job, workers develop a great deal of metaconscious knowledge of how to survive by knowing when they can count on whom for what, and how to

get around situations where official reporting lines or the division of labour between jobs simply do not work. As a result of a great deal of experience, the processes of differentiation and integration develop a great deal of metaconscious knowledge of how things work and the social roles everyone plays. Such knowledge gives workers a feel for the corporate culture and a sense of what will and will not fly. Workers develop a sense of how hierarchical or flexible the corporate culture is and how it deals with value conflicts.

8. Workers learn to mediate the gap that almost always exists between the values of the 'world' based on knowledge separated from experience and those of the 'world' based on knowledge embedded in experience. The former is preoccupied with the technology-based connectedness of the plant represented by such things as production schedules, job quotas, and minimum quality levels, all measured by what I have called performance values.[17] The 'world' of knowledge embedded in experience sees this connectedness as an intrusion on culture-based connectedness. Workers symbolize their experience in terms of its meaning and value for individual and collective human life, and the work-related experiences in the plant are no exception. For example, it is difficult to feel good about your job if you know that what is being made in the plant is poorly designed or of poor quality, thereby limiting its use value. In other words, workers constantly live the conflict between performance values and context values, corresponding to what is good for technology-based connectedness and what is good for culture-based connectedness respectively. The perceived corporate culture and a person's primary socialization influence how workers live such value conflicts. For example, sustaining informal patterns of cooperation does a lot more than simply keep a place running. It creates a happy workplace where people feel they count and are appreciated. A variety of rituals may affirm this. To celebrate someone's birthday, people may bring a cake or some alcohol to spike drinks. Within these informal structures, the metaconscious images people have of one another and how people *live* the organization chart, reporting relations and job descriptions, are extremely important. They form the basis for the shop-floor culture, including how it interfaces with that of the corporation. For example, a highly respected operator may get a quick response to a request for help from a maintenance mechanic, or a highly respected foreman may

close his eyes to a variety of things that are officially not supposed to happen but that he recognizes as essential to creating a happy workplace with good morale. People who are regarded as poor team players cannot expect other workers to put much effort into helping them out of a tight spot. As in daily life, interpersonal relations and informal patterns of collaboration depend on the metaconscious knowledge people acquire on the job. What other workers think of them may be much more important than what management thinks, particularly in jobs where there is little possibility for promotion.

In sum, workers acquire a great deal of technological knowledge embedded in experience, of the materials they work with, the machines they operate, and the organizational structure within which they work. When we speak of so-called unskilled or semi-skilled workers, what we are really saying is that they are so with respect to technological knowledge separated from experience. In our society, that is the only 'real' technological knowledge in relation to any other kind of knowledge. The situation is parallel to the relationship between scientific knowledge separated from experience and daily-life knowledge based on experience. Workers are not unskilled in relation to technological knowledge embedded in experience, but the value system of modern societies cannot recognize this. The technological knowledge of workers acquired from experience performs all the functions of a technological tradition, and it is based on a highly developed portion of the workers' structures of experience as opposed to relatively distinct substructures. It is the continuity with the world of daily-life experience based on a way of life (or culture-based connectedness) that distinguishes contemporary technological doing from technological knowing.

The above analysis also applies to highly skilled workers who have undergone an extensive apprenticeship program, such as tool and die makers. In these cases, technological knowledge embedded in experience is often complemented and refined by more extensive school learning. However, at no time will their experience lead to technological knowledge separated from experience. The analysis can equally be applied to low-level white-collar clerical workers. A study by Kusterer examines the knowledge acquired by bank tellers.[18] After some formal training, new tellers are teamed up with experienced ones, but it generally takes a long time before they reach the highest levels of the five-stage skill acquisition model. Almost all computer-based work systems are designed to extend the technology-based connectedness of a corporation or institution. No matter how well such a system is designed, it is incapable of anticipating all possible daily-life situations or how a business may evolve over the years the system is in use. Hence, gaps between the design and how things actually work will have to continue to be bridged by so-called unskilled workers. As we advance in our analysis, it will become increasingly evident that information technology significantly extends the technology-based connectedness of modern societies, with the result that the importance of technological knowledge embedded in experience is not likely to diminish.

The technological tradition embedded in experience maps what may be regarded as normal practice, that is, integral to the way the workplace normally functions. When anything disturbs or interferes with this normal practice in a manner that a worker perceives as being nontrivial, it becomes a 'problem' that he or she will try to solve. If unsuccessful, he or she may consult others. By means of these efforts, the problem is transformed and becomes a part of how things function, thereby modifying normal practice. At no time does the problem become an anomaly with respect to normal practice because the consistency of culture-based connectedness is not logical or mathematical in nature. This course of events is one in which the workers' technological knowledge embedded in experience can grow as they learn from problems. Unfortunately, this represents a minority of such situations. The more the technology-based connectedness dominates the culture-based connectedness in the workplace, the fewer are the opportunities for such learning and growth. These differences are crucial, since Karasek and Theorell have shown that healthy work requires a level of control to permit learning and growth.[19] Otherwise, workers are unable to react as human beings normally would by extending the culture-based connectedness of their work and lives. Events can then no longer have any meaning or value in their lives, creating a sense of helplessness and powerlessness and resulting in a variety of symptoms such as anxiety, nervousness, depression, and an alarming growth of mental illness.[20] Workers attempt to cope with these symptoms in a variety of ways including alcohol and substance abuse. In other words, a lack of control resulting from the technology-based connectedness excessively dominating the culture-based connectedness at work results in what, in the sociology of work, is referred to as alienation.[21] Marxism teaches that such alienation is the consequence of capitalism, but the industrial experience of socialist societies has shown, beyond any reasonable doubt, that socialism cannot make it go away. I contend that the dehumanizing and alienating character of a great deal of modern work is the result of an ever-extending technology-based connectedness increasing its dominance over culture-based connectedness.

Kusterer reports that, in interviews, workers said that their most important goals were to get along with everyone and to learn as much as they could so that they would be able to handle the problems that came up.[22] In other words, workers metaconsciously learn that the best way of coping with alienation is to attempt to expand the culture-based connectedness of work as much as they can. However, information technology in general and computer-based systems in particular have seriously diminished such opportunities, especially for white-collar clerical work. It is difficult to avoid the hypothesis that the rapidly growing expenditures in treating mental illness in the U.S. adult population reflect a growing incidence of mental illness associated with the extension of technology-based connectedness in human life.[23]

The third most important goal reported by workers was to produce the highest possible quality products.[24] Engineers and managers may scoff at these findings, particularly when they have a great deal of experience with quality circles or other attempts to improve quality. It should be remembered that they approach this issue on the basis of technological knowledge separated from experience. It frequently seeks to achieve higher levels of quality by extending the technology-based connectedness of work through organizing workers to this end. The difficulty is the gap that exists between this kind of quality and the quality that has a meaning and value in the culture-based connectedness of work. When someone is first trained on a job, a great deal of it makes little sense. It is like a person learning to drive a car. The instructor may give some simple rules to do this or that in these or other circumstances. With more practice, a metaconscious knowledge is gradually built up of how the job meshes with those of others and how what they do affects oneself and vice versa. Gradually, this creates a bit of room for personalizing one's own work procedures without jeopardizing their meaning and value relative to everything else in the workplace and the world beyond. This freedom allows for a little self-expression without undermining what others do. Ultimately, this is all about the meaning of what a person does within the daily-life world, which is closely related to the perceived usefulness of what is being produced. Workers begin to learn what poor quality means to an inspector and, more importantly, to the customer and the world beyond.

The less workers are able to understand how the trivial production step they perform fits into the broader scheme of things, the more difficult it is for them to assess which details are important in terms of use value and which are not.

Of course, such technological knowledge embedded in experience may quickly reach its limitations, particularly in the case of an extreme technical division of labour. Workers may then come to erroneous conclusions, for example, that a rough edge is detrimental to the use value of a part, but this may not necessarily be so. A metaconscious sense of the usefulness of what people produce is fundamental to any technological tradition embedded in experience. It helps workers to determine where to concentrate their efforts and how to prioritize choices because these are now a part of the meaning and value they have found in work. Some kind of commitment to the work a person does is essential to integrate it into the daily-life world of meaning and value symbolized by a culture. It helps transform work into a life activity. It makes people feel good about their work because they can see that they make some contribution to society. Unfortunately, with the advent of a mass society, mass production, mass consumption, and mass advertising, a process of desymbolization occurs that flattens all meanings and values, which, coupled to a growing technology-based connectedness, is aggravating some of the previously mentioned work-related problems. In sum, the technological traditions based on experience are under enormous pressure in modern societies, yet they are essential to limit alienation.

5.3 Contemporary Technological Knowing and Doing in Relation to Culture

When engineers undergo secondary socialization into their profession, the point of departure is not the daily-life world of experience and culture but a variety of mathematical abstractions such as those discussed earlier for physics and stress analysis. As a result, there are important similarities between technological knowing separated from experience and culture and its scientific counterpart, but there are also some very significant differences. The principal difference arises from something that is entirely obvious to any engineer or manager but whose consequences are almost entirely ignored. When I ask senior undergraduate or graduate engineering students how they would go about designing a new production facility, they immediately break the problem down into the kinds of components or aspects to which they can apply what they have learned. They begin to talk about how they would balance the line, optimize the material's handling system, adapt a database information system, and so on. They implicitly recognize that it is impossible to begin by writing down all the applicable equations corresponding to these various aspects of the production facility in order to jointly solve them to arrive at an overall design. I remind them that what they are really talking about is optimizing various details of an overall design without telling me how they arrived at the basic concept. How did they decide between existing alternatives such as a Fordist-Taylorist system, a lean production system, or a socio-technical design? Is it possible to come up with a better design? We fail our students by not helping them address the matter on this level. Similar experiences result when one asks civil engineering students to design a bridge, or mechanical engineering students to design a passenger vehicle. Students jump straight to a basic design concept that does not come from any of their courses but from their daily-life experience of factories, bridges, or cars, supplemented by site visits they make as students. Their engineering education provides them with a deep appreciation of the optimality of all the details and the ability to compare alternative designs in terms of particular aspects. In other words, a mechanical engineering student can apply what she has learned about mechanics, fluid mechanics, thermodynamics, heat transfer, and strength of materials to this or that aspect of a vehicle but not to the car as a whole. Her courses are based on knowledge separated from experience and culture, while a design concept primarily depends on knowledge embedded in experience and culture with its details optimized in terms of the former.

The situation is analogous to the one found in science. Because there is no science of the sciences able to integrate the findings of the many disciplines into a scientific world view, scientists are obliged to rely on knowledge embedded in experience and culture outside of their domain of expertise. Similarly, since there exists no technology of technologies capable of integrating the findings of the many engineering disciplines and specialties into a particular technology or the technology of a society as a whole, engineering specialists rely on knowledge embedded in experience beyond their area of competence. Engineering students have acquired concepts of factories, bridges, and cars from daily-life experience, whose details they refine from their engineering education. After graduation, they enter into what may be called normal practice, which frequently resembles the following pattern. When engineers encounter a technical problem, they rarely set out to invent or discover a new process, technique, or piece of hardware, but rather fall back on past technical achievements. Some of these achievements have become regarded as state-of-the-art practice and may be referred to as design exemplars for a particular engineering specialty. These must be distinguished from the analytical exemplars developed by the engineering sciences. The design exemplars are the models on which new solutions are based, so that engineers know in advance what the final result of their design work should look like. When engineers join a company, they quickly learn these design exemplars by examining recently sold units. There is no need to start from scratch unless the company is getting into a new line of products, and even these will reflect creative leaps from earlier design exemplars. Corporations thus develop technological traditions corresponding to normal practice. These correspond to evolving a technology based on interdependent design exemplars.

This is equally true for many other technical activities. Whether we observe an engineer designing a turbine to certain specifications, an architect designing a high-rise office building, an information systems specialist designing a new system, a planner designing a new subdivision, or a media specialist creating a series of broadcasts aimed at reshaping the public image of a politician, they all know what the end result should look like based on past achievements. From time to time, breakthroughs may occur and a new design exemplar may become recognized as state-of-the-art, displacing a previous one in some or all applications. Once a design exemplar is decided on as the model solution, the next step is to resolve the design task into a series of subproblems corresponding to other existing design exemplars. For example, the design of a high-rise office building thus becomes the design of its structure, exterior skin, heating and ventilation systems, electrical system, elevators, and so on. This process continues right down to the smallest detail commonly associated with a range of standard products. For example, the problem of transmitting power from one shaft to another is resolved by the use of a standard gear train, or two things are fastened together using standard bolts. These are some of the most micro-level design exemplars. The coexistence of several design exemplars usually means that each has its own unique advantages for certain applications. Where no standard products are

available, a design exemplar may have to be worked out and optimized using analytical exemplars.

It is rare for a design process to be limited to the routine application
of design exemplars, analytical exemplars, and standard products. In
the course of adapting these to a particular task, a creative practitioner
usually improves a number of details. The excellence of the design is
thus determined by the overall design exemplar and its constituents as
well as the analytical exemplars used to optimize the many details. The
whole process is highly complex and allows for a great deal of creativity. It can also be regarded as largely routine and thus best handled by a
computer-aided design software package.

The evolution of the state-of-the-art practice in a particular specialty
thus involves interactions between design exemplars and analytical
exemplars on many different levels. A breakthrough on any level may
lead to changes on other levels that can have repercussions even for the
design exemplar on the highest level. Similarly, a change in the overall
design exemplar has repercussions for all the constituent exemplars.
Analytical exemplars frequently lead to the modification of design
exemplars. Over time, many small changes may suddenly be recognized as paving the way for non-cumulative larger changes and vice
versa. In some cases, a design exemplar may derive directly from a new
analytical exemplar, so that the model of technology and applied science may appear plausible in this instance. In others, the reverse may
be the case. When performance values dominate design and decision-
making, variations in a product offered by competing firms are frequently due to minor aspects of the design that cannot be quantified.
The more the design process is guided by performance values, the more
technological designs will converge as choices between alternative details become simple quantitative choices based on precise calculations.

The functioning technology of a society is thus based on technical
accomplishments that are regarded by practitioners as being state-of-
the-art. This recognition converts these accomplishments into design
exemplars. These can be organized according to the socio-technical
systems they help constitute. For example, the electricity system of a
society is essentially conceptualized by a design exemplar as a one-way
system in which power plants transform energy into electricity that is
distributed to consumers by means of a grid. Similarly, industry is a
one-way system in which products produced in factories are distributed to consumers via a network of wholesalers and retailers. Engineer-

ing specialties correspond to some but not all of these design exemplars. For example, urban planners may decide how and where development is to take place. A specialist in power engineering may translate such a plan into requirements for new power plants and grid extensions. The overall design specifies the performance of the various functional components and how they are to be integrated into the whole. Such components are designed in terms of their constituent elements, including their joint integration, and these in turn are broken down further, and so on. For example, the design exemplar of a power plant may be based on a boiler that produces steam for a turbine, which turns a generator whose output of electricity is transformed up to a high voltage before being transmitted by power-lines to local grids for distribution.

The umbrella design exemplar is that of the technology-based connectedness of the functional components that, in turn, correspond to the constituent design exemplars, and so on. Collectively these design exemplars know the 'technological world' of the power plant from the 'outside in,' while the associated analytical exemplars know it 'inside out.' The technology-based connectedness generally correlates with the technical division of intellectual labour and the kinds of specialists involved. The technological paradigm of each specialist is characterized by one or more related umbrella design exemplars corresponding to the largest technological wholes, which the specialists help to design. The larger technological wholes of which these are a part will generally be known only from the 'outside in,' corresponding to, at best, a general knowledge of the analytical exemplars involved. In some instances, a few of the constituent technological wholes may be left to sub-specialties.

The differences between scientific and technological disciplines can be understood largely in terms of the presence or absence of design exemplars. The analytical exemplars are distinguished from their scientific counterparts by their close connection to these design exemplars. Because of these ties, the engineering sciences frequently use sets of approximations that are valid for certain kinds of applications and contexts, limiting their applicability to knowing certain portions of the 'technological world' of a society 'inside out.' Frequently, the ability of such analytical exemplars to specify the distribution of particular variables within the boundaries of a continuum or control volume is more important than understanding the actual physical phenomena, since this permits the precise adaptation and optimization of particular design exemplars to particular applications. In the extreme, one could argue that all that is required are the mathematical functions that interpolate and extrapolate the experimental data and a limited understanding of why these functions behave as they do. However, this latter aspect is not entirely necessary to proceed with the task at hand.

Analytical exemplars are thus distinguished from their scientific counterparts by their levels of generality. The former may be restricted to optimizing particular details of a class of machines, systems or processes, or pertain to several such classes that share similar phenomena. Exemplars in the natural sciences seek to expand the applicability to nature and the accuracy of the scientific paradigm to which they are integral. Scientific exemplars may be regarded as establishing the boundaries of the law-like behaviour of a certain group of phenomena in nature, while analytical exemplars in engineering are preoccupied with what is possible within such boundaries to achieve the optimal performance of specific technologies or their products. These may further constrain the range of some variables, so that limiting assumptions can be made for the purpose of simplifying the general equations governing their behaviour to obtain a general or approximate solution. The more this is possible, the more mechanical normal practice becomes, and the greater use it can make of computers. Given the very different contexts in which analytical exemplars and their scientific counterparts function, their use leads to different metaconscious values and beliefs. For analytical exemplars, predictions should be sufficiently accurate and preferably quantitative, and errors should be within certain limits over the ranges characteristic of the relevant applications. Solutions should be internally consistent but not necessarily externally consistent with others – a limitation frequently arising from the use of simplifying assumptions. The close relationship with design exemplars introduces additional values including those related to functionality, cost, health and safety issues, environmental standards, and professional codes of conduct. It should come as no surprise, therefore, that the engineering sciences quickly diverged from the natural sciences that gave birth to them. For example, the strength of materials, the theory of elasticity, hydraulics, and the theory of structures quickly distanced themselves from physics.[25] Other engineering sciences had their origin in design. Thermodynamics and kinematics are two obvious examples.[26] The former was based on Carnot's ideal heat engine, which was later translated into the language of science.

In conclusion, how do the practitioners of technology separated from experience learn to live with their bifocal mental lenses? When they look through the bifocal part of their lens, very little of the world catches their attention. For example, as aerodynamicists they may regard trees bending in the wind differently from most of us, or they may glance at a car and see certain aerodynamic 'facts.' The bifocal part of their mental lens filters out a great deal of the world of daily-life experience, leaving them with certain aspects that are significant in the technological specialty that includes the analytical exemplars of aerodynamics. In their professional capacity, they will concentrate only on those features that make it through this filter to become the 'facts' of a situation. The analytical exemplar relates these facts to the 'world' of aerodynamics. As technological practitioners, therefore, aerodynamicists do their work with minimal consideration of the context constituted by daily-life experience and culture. When they do consider this context, they need to switch back to the main portion of their mental lens. Hence, technological practice has an arm's-length contact with daily life and society. This makes it very difficult for these practitioners to effectively contribute to technological development that serves human values and aspirations. Instead, they tend to contribute to technological growth driven by performance values, with profound and detrimental effects to individual and collective human life.

We begin to apprehend the situation, although dimly, that when too many activities in a society are no longer based on culture but on scientific and technological approaches, collective human life is bound to become more fragmented because technology-based connectedness cannot provide the same support for human life as does culture. When scientific and technological practitioners look through the bifocal part of their mental lenses, they are convinced they are looking at the real world, but what then are they to make of what they experience through the general part of their mental lenses outside of their domains of specialization? If this is not the real world, how is one supposed to live? It is possible to bury oneself in one's work and to live as much as possible in that world, but this would require substantial changes in someone's personality, self-awareness, and awareness of the world. For some people, it can never be complete, making life more difficult as one grows older. Others may find this largely impossible from the beginning. In either case, scientific and technological practitioners face a variety of difficulties in living their lives.[27] To fully understand this requires a parallel examination of what is happening on the level of society and culture, because this affects their general mental lenses. What happens to a society when a significant portion of its members organize and reorganize a variety of activities using a bifocal part of their mental lenses acquired during secondary socialization into science or technology? What happens to the way of life of a society when significant portions of it are no longer evolving on the basis of culture? When this situation is internalized into the structures of experience of the members of a society, what will be the effect on how their structures of experience function as a mental lens or map? What happens to primary socialization and the ability of young people to make sense of and live in the world of their parents? What influence will this have on secondary socialization and the modern university? Again we face the issue that, when people change their technology, it also, inevitably, changes them.

chapter 6

Adapting to the New Technological Knowing and Doing

6.1 The Emergence of Universal Technology

Out of the struggle by the first generation of industrial societies to overcome the limitations of technological knowing and doing embedded in experience and culture, an entirely new and universal technology gradually emerged. This process began in the closing decades of the nineteenth century and came to full fruition during and after the Second World War. The separation of technological knowing and doing from experience and culture completed the transition from traditional technologies to modern universal technology. This came about by means of another chain-reaction-like process involving both technology and society. The principal modifications to technology included a restructuring of the technological cycle, the industrial corporation, and the way the structure of technology was embedded into that of society. The principal changes to society included the emergence of mass production, mass consumption, mass media, a large public sector in the economy, a social integration depending only partly on culture (erroneously referred to as a mass society), and political institutions partly or wholly based on new secular religions. All this added up to the emergence of entirely new kinds of societies, which were inadequately acknowledged in terms such as 'post-industrial society,' or 'post-capitalist society.' It is remarkable that discussions of globalization pay so little attention to what was radically new in the second half of the twentieth century, namely, that a growing dependence on a universal science and technology required that everything cultural be adapted to them and thus become unadapted to a particular people, time, and place. Such an evolution was impossible without a decisive transformation of the role of culture in human life, perhaps as decisive as the one that occurred with the emergence of societies. With hindsight, it is now apparent that the real significance of many events, particularly in the second half of the twentieth century, was related to their participation in what was becoming humanity's third megaproject: making sense of and living in the world by means of a desymbolized culture. Much of what was radically new during this period may well turn out to be symptoms of the emergence of a partial substitute for the role culture traditionally played in human life. In this chapter we will first examine the emergence of a universal technology, and the most immediate adaptations that had to be made to the technological, economic, social, and political dimensions of mediation. However, none of these adaptations would have been possible without much deeper and more far-ranging adjustments (to be examined in Part Three). These developments represented a further strengthening of the technology-based connectedness of a society, at the expense of its culture-based connectedness. After all, the new non-cultural technological knowledge was a response to the limitations of effectively symbolizing the technology-based connectedness of a society by means of experience and culture.

As we proceed in this analysis, we will encounter many familiar themes, concepts, and theories. Many of these emerged out of particular disciplines, and their validity flows from a core of what is genuinely new and important, while their weaknesses flow from not having seen this as integral to and inseparable from larger patterns that transcend each and every discipline. For example, as far as the economic implications of technological knowledge separating from experience are concerned, my analysis converges with that of J.K. Galbraith.[1] He examined the far-reaching economic implications of the advent of modern technology. With hindsight, I will reinterpret some of his key concepts, including that of a technostructure and the revised sequence in an economic and political democracy. Galbraith recognized, but did not develop, the fact that what he so well described was but part of a much larger phenomenon conceptualized by Weber at the beginning of the twentieth century as rationality,[2] and by Ellul fifty years later as technique.[3] Consequently, his conclusions were quickly upstaged by developments whose full impact was gradually being felt in the 1970s and '80s. The only other economic analysis comparable to that of Galbraith, but undertaken from a Marxist perspective, was stamped out as a secular heresy by the Soviet tanks that invaded Czechoslovakia in 1968. This analysis, published by R. Richta, concluded that modern science and technology had radically changed the patterns of evolution of socialist industrial societies, with the result that the struggle for genuine socialism would require different policies and strategies.[4] This was too much for communist orthodoxy.

Similarly, my analysis will converge with much of the literature on the emergence of mass societies in sociology. The birth of a different social personality, life in crowds as opposed to communities, the shift from private to public opinion, and the replacement of a traditional with a very different kind of morality are without doubt new and important developments, but once again their full significance cannot be grasped apart from a desymbolized culture. In the same vein, attempts to make sense of the changes in political life cannot be fully comprehended within the confines of conventional political science. To some extent this has always been the case because discipline-based knowledge treats all other disciplines as knowledge externalities. Nevertheless, these kinds of limitations become more acute if a particular social and historical transformation has great depth and scope. What could possibly be more decisive than a change in the way cultures help sustain human life and the world? This confronts us with the possibility that the contemporary university is largely irrelevant apart from its decisive role in evolving the new megaproject and the technical approach to life on which it is based. A non-cultural science and technology can emerge in human life only if everything else is adapted to them and the role of culture is diminished in the process. Very little of what has changed individual and collective human life during the latter half of the previous century can be fully understood without a cause-and-effect relationship to this central development.

In this larger context, I will now turn to an examination of the emergence of a universal technology. The emergence of modern technology that resulted from the separation of technological knowing and doing from experience can be understood in terms of how the technological cycle had to be restructured, from a process that was essentially linear and sequential to one whose phases had to be planned simultaneously. This planning could only be done by corporations that had undergone fundamental changes. I will use Galbraith's example contrasting the technological cycles of the Ford Model T and the Mustang, introduced in 1903 and 1964 respectively.[5] The Model T is an example of a product whose technological cycle involved mostly knowing and doing embedded in experience, while the Mustang was designed, produced, and marketed making use of knowing and doing separated from experience. Ford's idea for the Model T vehicle was easily converted into a prototype, using readily available materials and parts that could be purchased in the morning and used in fabrication in the afternoon. The parts that were unique to the Model T could be made in any workshop, using machines commonly found there. The people who had the skills to build the prototype were thus readily available and could be hired at short notice. Even Ford's partners were drawn from businesses other than automobile-making.

Little or no optimization of the prototype took place. As is evident from the previous chapters, when technological knowing and doing is embedded in experience, no optimization in the modern sense of the term is economic. The state-of-the-art is embodied in technological traditions constituted by the experiences of generations of practitioners. Imagine someone wanting to optimize the prototype, beginning with its wheels. A machine could be built that would rotate a wheel while it was impacted by a mechanism simulating a bumpy road. Without theoretical guidance, endless variations in the geometry of the wheel and in the materials used for its fabrication could be tested at great expense, with no assured substantial improvement. Even if there had been, there would still be many other details to be empirically investigated. Such a process would take years to complete, and the investment would almost certainly be uneconomic. To the best of my knowledge, no one has ever undertaken such a venture as long as knowing and doing was embedded in experience. Technological traditions took care of this in a much more effective manner. In the case of the Model T, the knowing and doing embedded in experience that went into it came from the technological traditions of that day, which had relatively little experience with cars. It should come as no surprise, therefore, that customers had many complaints about the production models. In some vehicles, the steering even turned the wheels the wrong way, and there were also complaints about the performance of the brakes, the carburetor and the cooling system.

Putting the Model T into production was a relatively straightforward affair.[6] It took only three months to set up the Ford Motor Company and another four months before the first cars reached the market. The authorized capital was $150,000. Production was essentially correlated with sales. When these grew steadily the business could be expanded, and when sales slumped standard machinery could be sold to reduce the production capacity of workshops, or entire workshops could be rededicated to making other products. In sum, the technological cycle of the Model T was a linear sequence of invention, minimal innovation and development, application, and diffusion.

A few statistics should alert us to the fact that the technological cycle of the Ford Mustang was a great deal more complex.[7] Its engineering and styling took three and a half years to complete and cost $9 million. Production plans included the number of units to be produced and sold. Tooling-up costs were $50 million. By the time the Mustang came on the market, Ford's assets had grown to $6 billion, and it employed on average 317,000 people, up from 125 in 1903. These trends continued, as demonstrated by the fact that the company invested $600 million to bring out the Zephyr and Fairmont models in 1977. Assets were then around $16 billion and employment 445,000 worldwide. As Galbraith puts it,

> Virtually all of the effects of the increased use of technology are revealed by these comparisons ... Technology means the systematic application of scientific or other organized knowledge to practical tasks. Its most important consequence, at least for the purpose of economics, is in forcing the division and subdivision of any such task into its component parts. Thus and only thus, can organized knowledge be brought to bear on performance.
>
> Specifically, there is no way that organized knowledge can be brought to bear on the production of an automobile as a whole or even on the manufacture of a body or chassis. It can only be applied if the task is so subdivided that it begins to be coterminous with some established area of scientific or engineering knowledge. Though metallurgical knowledge cannot be applied to the manufacture of the whole vehicle, it can be used in the design of the cooling system or the engine block. While knowledge of mechanical engineering cannot be brought to bear on the manufacture of the entire vehicle, it can be applied to the machining of the crankshaft. While chemistry cannot be applied to the composition of the car as a whole, it can be used to decide on the composition of the finish or trim.Nor do matters stop here. Metallurgical knowledge is brought to bear not on steel but on the characteristics of special steels for particular functions, and chemistry not on paints or plastics but on particular molecular structures and their rearrangement as required. Nearly all of the consequences of technology and much of the shape of modern industry derive from this need to divide and subdivide tasks, from the further need to bring knowledge to bear on these fractions and from the final need to combine the finished elements of the task into the finished product as a whole.[8]

Galbraith then goes on to describe the consequences that flow from the use of this kind of technology. In an accompanying footnote, he also makes it clear that 'the subdivision of tasks to accord with the area of organized knowledge is not confined to, nor has it any special relevance to, mechanical processes. It occurs in medicine, business management, building design, child and dog rearing and every other problem that involves an agglomerate of scientific knowledge.'[9] His recognition that the phenomenon he describes is not limited to what might be called industrial technology does not affect his subsequent analysis. He fails to see that technology becomes a system with far-reaching consequences, as we will see later. Nevertheless, his analysis of how the use of this new technology compels a restructuring of industry and its interactions with society clearly discerns a discontinuity with the kinds of patterns of economic growth described by neo-classical economics. It was this emerging intellectual division of labour, and not capital, that now constituted the new wealth of nations.

I will continue to follow my own course of analysis by showing how the use of technological knowing and doing separated from experience in the design and production of the Mustang required fundamental changes to the technological cycle. First, consider the 'invention' of the Mustang. It is likely that the concept for this car was partially inferred from market trends. Increasing sales of two-door hardtops, convertibles, and sports cars was reducing the sales of four-door family sedans. This trend, together with others including changes in demography and the family, the spreading suburban lifestyle, the ongoing consumer revolution, and the advent of a mass society, all pointed to the likely success of a new kind of car that would be a hybrid of the four-door sedan and its competitors. It would be similar in appearance to two-door hardtops, convertibles, and sports cars, while being able to accommodate the typical suburban family of two adults and two children and enough luggage for a long weekend. Its performance characteristics would also be somewhere between those of four-door sedans and sports cars. Its price tag would be affordable for the average suburban family and thus much closer to that of a four-door sedan than a foreign sports car. By a process of interpolation, some additional details of the concept car could be estimated, including its approximate size, weight, and engine requirements. Some preliminary sketches could now be made, and much else could be filled in with the use of one's imagination.

Next, imagine that the concept for such a car was presented to a meeting of executive vice-presidents, who agreed to proceed with working out the design for the purpose of eventually producing such a vehicle. It was now up to the vice-president responsible for research and development and her staff to proceed with the concept. These people were highly trained in technological knowing and doing separated from experience. Hence, it was no longer a question of building a prototype on the basis of a technological tradition using knowing and doing embedded in experience and then to 'optimize' it by driving it and carefully observing its behaviour under a variety of conditions. Formerly, if certain things did not work well, the solution was to take the car back to the shop and tinker with it in an attempt to remedy the problem. Engineering the Mustang was now a question of applying highly specialized technological knowledge separated from experience to work out and optimize every aspect of the concept. As pointed out in the previous chapter and also in the above quote from Galbraith, this could be accomplished only when the task of designing the car was broken down into ever-smaller components until these were amenable to the application of technological knowledge separated from experience.

The concept of the Mustang was the umbrella design exemplar, which now needed to be fully worked out. This process drew on a large number of other design exemplars corresponding to the chassis (complete with suspension and wheels), the power train, the electrical system, the body, and the interior. These design exemplars, in turn, were constituted of others. For example, the design exemplar for the power train could be broken down into those of the engine, transmission, drive shaft, differential, and rear axle. The design exemplar for the interior could be broken down into those of a dashboard, front bucket seats with optional console, rear bench seat, inside door panels with armrests, and so on. The automobile industry in general and Ford in particular had accumulated a variety of such larger and smaller design exemplars, so that choices could be made to proceed with this or that particular version. For example, it might have been decided to adapt the design exemplar for a dashboard used in another model. It should be emphasized that each design exemplar, large or small in scope, was presumably chosen for its ability to contribute to achieving the overall concept previously agreed upon. In sum, the umbrella design exemplar was broken down into several constituent design exemplars that, in their turn, were broken down into their constituent exemplars to form a structure like that of a tree trunk, dividing itself into ever-smaller branches. The metaphor is appropriate, provided that each branch is characteristic of the tree as a whole. At the very least, this means that they are all compatible so that the chassis fits within the body shell, the engine has enough room under the hood, the passengers have enough headroom, and so on. However, a good design requires a great deal more, namely, that every design exemplar is a microcosm of the overall concept. There must be an overriding unity that enfolds the overall concept into every aspect of the vehicle, as if it were designed by a single person. Most car designs are somewhere between these two extremes, giving evidence here and there of having been designed by committee.

In this way, the original concept of the car is roughed out by drawing on the extensive stock of design exemplars. The process proceeds on the basis of knowing this technological 'world' from the 'outside in' in terms of how the car will look and function. It requires breaking down the overall design task into a decision tree of choosing this or that design exemplar, while at the same time synthesizing all such decisions to achieve a high level of integrality. The choices of the design exemplars are simultaneously tested, adapted, and optimized using available analytical exemplars. The latter deal with the design exemplars from the 'inside out.' These complementary processes, based on technological knowledge separated from experience, were extensively discussed in previous chapters. I will, therefore, limit myself to a few examples.

Suppose that a decision had been made to offer customers a choice of two different engines, namely, a six-cylinder engine for economy, or a small eight for better performance. Also suppose that it had been decided to use an existing six-cylinder engine with an excellent track record, but that all the existing eight-cylinder engines were too large. A choice would then have to be made to scale down an existing engine or to design a new one from scratch to combine some of the latest advances. One end of the spectrum of possibilities would be to choose an entirely new power train concept. Today, it might be the hybrid design in which a small engine drives a generator that provides power to the electric motors in the wheels, which can also recover power when braking. On the opposite end of the spectrum is the choice to combine the best of several existing engines. For example, the flow of air and fuel into each engine, and the exhaust leaving it, may be examined using the analytical exemplars of fluid mechanics. In this way, the best features of each can be incorporated into the new engine. The same kind of comparative analysis may be undertaken using the analytical exemplars of machine design to examine the camshafts, piston design, crankshaft, or the engine block itself. Whatever course of action is followed, a design exemplar is worked out using various analytical exemplars. Their use is not limited to deciding on details, such as how strong a particular component has to be to resist the forces exerted on it, or the maximum allowable expansion of a material over the range of operating temperatures. Such are the requirements that constitute the envelope within which optimization can take place. The use of technological knowledge separated from experience to guide it can make optimization highly cost-effective. A mathematical function representing the behaviour of critical variables now points to optimal arrangements. For example, a metallurgist may suggest that, by slightly changing the composition of a particular material, its strength is barely affected but its machinability is greatly enhanced. According to fluid mechanics and heat transfer, the optimal position of the radiator is obvious: the stream of cooling air should form, as much as possible, a crossflow to the tubes containing the coolant. Optimization is a complex process because what may be optimal from the point of view of one analytical exemplar may not be so from those of others. What may be optimal in terms of strength of materials may not be so in terms of manufacturability.

It is worth emphasizing once more that the relationship between the design exemplars and the analytical exemplars is a complementary one. For example, it is impossible to solve the general equations of fluid mechanics to arrive at the optimal aerodynamic form for a car, the optimal shape for the intake or exhaust manifold of the engine, the best shape for the front windshield to avoid excessive drafts when the side windows are partly or entirely lowered, and so on. However, it is possible to use the analytical exemplars of fluid mechanics to comment on the aerodynamic characteristics of a particular body style, manifold, or front windshield. Similarly, metallurgists can comment on the suitability of choosing a particular material for a particular task, but they cannot design a vehicle. There is no technology of technologies whereby relevant analytical exemplars can be integrated to arrive at a design exemplar. Nor is it possible to resolve a design exemplar into various analytical exemplars relevant to it. Design exemplars deal with the vehicle from the 'outside in,' in functional and aesthetic terms belonging to the world of experience and culture; while the analytical exemplars deal with it 'inside out,' in terms of technological knowledge separated from experience.

What kind of organization would permit a corporation to make the fullest possible use of technological knowing and doing separated from experience wherever that is useful in its many operations? Such an organization is not found in science: neither within the university with its faculties, departments, and programs as clusters of cognate disciplines and sub-disciplines, nor outside the university with its invisible colleges and learned societies. The organization of science is ideally suited for the advancement of ever more specialized knowledge. The organization of the professions, either within or outside the university, is similar and also does not provide a suitable model. What is required is an organization capable of synthesizing highly specialized knowledge relevant to carrying out a particular task. It would have to facilitate a kind of local science of the sciences or technology of technologies relevant to the task at hand. But there is no formal and objective process for synthesizing specialized knowledge. It must be accomplished in another way.

Since the advance of scientific and technological knowledge separated from experience is based on a strategy that trades breadth for depth, there are very few decisions that corporations need to make for which a single person has all the requisite knowledge. The obvious solution is for most decisions to be taken by groups of specialists together possessing the required knowledge. This poses another problem, namely, how the quality of any decision is to be assured. If all the members of a team charged with making a decision represent different specialties, how can anyone judge the soundness and quality of the contribution of another? How then can these specialists function as a team? It may be more appropriate to regard them as a committee of sorts. Galbraith argues that such committees are at the heart of many industrial achievements:

> Thus decision in the modern business enterprise is the product not of individuals but of groups. The groups are numerous, as often informal as formal, and subject to constant change in composition. Each contains the men possessed of the information, or with access to the information, that bears on the particular decision, together with those whose skill consists in extracting and testing this information and obtaining a conclusion. This is how men act successfully on matters where no single one, however exalted or intelligent, has more than a fraction of the necessary knowledge. It is what makes modern business possible, and in other contexts it is what makes modern government possible. It is fortunate that men of limited knowledge are so constituted that they can work together in this way. Were it otherwise, business and government, at any given moment,

would be at a standstill awaiting the appearance of a man with the requisite breadth of knowledge to resolve the problem presently at hand.[10]

Galbraith also writes:

Association in a committee enables each member to come to know the intellectual resources and the reliability of his colleagues. Committee discussion enables members to pool information under circumstances which allow, also, of immediate probing to assess the relevance and reliability of the information offered. Uncertainty about one's information or error is revealed as in no other way. There is also, no doubt, considerable stimulus to mental effort from such association ... Committees are condemned by those who have been captured by the cliché that individual effort is somehow superior to group effort; by those who guiltily suspect that since group effort is more congenial, it must be less productive; by those who do not see that the process of extracting, and especially of testing, information has necessarily a somewhat undirected quality – briskly conducted meetings invariably decide matters previously decided; and by those who fail to realize that highly paid men, when sitting around a table as a committee, are not necessarily wasting more time than, in the aggregate, they would each waste in a private office by themselves ... Pooling and testing information is nonzero sum – all participants end with a larger score.[11]

In other words, a hierarchy of committees is the answer to the lack of a science of the sciences and a technology of technologies capable of synthesizing the findings of a variety of specialties relevant to the making of a particular decision. This organization makes the application of technological knowledge separated from experience a reliable process, since it is not dependent on individual genius but on scores of relatively ordinary men and women who have been deeply trained in their specialties. Great men and women do play important roles, but not in the way it is commonly supposed.[12] A special organization that applies technological knowledge separated from experience (and, as we shall see, also other kinds of knowledge separated from experience) to practical tasks Galbraith calls a technostructure.[13] Participation in such an organization does not depend, first and foremost, on rank and authority but on the specialty a person has mastered and its relevance to a particular task. It is very rare, indeed, that a decision taken by a group is overruled by a single person, regardless of their position in the formal organizational hierarchy. Such a decision can only be checked by a similar group whose members together possess the requisite knowledge. Any other basis for a decision would be seen as arbitrary and unreliable and could disturb the delicate and complex functioning of a technostructure. Once knowledge separates from experience and culture, power passes to the group with the requisite knowledge.

The concept of a technostructure is, of course, an ideal type. There are many sources of distortion, as illustrated by the following examples. Group members who pursue all the latest techniques for projecting a certain image are likely to have more influence than those who do not. Facial expression, body language, grooming, and dress may all be carefully calculated to create an impression of self-confidence, competence, and success. This is particularly true for the highest levels of management decision-making. Management fads succeed one another in rapid succession, and the cult of image is all-important. Further distortions result from the vanity of those engaged in the endless power struggles that rage at the top of many organizations. From my descriptions in this and the previous chapter, it follows that decisions about umbrella design exemplars are more vulnerable to these distortions than are decisions about lower-level design exemplars; and these, in turn, are more vulnerable than decisions that primarily involve analytical exemplars. Nevertheless, the concept of technostructure, referring to organized intelligence, is indispensable for understanding a world in which knowledge separated from experience and culture has gained the upper hand over knowledge embedded in experience and culture. The technostructure is that part of a corporation or government in which the former dominates the latter, while the reverse is true for the rest of the organization. It is by means of such a technostructure that the latest technological knowledge separated from experience was brought to bear on the design of the Mustang. This explains why it took a great deal longer to bring the Mustang to market than the Model T. It also explains the much greater investment, since, during that time, many specialists had to be paid and provided with support staff and offices.

My generic analysis of the design process remains incomplete in a fundamental way. Parallel transformations of the organization are imperative to improve the likelihood of successfully bringing a new product to market. This can be illustrated by the following hypothetical events. Suppose that, upon completion of the design for the Mustang, the executive vice-president in question reports back at a meeting with her colleagues. Let us also suppose that the decision is made to proceed with the development of a production strategy, and that this task is now turned over to the appropriate executive vice-president. He takes the engineering drawings and specifications to his staff, who begin to pore over them. As specialists in manufacturing, they examine every detail from this perspective. In many cases, the conclusion is likely to be that, had a particular detail been altered ever so slightly, the manufacturability of the material or part in question could have been improved without negatively affecting the design. To make such alterations is now much more difficult because adjacent parts might have to be altered as well. With the likelihood that there are many such details, the manufacturing cost could have been lowered provided that such considerations had been made during the design phase. To make the changes now would cost a great deal of time and money, and this could have been prevented. In sum, no corporation making use of technological knowledge separated from experience to bring to market a new product will first design it and then determine how best to manufacture it. The technostructure will simultaneously design the product and develop a strategy for its manufacture. These two phases of the technological cycle can no longer be considered sequentially.

If the enormous up-front investment required to bring a new product to market is to be economic, it must be recovered from the sale of a large number of units so that the price of each is only marginally affected. If this is to be the case, then there is no point in using general-purpose machines for the manufacture of any product. For example, why purchase a general-purpose lathe to dedicate it to a particular production task that requires only a few of the many operations it is capable of undertaking, thereby wasting the rest? During the postwar period, mass production based on long production runs to serve large and relatively stable markets was considered to be more economic if dedicated production facilities were designed and built as an integral part of the production strategy. The production capacity of such a facility was the key parameter in its design. All this further explains why bringing the Mustang to market required a great deal more time and a much larger investment than was required for the Model T.

Next suppose that the design of the Mustang and a manufacturing strategy had been jointly developed, and that once again the results were reported to a meeting of executive vice-presidents. When the executive vice-president responsible for marketing and her staff begin to examine the results, they might find themselves in a position where they have to ward off trouble ahead. Several years have passed since the original concept was agreed upon, and intervening events could have disrupted some or all of the trends from which the original concept of the car had been inferred. Possible examples might include a minor recession resulting in higher levels of unemployment; an overseas war compelling the government to raise taxes, thereby reducing disposable incomes; rising interest rates undermining consumer confidence; an oil crisis threatening a sharp rise in the cost of gasoline; or an opposition party that had a good chance of forming the next government, promising that it would pass tough anti-pollution regulations for motor vehicles. Such events could weaken the likelihood of successfully bringing a new product to market. Once again, these possibilities would be better dealt with by not proceeding through the technological cycle of a new product in a sequential manner. As the vehicle was being designed and as a manufacturing strategy was being worked out, specialists in marketing monitored the results and assessed them in the light of changing market trends. For example, a growing consumer preference for fuel-efficient vehicles could lead to a review of the engine options to be offered to customers. In other words, as the design and manufacture of the car were being planned, so must be its marketing strategy as well as a host of additional details. Arrangements with subcontractors had to be made, dealers' sales staff and mechanics had to be trained, and owners' manuals had to be written and printed.

Finally, there is also a compelling financial reason why the technological cycles of new products have to be planned in their entirety once technological knowledge separates from experience. The dramatic increase in the cost of designing a new product is economic only when it can be recovered from mass production and sales. In turn, mass production is most economic when dedicated production facilities are planned and built, further adding to the up-front investment that must be recovered. The output capacity of such a facility is an all-important design parameter; a factory producing Mustangs is very different from one producing Ferraris. Once this design parameter is decided, it eventually determines the number of units that have to be sold. It may be possible to add a shift if sales significantly exceed the plan, but this is often accompanied by serious problems. Generally speaking, production levels have to be kept within a narrow bandwidth if quality and factory maintenance problems, as well as financial losses, are to be avoided. Sales are sensitive to price; the price depends on how much of the initial up-front investment has to be recovered from each unit; and this in turn depends on sales volume. For these reasons, production planning in the automotive industry at the time of the first Mustang was based on a three-year cycle: after one year, minor design changes were made that required equally minor retooling; this was repeated after the second year, following which more substantial design changes were introduced. In effect, sales targets were set for at least three years, and the cost of model changes could be included. Hence, from a financial point of view, making use of knowledge separated from experience makes the phases of the technological cycle of any product much more interdependent: a significant change in one will have implications for some or all of the others and for the entire plan for bringing a new product to market.

The above comparison of the technological cycles of the Ford Mustang and the Model T illustrates the technological constraints that flow from knowing and doing separating from experience. When technological knowing and doing is embedded in experience, the phases of the technological cycle are largely sequential in time as well as being relatively independent from one another. When technological knowing and doing separates from experience, the entire technological cycle must be planned prior to a single unit being produced because of the high level of interdependence of its phases. Such planning involves many specialists and includes, but is far from limited to, engineers who collaborate together in a unique social process whose organizational locus is called a technostructure. There may well be other ways of accomplishing a synthesis of knowledge separated from experience for the solution of practical problems. Partial alternatives have been pointed out by observers.[14] The course of events might have followed a different path if the cultural context in general, and various socio-economic and political conditions in particular, had been different. It must be remembered, however, that these conditions were the result of 'people changing technology' and, simultaneously, 'technology changing people' during the first stage of industrialization. This greatly extended the technology-based connectedness in society. Technological knowing and doing separated from experience further extended the technology-based connectedness by transforming the technological cycle. Since this required organizational changes in the 'software' of technology, it led to the formation of technostructures. The changes in the 'hardware' and 'software' of technology mirrored and complemented one another.

There are both similarities and differences between bureaucracies and technostructures. Both social organizations are manifestations of technology-based connectedness, as opposed to culture-based connectedness. However, in the case of the former, participation is primarily determined by one's job description and its place in the organization chart. In the case of the latter, participation essentially depends on committees charged with making various decisions. Inclusion or exclusion from such committees is, first of all, determined by the relevance of a person's expertise to the decisions to be taken. The importance of the person's role on a committee is further determined by the significance of his or her expertise relative to that of the other members. The members of a technostructure usually participate in a variety of committees that require their expertise. In some they may play a minor role, while in others they may play a key role. This is true for a junior engineer just hired out of graduate school, a senior market analyst who has been with the firm for a long time, or a manager who has a great deal of experience in constituting and motivating such committees. As noted, a variety of factors can distort the operation of a technostructure. Some members may have more to recommend them in terms of their ability to climb the organizational pyramid or to seek to improve their status than their technical competence.[15] There will also be informal communication networks based on experience and culture, such as those resulting from acquaintances and friendships. The extent to which these are able to undermine this organization is not well known. The ideal type of this kind of organization is one that, before anything else, facilitates the synthesis of highly specialized knowledge separated from experience. In sum, a technostructure represents a constantly changing network of associations that form temporary hierarchies of committees, which, in industry, often correspond to the planning of the technological cycles of products. Since knowledge separated from experience takes centre stage in all these activities at the expense of knowledge embedded in experience, the corresponding technology-based connectedness dominates culture-based connectedness.

The emergence of technostructures within corporations had far-reaching consequences. I will briefly summarize the most important ones, and refer the reader to Galbraith's more extensive account.[16] It is worth emphasizing that the validity of my analysis is limited to the industrially advanced nations for the three or four decades following the Second World War, after which knowing and doing separated from experience in the remainder of society brought further fundamental changes. Within this time period, I will focus on what Galbraith termed the planning system, made up of large corporations employing technostructures. I am well aware that Galbraith's analysis has drawn a great deal of criticism, particularly from economists and both the political left and right. This was to be expected, since it essentially proposes a radical change in making economic sense of our world, which has far-reaching consequences for the conventional wisdom in a number of disciplines and in political life. Almost without exception, the critiques are of an ideological nature, rather than a critical assessment of the thesis that there has been a fundamental change in the knowledge on which modern economies depend. Whatever their merit may be, they shed no further light on the present analysis. There is little else we can draw on. There is no adequate sociology of technology distinct from the sociology of science. One likely reason for this lacuna is that it is next to impossible to gain access to corporations to empirically study a concept such as a technostructure. The fact that modern cultures assign a very high value to technology does not encourage a critical analysis. Whatever the case may be, the result is the same: Galbraith's underlying premise has drawn little critical scholarship. Subsequent to the time period analysed by Galbraith, some smaller companies (such as those involved in high technology in Silicon Valley) have evolved technostructure-like organizations to make use of the latest advances in science and technology.[17] In any case, the full weight of my analysis will not become apparent until Part Three.

Thus far I have shown that, in order to make use of knowledge separated from experience, corporations must plan the entire technological cycle of any new product. A new organizational locus for making use of this kind of knowledge emerges within the more traditional and bureaucratic corporate organization. These changes in turn produce others in an internal chain-reaction-like process of adjustments, of which the formation of technostructures is the most prominent feature, thereby creating a fourth stage in the development of the industrial corporation. The gradual increase in the use of knowledge separated from experience lengthened the time required to bring a new product to market and augmented the necessary capital investment. During the first half of the twentieth century, the operating capital of the largest corporations grew steadily, gradually falling beyond the reach of the wealthiest owner-entrepreneurs and their families. They were compelled to sell shares, and this changed their status into that of major shareholders. Such changes were frequently accompanied by battles over control. As the complexity of the corporation also grew, owner-entrepreneurs could no longer oversee all the operations of the firm, forcing them to rely more heavily on professional management teams.

This too did not come easily in many cases. The separation of knowing and doing from experience greatly accelerated these trends. Eventually, operating capital requirements grew to such levels that even major shareholders came to own only a very small fraction of the business.

Along with this change in the structure of ownership came a change in the way the legal owners were able to control their corporation. When the number of shareholders became very large, they could not exercise the direct control that owner-entrepreneurs could during the previous three stages in the development of the corporation. At shareholders' meetings, the legal owners elected a board of directors to act on their behalf. With the advent of technostructures, the individual shareholder was at a loss to understand, let alone suggest, alternatives to what management proposed. This would require access to another technostructure or at least a scaled-down equivalent in the form of a management consulting firm. Hence, although shareholders were the legal owners, it was difficult for them to exercise any direct control. As long as the corporation was doing reasonably well, it was likely, therefore, that they would accept the strategies and recommendations proposed by management.

Strangely enough, the conclusion that under most circumstances management in effect now controls the corporation remains controversial. How many shareholders are able to understand the balance sheet of a corporation the size of General Motors, especially as abbreviated in the shareholders' report? How many shareholders have the knowledge separated from experience required to understand the minutest fraction of the operations of such a firm? How many shareholders bother to attend shareholders' meetings? How many of these shareholders 'own' shares indirectly via a pension or mutual fund? It is, of course, possible, and indeed it occasionally happens, that the shareholders assert their legal rights, reject management's proposals, and elect a new board of directors, thereby hoping to change the direction of the corporation. The point is that this is exceedingly rare. I recognize that these assertions fly in the face of our present secular economic theology, particularly in the United States. However, the closing decades of the twentieth century have provided us with ample evidence of a problem. First, there are movements that seek to strengthen the rights of shareholders vis-à-vis those of management.[18] Second, there is the obscene skyrocketing of senior management salaries and benefits, which is clearly not in the interests of the corporation and its shareholders.[19] Third, a recent study has shown that senior executives of failed companies often walk away with staggering financial benefits of various kinds.[20]

The separation of ownership from control facilitated a necessary adjustment in the goals of the corporation. During the previous three stages of development, the only goal was profit maximization. The separation of knowledge from experience and its application by means of a technostructure led to a more complex set of goals. The key to understanding this change is the recognition that the technostructure, as the collective brain of the enterprise, must be kept intact at all costs. The implications go well beyond the impossibility of dismissing a significant part without damaging the brain of the corporation, on which everything depends. The complex and delicate decision-making processes made by hierarchies of committees are extremely vulnerable to outside interference, which can only be regarded as arbitrary by every committee. It would appear that this collective brain is as vulnerable as a human brain to outside interference by, for example, poisons or severe blows. Ample evidence is provided by the many 'doctors' attending to ailing collective brains, namely, consulting firms. In other words, the primary goal of the new large corporation is the protection of its technostructure (by far its most important factor of production). Its autonomy can be challenged primarily in two ways. If the shareholders are not receiving reasonable (which is not the same as maximum) dividends on their investments, they may challenge management and thus interfere with the long-term operations of the technostructure. If it becomes necessary for the corporation to borrow capital, it exposes itself to external influences. Although management does not own the corporation, it is nevertheless extremely interested in profits since in order to stay in control it must pay reasonable dividends to shareholders and finance its internal operations. Borrowing money from banks invites external scrutiny, demands for changes in the business plans, and possibly changes at the top.

In decreasing order of importance, the second goal of the new corporation (after protecting its technostructure) is growth. Growth means added responsibilities for the technostructure, which translates into promotions and salary increases for its members. At the same time, these corporations operate in societies whose cultures expect everything to grow. No matter how healthy a firm and how excellent its products, its shareholders and society expect it to grow. I recognize that, following the period I am describing, there was a frenzy of takeovers orchestrated by management, many of which could arguably be seen as not being in the best interests of their corporation and its shareholders.[21] This marked an era of unbridled corporate greed that met little resistance of a social, political, or moral kind from the mass societies in which they operated.

The third goal is technological virtuosity. By this, Galbraith means that when highly educated people in the technostructure are not provided with opportunities to make creative use of what they have learned, they have a sense of being underemployed. There is, therefore, a great deal of pressure in the technostructure to make use of the latest advances in knowledge separated from experience.

Finally, a fourth set of goals includes a variety of concerns. For example, to avoid interference with its strategies and plans, management may deem it preferable to make some compromises during labour negotiations at a slight cost to profitability rather than face a strike and slumping morale. Management may prefer to clean up environmental problems in plants located in communities or nations where there is a real threat of public opinion being aroused and politicians becoming involved. A more comprehensive set of motivations than profit maximization has emerged with the rise of the planning system. This development, together with the new technical means at the disposal of management, is transforming the role of the corporation with respect to mass societies. The autonomy of the technostructure has been extended by a series of developments: the ability of corporations to monitor public opinion and to influence it by means of public relations and advertising; the possibility of optimizing donations in order to improve the corporate image; the effective use of political lobbying; the power to negotiate tax breaks and bail-outs; the use of free-trade agreements to undermine national sovereignty and local democracy – all these and more are transforming the autonomy of the technostructure into what (in Part Three) will be described as the autonomy of technique. For now, this can be understood in terms of the growing dominance of the technology-based connectedness over the culture-based connectedness of human life.

The pressure for constant innovation and development has increased steadily. Technostructures cannot be kept idle waiting for new inventions to happen. They are constantly innovating the product lines of corporations. It would be too simplistic to believe that this pressure for innovation matches the needs generated by the way of life of a society. In planning the technological cycles of all of its products, a technostructure also plans their obsolescence. This is a serious concern at a time when we recognize that the world's resources are finite, that the small portion of humanity living in the industrially advanced nations consumes a disproportionate share of these resources, and that we are seriously damaging our life-support systems. In other words, the autonomy of the technostructure also extends to the needs and values of modern societies.

Once knowing and doing separates from experience, a great deal of planning becomes necessary. Although Western societies once regarded planning as being incompatible with democracy, the necessity for planning that accompanied the process of knowledge separating from experience has led to its widespread acceptance. The markets for many goods and services have become dominated by a few producers, whose internal operations are largely planned, thereby pushing back the role of the Market, as we shall see. Today the scale of this planning is so large that among the one hundred largest economies in the world, fifty-one are corporations.[22] This fact appears to bother no one, particularly economists and politicians, who mostly extol the virtues of the Market. I suppose most people feel this is a small price to pay for having the new technology. It is ironic that central planning virtually disappeared with the collapse of the Soviet Union while transnational corporations became ever-larger and more powerful domains of planning. How this is compatible with the secular professions of faith in democracy made by many politicians defies all understanding.

The separation of knowing and doing from experience greatly increases the risk involved in bringing new products to market. The planning of the entire technological cycle of a product requires an enormous up-front investment, which cannot be recovered until several years later when it begins to be sold. This investment is not flexible and cannot be transferred to other products. For example, the facilities designed to produce a particular automobile can be converted to the production of another vehicle or truck only at great expense. This was not so before knowing and doing separated from experience. To manage this risk, corporations had to diversify their product lines to make the risk acceptable. With such a strategy, it became possible to survive a major failure (recall the Ford Edsel and GM's Corvair) without risking the company. For example, a substantial loss in the automotive division may be offset by healthy profits in the truck division, the aircraft engine division, the diesel locomotive division, the refrigeration division, and so on. It will not do, therefore, to explain the enormous growth in the size of the modern corporation during the decades following the Second World War simply in terms of greed or the drive for profits. There is a certain logic inherent in the structural changes associated with the application of technological knowledge separated from experience. Risk management through product diversification is one factor in this growth. Despite many concerns over globalization and the expansion of free trade without social and environmental standards, there is little understanding of the deep structural roots of the above trends.

Along with the restructuring of the technological cycle and the corporation, the separation of technological knowing and doing from experience further changed the structure of technology and the way this structure was embedded into that of society. From the classification of technology developed in chapter 3, it follows that this separation caused technology to behave more and more as a system with respect to its host society during the decades following the Second World War. Superimposed on the ongoing linking together of the 'hardware' of technology was the linking together of its 'software.' This took place on two levels. Technological know-how was split into two interacting activities that were more conscious and explicit in nature, namely, design and analysis, which each had their own communities of practitioners, organizations, and institutional loci. Insofar as design and analytical exemplars could now be expressed in abstract terms, they could be detached from their original contexts to be applied elsewhere, thereby converting each area of specialization into a transmitter and receiver of technical information. The internal relations were strengthened with respect to the external ones. Technology acquired its own internal mechanism of development based on the permutations and combinations of a multitude of large and small advances. Each invention, innovation, or theoretical advance triggered others. The growing globalization of the flow of technological information helped create the first universal technology.

The new internal mechanism of development resulting from the flow of information was not sufficient to cause the emerging universal technology to behave as a system towards its host societies. Another transformation was required in the cultures of the societies involved, to ensure that much of what became technologically possible by means of this new internal mechanism of development would be regarded as socially beneficial or necessary. That such a change was forthcoming is evident when we apply the theory of culture once more. The meanings and values assigned to technology by the cultures of the most industrially advanced nations kept on rising. After all, life without technology was becoming unthinkable. Many daily-life activities depended on one technology or another, particularly with the diffusion of mass production, mass consumption, and the mass media. Culturally, attributing an increasing importance to technology had a double effect on how society dealt with it. First, whatever became technologically possible had a very high value and should therefore be adopted as soon as possible. This ensured that the internal mechanism of technological development was minimally affected by traditional (context) values and that performance values rose in a society's hierarchy of values. There was an ever greater preoccupation with efficiency. Second, whatever was (metaconsciously) declared as very good or even the highest good in human experience by a culture could not be associated with any serious problems it might cause. These had to be the result of something else, such as the misuse of technology or its use by some to advance their wealth and power at the expense of others. This development further reinforced the separation of technology and the economy from the remainder of society and from what was increasingly regarded as a separate environment. To put it somewhat simplistically, technology and the economy were the domain of what had been declared as very good. In contrast, society and the environment were the recipients of most of the negative consequences and hence had to be distanced from the good. Whether this was worked out by means of a political ideology on the left or the right made no essential difference. The result was that everywhere the desired results of technology became dissociated from the undesired ones. The parallel crisis in society is only now beginning to get some indirect attention, as is the case, for example, in discussions about the long-term viability of public health care.

The dazzling accomplishments of technology in the domain of performance values began to be dissociated from its equally spectacular failures in the domain of context values. For example, the environmental crisis that began to attract attention during the 1960s was not generally regarded as a product of technology, but the result of population increases and rising levels of affluence. It became next to impossible for people to believe, let alone act on, the research of people such as Barry Commoner, who showed that the contribution of technological development to the environmental crisis from 1946 to 1968 outstripped that of population growth and rising affluence in the U.S. economy.[23]

The universal character of the new technology separated from experience and culture is particularly evident when we examine the intellectual and professional division of labour on which it is based. Specialists of all kinds relate to the world on the basis of their substructures of experience, as opposed to their structures of experience. As a result, their attempts at making things better are based on their specialties, regardless of their cultures. Technostructures coordinate the application of their specialties to the solution of complex problems or to the formulation of strategies to deal with particular issues, in a manner that breaks their tasks down to be coterminous with this specialized knowledge. Once again, culture plays little or no role. When technological knowing and doing become separated from experience and culture, it follows that the technological cycle, the corporation, and the structure of technology must also be separated from experience and culture. We could not possibly acknowledge this in terms of technology becoming more autonomous with respect to human life and society. Yet this is exactly what its separation from experience and culture implies, namely, that it is no longer adapted to local ways of life (including the ecosystems in which they are embedded). Technology now manipulates the world 'inside out,' which is autonomous from doing so from the 'outside in.' Surely the arms race and the environmental crisis so much debated during the decades following the Second World War are unthinkable without an autonomous technology, unless one believes that it was a deliberate and calculated strategy to destroy life. The alternative to an interpretation based on an increasingly autonomous technology is one based on the autonomy of a people. If people had a great deal of autonomy with respect to their technological creations, we would not have witnessed the emergence of a universal technology, but instead the emergence of a plurality of contemporary technologies, each perfectly adapted and integral to their ways of life and cultures, as was the case for all earlier peoples and times.

6.2 Living with a New Economy

Once again, a fundamental change in the technologies of the first generation of industrial societies was accompanied by a chain-reaction-like process of accommodation by these societies, beginning around the time of the Second World War. Three to four decades later, the separation of knowing and doing from experience in other dimensions of mediation began to make its full impact, and it profoundly marked individual and collective human life during the closing decades of the twentieth century. These time periods are approximate since, for reasons examined earlier, the separation of technological knowing and doing from experience did not occur uniformly across industry. The accommodations described in the rest of this chapter varied in terms of pathway and time, but in the end they yielded much the same result.

I have already described the adjustments the first generation of industrial societies had to make to their scientific and technological dimensions of mediation. The remainder of this chapter will describe the economic, social, and political adjustments, deferring the remaining transformations to Part Three.

Three main adjustments to the economic dimension of mediation were required as a consequence of knowing and doing separating from experience. The first resulted from the fact that it was impossible for the new corporations equipped with a technostructure to rely on the Market for the procurement of all necessary inputs and to sell the goods and services they produced. The role of the Market increasingly had to be complemented by technical planning. The second adjustment resulted from the need for governments to become involved in the economy through the creation of a large public sector to complement the planning of the new corporations in the private sector. The third adjustment resulted from the separation of knowing and doing from experience becoming the prime factor of production, pushing capital to second place. All of these adjustments were entirely contrary to the reigning secular economic theologies, and indeed (as Galbraith showed so well) against economic theory. It may be supposed, therefore, that these adjustments did not flow from human values and aspirations, but from the constraints imposed by the decision to take advantage of the new kind of knowing and doing separated from experience. They therefore belong to the domain of 'technology changing people' as opposed to 'people changing technology.'

As for the first stage of industrialization, rationalization, modernization, and secularization, I will begin my analysis using the qualitative analogy of input-output methods. From this perspective, the new corporation may be represented as a 'black box' containing the locus of activities that transform a variety of inputs into outputs by means of the technology and business practices of the time. Inputs include material inputs that are the outputs of other activities such as raw materials from mining, processed materials from steel plants or chemical installations, and components or sub-assemblies from suppliers; energy inputs in the form of fuels or energy carriers, which are again the outputs of other activities; capital goods for producing the material outputs; plant and office facilities; land for these facilities (or rather 'pieces' of local ecosystems); capital; and people with the requisite skills based on knowledge embedded in experience or knowledge separated from experience. Outputs include goods and services sold; wages and profits; people whose abilities and health may have been positively or negatively affected; patents and knowledge; waste materials, pollutants, and eventually worn-out or outmoded capital goods and facilities. Some inputs are transformed into outputs on a more or less continuous basis, while others remain in the 'black box' for a much longer duration.

The entrepreneurial firm of the first generation of industrial societies relied on the market to purchase its inputs and sell its outputs of goods or services. The consequences of the undesired outputs were mostly borne by third parties, thus creating the well-known market externalities. Reliance on the Market became more difficult as the entrepreneurial firm grew in size relative to the markets in which it bought and sold. We have seen that during the third stage in the development of the entrepreneurial firm, there were strong pressures on it to grow in order to gain a competitive advantage from the economies of scale. In the United States, entrepreneurs such as Chrysler, Du Pont, Guggenheim, and Rockefeller began to consolidate large enterprises. The next major impetus to growth came from technological knowledge separating from experience. Recall our discussion of the technological cycle of the Mustang. Suppose it had been decided to slightly modify one of the alloys used in the fabrication of one or more parts, and that this in effect created a specialty item not produced by any steel mill. In such a case, Ford could not rely on the market for this new material it had devised. The same was true for all the parts designed for the Mustang that Ford did not produce itself. It was also true for specially designed capital goods. There simply was no market for highly specialized materials, components, or subsystems designed on the basis of technological knowledge separated from experience for a particular product. Even when the decision was made to rely on materials, components, or subassemblies available in the market, it could not be expected that large quantities would be forthcoming at prices and delivery dates specified in the plan for the technological cycle of a new product.

The inability of the large corporation to rely on the Market for its material inputs was overcome in one of three ways.[24] The first was through vertical integration, which refers to a large corporation's acquisitions of another company capable of producing a desired input. The planning of the entire technological cycle of a new product could thus be extended to include the production of the necessary inputs. This did not mean that the uncertainty of the market was entirely eliminated, because these production units depended on others for their inputs, and so on, forming long chains all the way back to the biosphere as the ultimate source. One or more links in these chains might be affected by market uncertainties, but the risks could usually be reduced in this manner. A second solution was to work with small groups of potential suppliers for the purpose of selecting one of them as a business partner by signing a long-term contract for the supply of a particular input at agreed-on quantities, prices, and delivery dates. Even when these first two ways were not followed, the new corporation had a third in the tremendous influence it exerted on a particular market because the orders it placed were so large relative to that market. For example, despite the appearance of competitive buying and selling, the fact of the matter was that when a corporation the size of Ford decided to purchase or not to purchase, it could mean years of prosperity or hardship for a supplier or, in the extreme, its very survival. Needless to say, such suppliers would be exceedingly cooperative. The first two solutions eliminated the market altogether, while the third severely constrained it.

The same argument applies to all but one of the other inputs. For example, the spread of so-called wholesale competition, by which consumers of electricity can select their own suppliers, opened up the above three options for procuring this input.[25] The new corporations were generally able to provide themselves with the required operating capital from retained earnings, and it was only in unusual circumstances that they needed to borrow capital in financial markets. In terms of people, the labour market did not perform reliably with regard to technical experts having a great deal of education in knowledge separated from experience. Suppose a shortage occurred in a particular type of expertise. Salaries would rise, job opportunities would increase, and more excellent students would probably line up to get into the professional schools or university faculties providing the required education. Pressure would build to admit more of them, but even after the financial resources had been allocated, it took time to hire new faculty and build new classrooms and laboratories. Even then, it generally took another four years before the first graduates became available. It was for this reason that governments, in cooperation with industry, undertook educational planning to ensure the mix of expertise the economy required.[26] (It is worth noting that GM used vertical integration to secure an adequate supply of professionals by means of its General

Motors Institute.) The market was relied on only for the supply of employees whose work was based on knowledge embedded in experience, so that, following a short basic education or apprenticeship program, they could learn on the job, such as assembly-line workers, secretaries, and data-entry clerks.

What I have shown thus far is that the separation of technological knowing and doing from experience replaced the contents of the 'black box' with a different set of technological and business practices that connected the inputs and outputs of the new corporations. Technology now required the planning of the technological cycles of the products produced by these corporations. Such planning was incompatible with fluctuations in the markets for the inputs on which the modern corporation depended. The above three common strategies extended the influence of the corporation, thereby pushing back the boundary between the planning system composed of large corporations and the remainder of the economy, where much smaller entrepreneurial corporations operated in markets. Such corporations often relied less heavily on technological knowledge separated from experience, which limited the kinds of products they could competitively develop, manufacture, and sell. The new large corporations surrounded themselves with suppliers that tended to be much smaller in size and with whom they dealt mostly outside of the market.[27]

On the output side of the large corporation, we can observe a similar pattern of change. Significant economic accommodations had to be made by society if the planning required to make use of knowing and doing separated from experience was to have a reasonable chance of success. The corporation was vulnerable to the following uncertainties. The concept for any new product, if not directly inferred from market trends, would in any case be carefully assessed in terms of consumer attitudes and preferences. Such analyses, although meticulous and detailed, were by their very nature speculative because the new product did not yet exist. This introduced a first source of uncertainty. Because making use of knowing and doing separated from experience substantially increased the time required to bring a new product to market, consumer attitudes and behaviour might change in the meantime in response to changing conditions in society. During the planning of the technological cycle of a new product, early detection of such changes permitted adjustments to the basic concept, but this became increasingly difficult as the planning process advanced. Hence, there was a second inevitable uncertainty about how well the new product would

264 Disconnecting from and Reconnecting to Experience and Culture

fit the market when it was finally launched. The third uncertainty involved its planned price tag, its sales volume as constrained by the dedicated production facility, and its material and labour costs, although planning would have made them as insensitive to supply markets as possible. Finally, there was the inherent uncertainty of the market for the product and its competing alternatives as a result of, for example, business cycles and recessions. Together, these uncertainties could overwhelm even the best of plans.

It must be remembered that before bringing a new product to market, the large corporation was locked into an enormous up-front investment in the application of knowledge separated from experience to the entire technological cycle, including the building of a dedicated production facility, all of which could not readily be transferred to another product. According to modern economics, we are to believe that, at this point, the company would simply fold its arms and wait patiently for people to decide whether or not they would buy the new product. Was this not a society in which the customer was king (or queen)? This commonplace made us all kings and queens that ruled society by means of the Market, in the supposed absence of any means of persuasion or exercise of power. It implied that advertising, on which our society spent a considerable portion of its wealth, merely provided us with a few (mostly trivial) facts to make us more informed consumers. What was the economic reality of individual and collective human life in a modern society, and how well did our economic beliefs correspond to it?

Many things taken for granted by modern economics and popular belief stem from a time in which they were largely self-evident. Once again we come face to face with the failure of modern science in general and the university in particular. As I will examine shortly in more detail, the spectacular success of modern science is rooted in a knowledge strategy that simplifies the task by abstracting what we wish to know from the unmanageable complexity of reality, to place it into the more manageable intellectual context of a scientific discipline and the physical context of a laboratory, carefully designed so that a very small number of variables can be studied, preferably one at a time. Generally, the founders of a discipline abstracted a family of phenomena in such a way that the connections to the remainder of the world, severed at the boundaries of these phenomena, had a minimal significance. The subject was then advanced with minimal reference to other disciplines. In other words, if the significance of the connections intellectually severed by a process of abstraction changed in the course of the evolution of a society, this might or might not receive critical attention; and in any case, there was no established part of any scientific discipline assigned to this task. As a consequence, each discipline advanced by treating the findings of all others as knowledge externalities.

In the social sciences and humanities, the choice of boundaries between what was to be abstracted for study and the rest of the world generally remained in dispute, giving rise to parallel schools, with the result that no single approach to the subject emerged. Because each discipline and school treated the findings of all others as knowledge externalities, they opened themselves up for what elsewhere I have referred to as a professional and intellectual fundamentalism.[28] By this, I mean that the implicit and explicit assumptions about the world from which a discipline or school abstracts a family of phenomena may have been good enough for the world of our intellectual fathers and mothers, but it may not be good enough for us. To break this intellectual and professional fundamentalism would require internalizing the findings of other disciplines and schools to broaden our understanding of the world, in order to critically assess the loss of understanding that has occurred from severing the relationships between that portion of the world corresponding to the subject matter and the rest of the world. Since all other disciplines and schools are equally at risk, the relevance to our world of science in general and the university in particular may be very limited.[29]

What concerns us here are the intellectual fundamentalisms that reign over much of modern economics and popular economic belief. These have been exposed, but without effect. Surely it is impossible to understand the world of the second half of the twentieth century without taking into account the substantial influence of science and technology. How have they transformed the context within which economic activities occur and on which they depend: society and the ecosystems in which it exists? How have these changes influenced economic institutions and their place in the institutional framework, way of life, and culture? In turn, how have these changes individually and collectively affected human lives in general, and economic behaviour and attitudes in particular? When we examine a representative sample of introductory textbooks on modern economics for answers to such questions, and critically assess them in the light of the findings of other disciplines and schools, we quickly isolate a number of implicit and explicit assumptions about the world and the place of the economy in it. It is almost as if our world is only quantitatively different from the world of Smith, Ricardo, or Mill, but not qualitatively so.[30] In it, we find *homo economicus* maximizing the utility derived from disposable income. This is accomplished by ensuring that roughly equal satisfaction is derived from the last dollar spent on things such as food and drink, clothing, shelter, and entertainment. The demand for any good is determined by the marginal utility derived from the last dollar spent on it (being more or less equal to a dollar spent on another good). Otherwise, satisfaction can be increased by spending less on a good with a lower marginal utility and spending more on one with a higher marginal utility. Such a reallocation would increase the marginal utility of the former and decrease that of the latter, until an equilibrium is re-established. Any outside interference with this process of translating needs and wants into demands would reduce the happiness a person derives from an income. The demands created by *homo economicus* are received by the Market and transmitted to producers of goods and services, who respond to them by maximizing their profits in the allocation of factors of production. In this way, the great invisible hand of the Market creates the best of all worlds for most people. This economic vision of things is elegant in its simplicity, seductive in its explanatory power, and reassuring in its political implications. It portrays a world of economic democracy in which the consumer reigns over the Market, to which producers respond.

The first thing to notice about this description is the complete separation of the actions of *homo economicus* from experience and culture. This model regards the complexity of daily life from the 'inside out.' Underneath this complexity, subject to influences from within (including those of the unconscious and of primary socialization) and from without (society), lies a mathematical algorithm for maximizing utility. It makes the economic model acultural and ahistorical, with the consequence that the findings of any other discipline or school become knowledge externalities. Such an economic vision of human life can only approximate reality under social and historical conditions in which there is a very modest material standard of living, compelling each person to give careful consideration to every penny earned and spent. Under such conditions, internal and external influences will have little or no effect. In a society with a high material standard of living, this is not the case. Constantly changing fashions and fads, the encouragement of impulse buying, and the ability to make a big fuss over the most trivial characteristics of goods are but some indicators that external influences are bound to be significant. However, since the above description of economic life is ahistorical, it apparently is equally valid for the early part of the nineteenth century with its relatively simple technology, small businesses, and modest material standard of living as it is for the second half of the twentieth century with its mass production, mass consumption, mass media, and the 'lonely crowd.'[31] However, life in a mass society is substantially different from life in a traditional society. The economic implications are not trivial, because individuals in a mass society are much more susceptible to external influences than individuals living in traditional societies. In the next section, it will become evident that the advent of modern mass advertising has made the above economic description of things increasingly irrelevant.

The methodological issue at stake parallels one that we have examined earlier: as a result of the cultural cycle, when people change technology, technology simultaneously changes people. When the members of a mass society live in a life-milieu that bombards them with images of all kinds, and when the function of their structures of experience as a mental map is greatly weakened because of the absence of an internalized viable tradition and morality, the result is that consumers influence producers while simultaneously producers influence consumers. The former direction of influence is the accepted sequence that dominates economic thinking, while the latter has been referred to by Galbraith as the revised sequence.[32]

Mass advertising is an important economic phenomenon. In the remainder of this section I will examine its modus operandi, and in the next section I will estimate its influence. The focus is on the advertising of the planning system, because the advertising of entrepreneurial firms is often somewhat different in character. Mass advertising was developed in close conjunction with mass production and mass consumption, while its form and content reflected human life in an emerging mass society. Already during the decades following the Second World War, advertising had little to do with creating informed consumers. This continues to be the case. Suppose we develop a simple checklist made up of the kind of information that may have been transmitted by an advertisement of that time: Is the function or purpose of the product explained? Which of its attributes are described? Are these compared to those of alternative products? Are we told how much it costs and where it may be purchased? What is the warranty? How green is the product in comparison to others? Are there any health risks associated with its use? Is it safe for children? Is the product made with child labour? Is it manufactured under adequate labour and environmental standards? Is the manufacturer a company in which responsible mutual funds invest? If we next score a number of ads from that period in terms of their information content, it will become evident that they contained little such information. In fact, the structure of these ads appears to have much more to do with associating a product with the symbols of a society of that time. These symbols operated on the metaconscious level and evoked an emotional response associated with a large body of experiences. If a symbol did not evoke such a response or required an explanation, it had ceased to be the living symbol of a culture. The advertisements of a previous time period may not work today, as the symbols of a culture constantly evolve and decline and because of the effects of desymbolization. (This will be examined in Part Three.)

For example, an advertisement for a car of the 1960s or 1970s frequently associated it with the stereotypical 'sexy blonde' posed along with it.[33] A beautiful young woman symbolized two things. To mostly male car buyers, the car became 'sexy,' even though cars can only get you from A to B with more or less comfort, safety, and fuel efficiency. Apparently, some of the characteristics of the woman had become associated with the car – something that was hardly the result of chance but was carefully developed and tested on consumers with a battery of well-known techniques. Sex had become a powerful symbol in the kind of society that emerged following the Second World War. This had a great deal to do with technology, as will become evident in Part Three. The fact that the women in the ads were young was also highly symbolic. In a traditional society, it was the elders and their wise counsel who were revered because they had spent a lifetime living out of a tradition and because they were closer to its source, be it a mythical hero ancestor or its reception from the gods. Even though events in the first half of the twentieth century demystified progress, the rapid technological and economic growth following the Second World War led to a body of daily-life experiences that implied that today was better than yesterday and tomorrow was likely to be better than today, and youth symbolized this.

Advertisements reflected the condition of human life in a mass society. The disappearance of a strong tradition and morality, the weakening of social ties to family and friends, and the accompanying weakening of traditional social groups, neighbourhoods, and communities created 'loneliness in the crowd.' Television advertising in particular introduced consumers to a world in which products were endowed with magic powers, necessary to live in and succeed in that world. Toothpaste could now do a great deal more than get your teeth more or less white: it could bring friends and even sex. Beer became associated with wonderful parties, beautiful people, and the promise of an end to loneliness – and more sex. The examples could be multiplied almost indefinitely as consumer goods were associated with powers that were deeply needed by people in a mass society. By bonding powerful symbols to consumer goods, an attempt was made to satisfy deeply felt non-material needs with material things. Technology, having become central to individual and collective human life, was assigned a very high value by the cultures of that time, with the result that it was difficult to see any limits. The mass media in general and advertising in particular played out this drama of modern existence.[34] For the individual, these ads portrayed the cultural unity of their lives.

Because symbol-based advertising operated on the metaconscious level, consumers were unaware of its influence. They could sit in front of their television sets and ridicule the content of advertisements, but this provides no clue as to their influence. On a daily-life basis, many might find the concept of a revised sequence hard to believe: Did I not read a great many test reports on cars and did I not drive the most promising ones before deciding which car to purchase? What, other than a trivial role, could advertising have played in my decision? Advertising simply did not reach people on this level, and its metaconscious influence could not be detected by asking these kinds of questions. It left intact people's convictions about consumer sovereignty and economic democracy. It left unexplained how the extensive planning of the large corporations interfaced with this supposed consumer sovereignty, why these corporations invested large sums in advertising, and how, in a mass society, people found their bearings in the absence of traditional reference points for life. A theory of the revised sequence resulting from symbol-based advertising can answer these and other questions much more satisfactorily than can the theory of the accepted sequence. The conclusion to which we are building is that, on the output side, the large corporation did not rely on the market but on the management of consumer demand to complement its planning. This claim must be carefully nuanced. It is not true that corporations could sell anything by advertising. Recall that the planning of the entire technological cycle of a new product began with the most careful examination of consumer behaviour. The revised sequence allowed large corporations to manage consumer behaviour to a sufficient extent to keep their plans on track by dealing with slight deviations in this behaviour.

To sum up, in order to make the fullest possible use of knowing and doing separated from experience, corporations had to grow very large, acquire a technostructure, and plan the entire technological cycle of their products. Together they constituted the corporate planning system, distinct from the remainder of the economy, which was composed of much smaller firms whose central feature was a less exhaustive use of knowledge separated from experience. The planning system could not rely on markets for its inputs and outputs, and minimized its dependence on them by means of strategies such as vertical integration, long-term contracts, and advertising. The time frame of the analysis is the three or four decades following the Second World War, after which information technology and the separation of knowing and doing from experience throughout society compelled additional changes.

Consumer sovereignty and economic democracy were not the only casualties of corporate planning. A second major economic transformation was required when technological knowing and doing separated from experience – the role of the state in the economy had to change. With hindsight, the reason is obvious. Symbol-based advertising could not protect corporate planning from changes in consumer attitudes and behaviour that stemmed from fluctuating economic conditions affecting their purchasing power. Recessions and declines in employment negatively affected the aggregate demand for goods and services. For corporate planning to work, purchasing power had to be adequate to ensure that goods and services were acquired at quantities and for prices foreseen in corporate planning. Here we encounter a crucial problem that industrializing societies were unable to resolve between the two world wars. Unemployment and inflation had been the social scourges of these societies most of the time. We do well to recall that the leading industrial nation making the most extensive use of knowledge separated from experience at that time, Germany, suffered so severely from these problems that Hitler and his party were able to convince the German people that extreme circumstances necessitated extreme solutions. An economic miracle was wrought, albeit with the sacrifice of 50 million lives.[35]

The Keynesian revolution implemented after the Second World War temporarily solved the problem, but it had the potential to cost even more lives. The basic concept of Keynesianism is simple and elegant. Imagine plotting the total output of goods and services of an industrial economy over time. The curve would fluctuate significantly enough to make corporate planning as described thus far less successful than it had to be to make the 'system' viable. Next imagine superimposing a public sector large enough to be expanded or contracted to offset fluctuations in the private sector, to ensure that total economic output will grow at a stable and fairly predictable rate. However, not all public expenditures are readily adjustable in this way. Consider a rapid decline in the private sector of the economy. A government may attempt to offset this by means of public expenditures that add to the standard of living, as is the case for schools, hospitals, roads or parks. Their planning and construction often take a long time, with the result that the impact of such expenditures comes too slowly. Furthermore, once the private sector of the economy begins to expand once more, it is not politically feasible to halt their construction. The only possibility is to delay new construction. No such problems occur with public expenditures on scientific and technical research, defence, or space exploration. To most people it makes little difference whether a space mission is launched a year earlier or later, whether the power of our weapon systems is improved by 5 or 10 per cent, or whether a supersonic transport project is abandoned. Public expenditures on projects of this kind tend to be very large and involve relatively few producers, thereby facilitating the management of a large public sector. This was the road generally taken.

What turned out to be even more important in the long run was the ability to underwrite at public expense the increasingly large and risky investments required to advance the technological frontier. Many such advances would be uneconomic otherwise. However, in the name of national security, the advancement of democracy, a life-and-death struggle against an evil empire, the prestige of the nation, and economic survival in an increasingly global technological race, almost any sacrifice could be justified, including each citizen having to work for the state for nearly eight months of the year.[36] Spending money on the poor, on hospitals, or on roads would not accomplish any of these things, so vital to the new universal technology and its economic use. People had tremendous difficulty admitting that the running of modern economies was so dependent on military expenditures.[37] This would undermine all their cherished images of themselves and their technological and economic accomplishments. The Cold War filled a deep need of communism and democracy alike. The Keynesian revolution was a triumph in overcoming possibly the most acute problems of industrial civilization, but the price was high. Despite the doctrine of security through mutually assured destruction, the nuclear powers kept getting closer to an accidental nuclear war due to frequent computer errors indicating an attack.[38] The impact of a growing military-industrial complex directly and indirectly undermined the private sector of the economy.[39] The success of the Japanese economy relative to the American one for decades must in part be attributed to Japan concentrating all its resources on developing the private sector of its economy. The negative impact of the military-industrial complex in the United States on the private sector of its economy has been extensively debated.[40] In any case, defence based on modern weapon systems was the ultimate end-of-pipe solution to the more fundamental problems of industrial civilization. The risk of war is always better dealt with by preventing the kinds of conditions that back nations against the wall with no choice but to fight. The separation of knowing and doing from experience thus had many unintended and far-reaching consequences.

For several decades, the large public sector created by most governments in the industrial nations following the Second World War was financed by a progressive tax system. When the private sector of the economy grew and incomes with it, so did taxes, thereby progressively curtailing demand. When, during a time of recession or growing unemployment, incomes fell, so did taxes, thereby progressively supporting demand. This self-adjusting level of taxation functioned for both individual and corporate incomes. After a number of decades, this began to work less well and national debts began to rise. Technology was an important factor contributing to this situation.[41] As the influence of performance values on technological development strengthened following the Second World War, the desired results (the 'signal') were increasingly undermined by the undesired ones (the 'noise'). Net wealth creation (as opposed to gross wealth creation measured by the GDP) was estimated to be in decline as the 'noise' of technological and economic growth engulfed societies. In the interim, the Keynesian revolution permanently established a large public sector and thus an expanded role for government in the economy. The economic planning of the state involved in running such a large public sector complemented the planning in the private sector. It also began to make increasing use of knowledge separated from experience, compelling, as we shall see shortly, some fundamental political changes.

Galbraith showed convincingly that the revised sequence also applied to government procurement.[42] For example, the technostructures in the military-industrial complex (one of the major components of the planning system) were in a strong position to provide advice on what was technologically achievable in developing the next generation of a weapon system. At the same time, the military would interpret this in the context of defence strategy; and this information, in turn, permitted these technostructures to plan possible avenues and priorities. In this way, the planning of the military-industrial complex affected military planning and vice versa, with the result that a revised sequence operated between the government as military consumer and the arms industry. It was this relatively seamless partnership in the name of ultimate causes that helped technology to unshackle itself from its traditional economic chains. Current economic ideologies are embarrassingly silent about this all-important part of a modern economy.

The above provides further evidence for the previously developed claim that knowledge separated from experience had become the prime factor of production. The relationship between technology and capital was gradually reversed in the course of the twentieth century. Corporations making intensive use of knowledge separated from experience were usually able to generate their working capital. The underwriting of the most advanced and risky technology was achieved through the large public sector, which in its turn was underwritten by taxes. This development should not surprise us. It was the use of industrial technology that created the system appropriately named after its prime factor of production, namely, capitalism. It was organized for the renewal and accumulation of capital as the lifeblood of the first generation of industrial societies. From a historical perspective, the rise of capital as the prime factor of production was relatively sudden, and so was its downfall. Its rise transformed the social structure of a society and its relations of power. The rise of knowledge separated from experience as the prime factor of production required once again a fundamental restructuring of society.

With hindsight, all this might have been expected. Did Marx not give a prominent role to technology in the forces and relations of production?[43] Can Weber not be interpreted as implying that rationality is in the process of becoming the prime factor of production?[44] Did Ellul's description of technique not include its role as the new factor of production?[45] Their works are landmarks in describing what, during their lifetimes, were the embryonic beginnings of major developments that continue to shape our world. Despite vast ideological differences, it is these embryonic beginnings at the core of their separate visions that continue to be of such fundamental importance for scholarship. And in all of this, economics served as the 'secular theology' of the system, as opposed to providing us with a scientific description of economic life.[46] It veiled a new universal technology that was gradually imposing itself on economic reality by transforming people's wants and needs to more accurately match what the 'system' could produce, by imposing itself on the Market, by becoming the new prime factor of production, by requiring the state to play a new economic role, by harnessing economic production to war, and by compelling citizens to work the better part of each year for the state to pay a for all this. Since this did not correspond to any political or religious tradition or to any human values or aspirations that I know of, it should be considered as evidence of the growing autonomy of a universal technology with respect to human life and society – an autonomy matching its previously described separation from experience and culture.

The role the economy played in all of this was to count commodities and to discount life. However, this no longer resulted from the Market force of commoditization but from a growing technical approach to the world that reified everything it touched, thereby strengthening the technology-based connectedness of human life and correspondingly weakening the culture-based connectedness. This helped to create a new wealth of nations, this time based on an intellectual and professional division of labour, as well as a new 'poverty of nations.' The invisible hand of the Market was left to preside over a growing number of market transactions that were purely speculative and corresponded to little or nothing real. Noam Chomsky reported that 'In 1971, 90 percent of international financial transactions were related to the real economy – trade or long-term investment – and 10 percent were speculative. By 1990 the percentages were reversed, and by 1995 about 95 percent of the vastly greater sums were speculative, with daily flows regularly exceeding the combined foreign exchange reserves of the seven biggest industrial powers, over $1 trillion a day, and very short-term: about 80 percent with round trips of a week or less.'[47]

6.3 Living in a Mass Society

Making the fullest possible use of the new prime factor of production for economic development required a further social transformation, and one which superimposed itself on those undergone by the first generation of industrial societies. The result was an entirely new kind of society that had little in common with its precursors. These so-called mass societies are characterized by an accommodation to a universal science and technology corresponding to a greatly weakened role for culture. The deficit had to be made up by technical means using the mass media as their infrastructure. An entirely different way for establishing a social hierarchy and for weaving the social fabric had to be found in order to remove significant barriers to the fullest possible use of knowing and doing separated from experience and culture. Social mobility could no longer be restricted. Until then, the members of each new generation occupied the social positions of their parents. The moral and religious justifications varied considerably. For example, it was thought that people were assigned their station in society at birth according to the merit or lack thereof of their previous lives, or according to the sovereign decision of a god whose wisdom could not be challenged. The universality of this system of assigning social position by birth surely reflects a deeper underlying reality. The transmission of a way of life from one generation to the next is best assured in this way when it is based on a knowing and doing embedded in experience. Because work was mostly carried out in the context of the extended family, children grew up watching their parents work and quickly learned to help out as they were able; so that it was not long before they could do a great many things if, for example, one of the parents fell ill. In this way, the restriction of social mobility ensured the effective transmission of a way of life from one generation to the next. Intelligence (understood in the modern sense) was not a significant factor in this process of transmission, since all children are able to acquire what continues to be the most complex symbolic system human beings learn, namely, their mother tongue and their culture, to which a knowing and doing embedded in experience is integral. Intelligence did not prove to be a decisive factor until ways of life began to make extensive use of knowing and doing separated from experience.

A formal education was not strictly necessary for most social positions. For example, growing up with a father who was a craftsman meant that, by the time you reached your early teens, you were ready to finish up what you had been learning at home with a formal apprenticeship, usually under another master. An elementary education might be beneficial but was not strictly necessary. A higher education was restricted to the professions, and to the members of the social elite as literate people of leisure. As long as ways of life primarily relied on knowing and doing embedded in experience, it appears that this system of transmitting a way of life from one generation to another was almost inevitable.

When ways of life began to make extensive use of knowledge separated from experience, fundamental changes had to be made. If higher education had been restricted to the upper strata of the social hierarchy, there simply would not have been enough people qualified to participate in the ever-larger number of technostructures. More important, the abilities of people to acquire and use knowledge separated from experience appeared to vary a great deal more than the abilities associated with knowledge embedded in experience. Hence, the social structure enabling a society to make maximal use of knowledge separated from experience was one in which everyone was given an equal chance at an education, but where the educational system identified, streamed, and rewarded those who excelled in mastering knowledge separated from experience. These students were encouraged to continue on to higher education, which again selected the best for the more advanced levels. Ideally, their level of achievement should have been the determining factor in obtaining positions within technostructures, which in turn would open the door to positions of leadership in industry and government. Alternatively, those students who were identified by the school system as having an inadequate ability in mastering knowledge separated from experience were directed to training for positions in which knowledge embedded in experience dominated, although it may have been supplemented with various levels of knowledge separated from experience. Such would be the ideal type, for which societies strove with varying degrees of success. The result of these efforts was to essentially divide the social structure of a society into two. The top of the social hierarchy was reserved for those who could participate in the development, application, and transmission of knowledge separated from experience, while the bottom part was reserved for those who could not do so effectively.[48] This invisible division in the social structure of society corresponded to one in the corporate world, as the dream of starting at the bottom of an organization and working one's way to the top vanished with knowing and doing separating from experience.

I have constructed this ideal type of a modern educational system and its corresponding social structure with the benefit of hindsight. Reality was never so tidy, as societies sought to make sense of many new developments whose long-term implications were often far from clear at the time. As such, the ideal type frequently deviated from reality in many ways, but thus far these deviations appear not to be decisive. The ideal type is designed to describe the mainstream and not whatever developments may occur on the fringes. As in the past, societies developed ideologies to make this new assignment of social positions morally compelling. Concepts of equality, liberty, democracy, and human rights received a strong boost. Instead of being radical ideas held by marginal groups, they became compelling to entire societies. In fact, they now appear so self-evident to us that we do well to recall the reservations expressed by earlier thinkers. Some saw the values espoused by the French Revolution and the Industrial Revolution as threats to social order. They conjured up images of agitated mobs whose members were collectively unable to make reasonable decisions. De Tocqueville's reservations about American democracy are well known.[49] How can an egalitarian society recognize and respect leadership based on the ability to creatively unfold the potential of a way of life, especially when this runs contrary to public opinion? How can the general public in a nation state make reasonable decisions that command respect? Must individual expression be subordinated to that of the majority? In Europe, the bourgeoisie believed that the people submerged in the urban masses could not behave as responsible citizens, and consequently opposed extending a vote to them. Marx made the proletariat the revolutionary class, but acknowledged that both the bourgeoisie and the proletariat were alienated by the capitalist system.[50] Simmel recognized that the individual found new liberties in the city at the price of becoming more calculating and blasé.[51] Many people had difficulty making up their minds as to whether the increased rationality of the age of science and technology would carry the day, or whether it would be the increased irrationality of the crowd. Public education was seen by many as a way of ensuring that the balance would tip in favour of the former. Others were less sure because science itself undermined the beliefs and religious certainties that had held society together. Marx held that science was part of the superstructure of a society.[52] Freud showed that science was not free from the unconscious.[53] In one form or another, many such ambivalences continue in individual and collective life today. Aron has explained them in terms of the dialectics of modern societies.[54]

As societies struggled to enhance their capacity to make use of knowledge separated from experience, many traditional prejudices came under pressure, and in some cases were made illegal. If, as a society, you face a shortage of people who can help industry and government make full use of the latest scientific and technological knowledge, and if this is seen as absolutely vital for economic success and for the survival of the nation state, then refusing highly capable people simply on the grounds that they have the wrong skin colour or creed may find little support. If, on the other hand, the economy is not growing much and the need for such people diminishes, then some of these prejudices quickly rear their ugly heads again, although probably in a somewhat oblique form. In other words, this is not a great triumph of humanism or of a new rational spirit but a response to necessity. I am not suggesting that the people who struggled in the civil rights movement, the women's liberation movement, and the human rights movement were opportunists. What I am saying is that, in the past, such movements were on the fringe because they threatened society; society now stands to gain from, at least in part, acting on their message. No egalitarian societies have emerged. Everyone was a little more equal until the educational system began to classify a person's talents in the domain of knowledge separated from experience, thereby creating a new basis for social inequalities.

The separation of knowing and doing from experience necessitated a public education system accessible to all citizens in order to graduate the required number of educated people. The general pattern appeared to be that barriers were less well tolerated when there was a shortage of such people, while lip service was paid to equality when there was a surplus. In every nation where knowing and doing separated from experience became essential for technological and economic growth, the state intervened in education, much of which had traditionally been sectarian. Here we encounter a second reason (in addition to the economic one examined in the previous section) why the state inevitably grew when knowing and doing separated from experience became the prime factor of production.

The separation of knowing and doing from experience provided the final impetus towards the development of an entirely new kind of society, commonly referred to as a mass society. Recall how, in western Europe, the agricultural revolution reduced the requirements for peasant labour. The resulting migration to the growing industrial centres and the colonies did not lead to the recreation of a traditional society. In the case of the former, individuals were increasingly swallowed up in crowds in the factories and the tenements of working-class neighbourhoods. Labour unions and political parties sought to mobilize and organize these individuals. As time went on, many services traditionally provided by the extended family and neighbours were taken on by large anonymous bureaucracies. Those who migrated to the colonies of the New World experienced the frontier and its social dislocations. Again, no traditional societies were re-created. The move towards a mass society took slightly different roads, but the end results were very similar.

The individual was thus increasingly trapped in organizational webs that, unlike a traditional social fabric, had little meaning or value for life. The relationship between the individual and society was transformed, as manifested by the three successive personality types identified by Reisman[55] – tradition-directed, inner-directed, and other directed – which can be explained using the previously developed model of culture. Before the migration to the industrial centres, the rural peasants used their structures of experience as mental maps symbolizing an evolving tradition. Life as factory workers in these new centres led to many new experiences that could not easily be interpreted by and located on these maps. All that people could do was to cling to the deeper metaconscious patterns containing the familiar values and myths, thereby rescuing some direction and reference points for life in the midst of all the upheavals. Their structures of experience were thus no longer used as mental maps but as gyroscopes, allowing them to steer some kind of course. This represents the transition from what Reisman called the tradition-directed personality type to one that was inner-directed. As generations succeeded each other, the ability to use structures of experience as gyroscopes lessened because their basis had become too remote and distant in the past, causing gradual weakening of the cultural unity. All traditional values, morality, and reference points for life had been undermined. The only option left open was to turn to others for guidance. Individuals began to use their structures of experience less like a gyroscope and more like a radar. They constantly scanned what others were doing to ensure that they were going with the flow, so as not to be left out and given over to loneliness. The shift was now from the inner-directed personality type to the other-directed personality type. These individual transformations went hand in hand with those of society.

What did the predominance of other-directed personalities imply about a society? It would suggest that growing up in such a society no longer provided individuals with clear reference points by means of which they could establish meaning and value for their individual and collective lives. The role of culture had been undermined, including a desymbolization of the cultural unity. Individuals now relied on others to help them find their way, but these others were in the same position. Individually and jointly they had to be fed information from the outside by means other than cultural. Presumably such means were provided by the society they lived in. Their dependence made them less self-reliant, with the result that the unique dialectical tension between individual diversity and cultural unity that was one of the hallmarks of traditional Western societies was disturbed, at the expense of the individual and in favour of society. This is one of the characteristics of a mass society.

To shed further light on this question, we need to shift our level of analysis to that of society, and examine how the transition from traditional societies to mass societies changed the relationship between the individual and society. It may be regarded as a progressive liberation of individuals from their bonds to traditional families, neighbourhoods, and communities. Their newly acquired freedom allowed people to choose those social relations that best satisfied their goals and aspirations, to encourage a fuller development of their potential. They were greatly aided in this quest by the proliferation of technologies that helped them overcome the limitations of space and time, including telephones, cars, and, most recently, the Internet. These permitted all kinds of new 'relations' and 'communities' of people's own choosing. One could now have 'friends' anywhere and 'talk' with them at any time. Many saw this as a positive, liberating experience, a further step along the road of progress, modernization, and secularization.

Alternatively, the transition process may be regarded as a loss of how traditional societies sustained individuals by means of reference points for life, such as a tradition, morality, and religious direction. People were now joining the 'lonely crowd,' from which they sought some guidance by means of their other-directed personalities, which signified an entirely different role for culture. The proliferation of transportation and communication technologies indeed brought people closer together in space and time, but only via a multitude of shallow, impersonal, and short-term relations that increased the influence of the crowd on the individual.[56]

Both these interpretations are unsatisfactory because they are incomplete.[57] If a mass society were composed of liberated individuals, people would be exposed to all manner of external influences from which they were previously shielded by being integrated into a local group with a well-structured material, emotional, and spiritual life. Such liberated individuals would now have to find their own way in the world, which almost amounts to having to invent their own culture, a task which they previously shared with their local group and the many preceding generations that brought the group into existence. They would now have to determine everything for themselves, with the result that their individual lives would have to become the measure of all things. Surely, such a situation would be unbearable, leaving people defenceless against external influences that would overwhelm and alienate them. There would be no society at all, at least not in a form resembling our own. Some social support had to be forthcoming from somewhere, and, presumably, this is integral to a mass society.

Neither can a mass society be simply a collection of crowds that together form an unstructured mass. Such a mass would be without any direction and thus without a way of life and history, as individuals using their other-directed personalities look in vain to others for direction. Individual differences would weaken, making the mass a random movement of lives without any coherence, so that a mass society would simply be a traditional society in the process of disintegration and disappearance. Again, this was not where mass societies appeared to be heading. Hence, it was wrong to lament the loss of points of reference for human life resulting from individuals no longer being integrated into local groups. What was really happening was a transition to a different set of reference points. In a traditional society, the local group interposed itself between the individual and society. Changes in society at large came to the individual via his or her group, and were responded to as a member of that group in a way that was individually unique yet culturally typical. With industrialization, a technology-based connectedness increasingly imposed itself on the culture-based connectedness. Is it possible that in a mass society a new set of reference points for individual and collective human life was derived from this technology-based connectedness? Have organizations such as unions, corporations, political parties, and the state rushed in to fill the vacuum left by a vanishing way of life?

The problem goes deeper than having to account for a new kind of social cohesion that distinguished mass societies from their traditional predecessors. Such social cohesion did much more than provide social support, fundamental as that was. Something equivalent to the role of the cultural unity of traditional society had to be established. People's lives had to do more than make sense individually and collectively. People had to be able to orient themselves in reality in a way that justified them. They needed assurances that what they were engaged in did not merely make sense but was something they ought to be doing. They wanted to be justified in the eyes of their families, friends, neighbours, and colleagues, and to any groups and organizations in which they participated. In turn, they wished these groups and organizations to be right, including the nation. This had been the well-known function of traditional morality and religion that, at the cost of alienation, once veiled human reality so as to make it right and just. Having lost this, some other way had to be found.

In Western civilization, the transition from pre-industrial traditional societies to mass societies went hand in hand with the transformation of the 'hardware' of technology, on which the transformation of the 'software' was superimposed a century and a half later. The separation of technical knowing and doing from experience sharply reinforced many trends that emerged during the first phase of industrialization. The extended family was dealt a severe blow when it ceased to be the social context for many daily-life activities, beginning with work. In turn, local communities, as the social context for these weakened extended families, could not re-establish themselves in the new industrial centres. Individuals now carried out more daily-life activities in crowds, factories, tenements, and the streets. As a result, the influence of local customs and traditions on individuals weakened to make way for the more encompassing influences from society, which were mediated via pressures from crowds. This condition was both cause and effect of broad ideological currents that were beginning to have a substantial influence in the nineteenth century. Ideologies fed individuals information about their society from the outside, in the sense that this did not derive from personal experience and participation in a culture. Two well-known facts ought to alert us to the change in the relationship between the individual and society. We know that crowds are easily manipulated by outside influences and that their members are inhibited in the use of their critical faculties and may even engage in activities that would normally not enter their minds. We also know that totalitarian regimes, such as those of Lenin and Mao Tse-tung, deliberately broke up peasant communities in order to permit their propaganda to take hold of individuals, who were now much less protected.[58]

When knowing and doing separated from experience, the nuclear family and the neighbourhood community were further weakened. Imagine a family reunion. One of the children might have done very well in university and gone on to a highly successful career, having reached a senior corporate position. A second child, who was not interested in school but loved working with her hands, became a tool-and-die maker and eventually started a successful business. A third child dropped out of high school and was currently surviving on two or three part-time jobs. As a result, the social 'worlds' of these three people were very different, and may also have been different from that of their parents. Their material standards of living, lifestyles, religious affiliations, and political commitments may have become so different as to estrange family members from each other, a circumstance which weakened many nuclear families. The same argument could be made for the neighbourhood community. Although it may have been relatively socio-economically homogeneous, the social 'worlds' of the children who grew up there would almost certainly have become much less so, not to mention the fact that some of them would not have spent much time in any one neighbourhood. Hence, there was an estrangement from childhood playmates and teenage friends as well.

The separation of knowing and doing from experience accelerated the trend towards mass production, mass consumption, and mass advertising. Material standards of living rose dramatically, with the result that the preoccupation with one's daily bread gave way to that with consumption. Such a shift must be understood in the context of the fact that people living in these mass societies had less well developed metaconscious images of their social selves, since the majority of daily-life contacts were too numerous, impersonal, and discontinuous to function as the 'social mirror' in which people could clearly see themselves. Coupled with the weakening of traditional reference points for life, the result translated into people less sure of themselves and thus more vulnerable to outside influences. Lacking cultural guidance, they were ready to be informed of what they must do to belong and to be justified: What should I look like to appeal to others? What should I wear to make a good impression? What must I own to get along? How must I live to fit in? What must I be informed about to participate in conversations? People needed and wanted substitutes for the kinds of things customs, tradition, morality, and religion used to provide, and this had to work in the kinds of relationships common in a mass society where people rarely knew each other well. Did information about the latest fads, fashions, and trends provide them with these clues? Whatever merit *homo economicus* may have had as a model for people's economic behaviour in nineteenth century industrializing societies, it has no relevance for mass societies. People's supposed preoccupation with shiny hair, whiter teeth, and fresh breath, enticing producers to satisfy these needs, is simply a contemporary economic fairy tale. Advertising plays on people's needs to belong and soothes their anxieties and aimlessness. This is particularly evident among the young.

The other-directed personality that had to receive information from the outside was closely tied to the appearance of what has been referred to as a statistical morality.[59] Its significance can readily be grasped from the following common occurrence. Suppose a young teenager approaches her parents for permission to go and see a movie. The parents try to dissuade her from going by warning her that the movie is needlessly violent: 'It is one thing to depict the violence of war or of gangs, but it is quite another matter to almost celebrate violence for the sake of violence.' Their teenage daughter is not the least convinced: 'Everyone in the class has seen it, so why shouldn't I?' The moral argument of the more traditional parents makes little sense in the world of their teenage daughter, in which what everyone does is normal and what is normal is normative. She cannot understand that in terms of any traditional morality or religion this makes no sense at all: just because most people do something does not make it right. The other-directed personality scans what everyone does, which has now become the norm. Statistical morality thus entered on the human stage. This development further confirmed that the cultural unity of a mass society was weak compared to that of a traditional society and incapable of supporting a morality of a traditional type.

A similar transformation occurred in the formation of opinions. A change from private to public opinion went hand in hand with the emergence of the other-directed personality and mass society. It involved a change in the process by which individuals formed opinions and morally coped with the world. This is still true today. Most of us have seen television programs in which a reporter randomly interviews passers-by on the street regarding a recent event on the news. We have also listened to call-in radio shows soliciting people's opinions on various subjects. It ought to come as a surprise that virtually everyone has an opinion, yet the source of that opinion is hard to pin down. Did the opinion result from speaking to someone who witnessed the event, from speaking to someone who has spoken with witnesses, from reading an article or book on the event, from listening to a radio broadcast, from watching a television documentary, or from participating in a group discussion? Even in the relatively rare situation that one of these is the source, the information about the event is usually so piecemeal and limited that the most elementary critical scrutiny would lead to the recognition that these people are not really in a position to form an opinion. In other words, many of these opinions appear not to involve their faculties in any usual sense. This corresponds to the extensively studied phenomenon of a public opinion that has mostly displaced private opinion in mass societies.[60] In traditional societies, private opinions resulted from people's direct or indirect experience of mostly local events that were reasonably accessible. The meaning and value of such events were determined by cultural processes that symbolically located them in individual and collective human life, understood as a psychosocial process involving others. The other-directed person appeared to receive ready-made public opinions from the 'outside.'

Finally, the separation of knowing and doing from experience led, as we have seen, to structures of experience beginning to function as bifocal mental lenses. Because of the concomitant devaluation of knowing and doing embedded in experience relative to knowing and doing separated from experience, the general portion of this lens was less well trusted, with the result that the other-directed orientation of intellectuals and professionals turned into an information-directedness. Rather than relying on their cultural resources, such people preferred information fed to them from the outside. This, as we shall see, had some counter-intuitive effects.

The picture of human life in a mass society that is beginning to emerge is one of a decline of the role of culture in individual and collective human life. The other-directed personality was complemented by a society that had to compensate for a loss of 'internal' symbolic functions by providing 'external' symbolic functions that were not cultural in nature. These functions integrated individuals into a society via mass social currents of information. These were instrumental in weaving a social fabric of an entirely different nature from the one found in a traditional society. Not surprisingly, the infrastructure for these mass currents of information was provided by the mass media. We are now in a position to understand the role of advertising, movies, documentaries, and the news, which is still valid for our own society. Collectively, they provide us with images of how we must live in our world. No detail is left to chance, including our appearance, clothing, behaviour patterns, food and drink, things we must own to belong, opinions about events and organizations, what is 'good' and 'evil' in the world. Much of this is carefully structured to correspond to the needs experienced by a person in the crowd. Advertising, by associating metaconscious symbols with consumer goods, endows the latter with the kinds of magical powers capable of satisfying these needs. What the mass media portray is first and foremost what traditionally was provided by culture, internalized as a mental map for making sense of and living a good life in the world. It is this information flow, produced by a mass society, that Ellul has called integration propaganda, as distinct from totalitarian propaganda.[61] Its primary components are advertising, human relations, and public relations. Another significant component is what Ellul calls 'the technological bluff'[62]: a kaleidoscope of images portraying the miracles technology will soon be performing in domains such as science, medicine, space, and biotechnology. Integration propaganda is an essential and necessary compensation for the greatly weakened role of culture. Without them, a mass society would not be a viable social entity. I have only begun to outline the tip of the iceberg. Much of the analysis in Part Three is required to complete the picture. For now, I will limit myself to pointing out one important consequence. As will become evident, a mass society seeks to meet many non-material needs with material things. It creates a society with a very high standard of living, but one which leaves many non-material needs unsatisfied and undermines the capacity of the biosphere to provide life support – a poor bargain indeed.

As societies began to take full advantage of a new universal technology separated from experience and culture, their social fabric continued to weaken. Once again these social adjustments were a manifestation of a deeper underlying phenomenon, namely, the ongoing strengthening of the technology-based connectedness of human life and society and the corresponding weakening of the culture-based connectedness. The outcome was not a further disintegration of society, but the beginning of its recreation as an entirely new kind of society, commonly referred to as a mass society. Social relations, groups, and institutions continued to be sustained by a greatly weakened culture that had to be complemented by flows of information introduced into daily life from the outside. The mass media immersed people into a bath of images, which performed a social role analogous to those of customs and traditions in earlier societies. One of the primary components, namely, advertising, combined the planning and strategies of technostructures with culture-based symbols in order to bond products and services to daily life. The increasingly essential role of information flows was to make a relatively autonomous technology livable while recreating social cohesion. One of the price tags was a weakening of the dialectical tension between the individual and society in favour of the latter. This reversed a great deal of effort Western civilization had put into strengthening this tension in the opposite direction.[63] In the process entirely new social entities were created, the nature of which will become more fully apparent as we proceed with our analysis.

6.4 Living with a Limitless Politics

Particularly during the decades following the Second World War, the political framework also had to be accommodated to the growing use of knowing and doing separated from experience.[64] Once again, we are confronted with the problem that much of contemporary political science has not internalized the many changes accompanying an extensive use of a universal science and technology.[65] The size of the state grew enormously, as it was compelled to create a large public sector in the economy and organize a mass education system. These were not the only causes for its growth. As knowing and doing separated from experience throughout society, many spheres lost their ability to regulate themselves as the functions of knowing, doing, and controlling were separated, thereby creating long and ineffective negative feedback loops. In many cases, the state had no choice but to step in, making it the locus for organizing an ever-growing number of activities. This trend quickly reached a point where it was next to impossible to identify areas of individual and collective human life not directly or indirectly affected by the state. Nothing makes this more evident today than the widespread belief that ultimately everything is political. This amounts to an assertion that in principle the state is omnipotent and that its role in human life is without limits. Any ineffectiveness of the state is thus blamed on ineffective or corrupt politicians or the political party of our choice not being in power. These developments ought to alert us to the possibility of the state increasingly contributing to a new secular sacred, emerging along with a universal science and technology. This in turn would open the door to politics taking on a secular religious character, as will become evident in Part Three. For now I wish to stress once again how unacceptable such a situation once was to Western beliefs and values. Political ideologies without exception saw the growth of the state as undesirable. For democrats, it was clearly impossible to have a large state *and* have local grass-roots control over things. Marx predicted that the state would wither away in a socialist society; and even Lenin, at the end of his life, was troubled by the fact that the Soviet state kept growing. The Jewish and Christian traditions long regarded empires as oppressors that enslaved people in one way or another, beginning with the prophetic warning about the consequences of Israel wanting a king, and later, the pronouncements against the empires of the time. Anarchists have, of course, always been against the state. It would appear, therefore, that the growing role of the state that accompanied industrialization was an accommodation to necessity as opposed to the outcome of political aspirations.

Following the Second World War, as the separation of knowing and doing from experience obliged the state to become the 'nervous system' of society, the scope and complexity of the problems it had to deal with increased. It therefore had every incentive to bring the latest knowledge separated from experience to bear on these problems, and to do so it had to reorganize its departments or ministries into technostructure-like entities. This development, which paralleled the one we have examined for the modern corporation, led to similar results. Just as the board of directors, representing the shareholders, found it almost impossible to exercise effective control over 'their' corporation, so also members of cabinet, elected to carry out the wishes of the people, found it more difficult to control 'their' ministry or department. This has been humorously portrayed by the BBC television series *Yes, Minister!*

Suppose a cabinet decision was made to prepare a national energy plan. The minister responsible for doing so called a meeting with her deputy minister to set the process in motion. A year later, the deputy minister returned with a plan. The minister took it home to read, and discovered to her surprise that it recommended substantially increasing the nation's reliance on nuclear power (as was common during several decades following the Second World War). Returning to her office the next day, she called in the deputy minister to explain that she had campaigned for a moratorium on nuclear power until the problem of radioactive waste could be solved. Let us assume that the minister happened to be an electrical engineer who had specialized in power. It would then be possible for her to professionally assess a few pages of the report that coincided with her area of expertise, but much of it would be beyond her. The minister made several suggestions to the deputy minister, with the request that the proposed plan be revised. After a few days, the deputy minister requested another meeting, and politely but firmly pointed out that these suggestions were not feasible for one reason or another. In fact, the minister was advised that all these things had been carefully looked at by experts, and if she would like additional details, these could readily be provided. The point was obvious: the plan represented a synthesis of a great deal of knowledge separated from experience, requiring many specialists to cooperate in a process similar to the one we have described for the technostructure of a modern corporation. No individual could master all this knowledge, and it was not easy for anyone to bring to the surface the many implicit assumptions and value judgments that it inevitably contained. I am not suggesting that this is impossible. Elsewhere,[66] I have shown that, during the energy planning following the Second World War, a few voices of common sense (including Schumacher and Lovins) put forth sound arguments that eventually carried the day, but not until after one of the most monumental failures of energy planning. The minister would probably find that the arguments worked out by the ministry and defended by the deputy minister appeared so rational and technically self-evident that, given the current state of knowledge, theirs had to be the most reasonable plan that could be devised. Eventually, the minister would have little choice but to recognize that her election promises, although made in good faith, could not be defended by technical and economic rationality. Hence, as government ministries began to use knowledge separated from experience to the fullest extent possible, governments had increasing difficulty imposing their will in the name of the people. They were unable to discern the limitations of bringing the latest knowledge separated from experience to bear on the complex issues that the state faced (by exposing implicit values) and thus to demonstrate that alternative solutions were not only politically desirable but also feasible. A government could not easily interfere in the complex planning process required to bring the latest knowledge separated from experience to bear on an issue. There was now a high level of autonomy of the state's machinery with regard to its political masters. Of course, governments regularly overruled their bureaucracies, but this frequently resulted in actions that were neither technically sound nor politically astute. It also explains why opposition parties often felt compelled to moderate their policies once they took power and gained access to the state's technostructures.

The situation was further complicated by the fact that, in a democratic mass society, public opinion could not be ignored. Governments elected to carry out the wishes of the people could not act against public opinion without undermining their legitimacy. At the same time, it was impossible to govern on the basis of public opinion, which, as all studies showed, was incapable of anticipating the problems governments must attend to and of sustaining the planning and execution of actions to deal with them. The result was a transformation of the relationships between citizens and elected politicians on the one hand, and between citizens and the state on the other. One of the principal causes of this political dilemma was the transformation of the relationship between the individual and society. Prior to the emergence of mass society, the individual's 'world' was that of the local group, and this 'world' was largely self-regulating thanks to culture.[67] Events only occasionally had effects felt outside the local group. Actions initiated beyond it, such as those of the state, rarely had much effect on local affairs for lack of means. As technology-based connectedness imposed itself on culture-based connectedness, the opposite situation emerged. It undermined the self-sufficiency and self-regulatory character of the local group, gaining considerable autonomy. This technology-based connectedness was not self-regulating, and led to what I have called the 'labyrinth of technology.'[68]

As a first example, consider the transformation of work in England. Before industrialization began, any economic initiatives of the government in London had little effect on people's work, and vice versa. At the time of the Industrial Revolution, work rapidly moved centre stage in both local and national affairs. Adam Smith associated the division of labour with the wealth of nations.[69] Marx built his entire economic interpretation of capitalism on the surplus value of labour as the source of wealth.[70] In the eyes of these and other nineteenth century observers, technology had not yet become the prime factor of production, and this was certainly the case. For people of the nineteenth century, work increasingly took a central place in their structures of experience, as shown by the fact that work became a myth along with progress and happiness. In the ideologies of the time, work was the source of all virtues, and laziness the source of all vices.[71] On the collective plane, work was all-important, and governments paid ever more attention to it. A work stoppage could now have national implications; at the same time, decisions made far away could significantly affect the work available to a local community. It is not surprising, therefore, that citizens took a growing interest in any government initiatives that might affect their work and thus their livelihood. However, they were entirely unprepared to understand the complexity of the economic system and all the factors that shaped their work. This might have developed into an intolerable situation because people could make no sense of what was happening to their work, even though it affected their entire lives as wage-earners. They had to become informed so that they could form opinions on government initiatives or other events affecting work.

Making sense of their world, centred as it was on work, was not enough. Popular stereotypes to the contrary, industrial civilization demanded more work from people than almost any other society. In addition, work became more stressful because of machine pacing and other technological factors, coupled to a loss of control.[72] Such efforts could not be sustained without a deeply founded motivation. In the nineteenth century, this was assured by the reigning ideologies, the myth of work, and an adaptation of the religious traditions. Later, this was reinforced by rising standards of living and reductions in working hours. When this was not the case, as during a time of war, massive propaganda efforts were required to sustain workers' morale. Following the Second World War, the further decline of the role of culture substantially weakened the socio-cultural support for work. Governments and corporations turned to public relations to feed people information about the importance of work for the nation and 'their' company. As a result, work and any initiatives affecting it became important political issues.

Taxes followed a similar development. The historically unprecedented growth of the state was cause and effect of the fact that citizens now worked for the state for the better part of the year. For this to appear legitimate and not as some new form of enslavement, people had to become convinced that the payment of taxes served their interests and aspirations. Politicians and the state were compelled to inform citizens of how their taxes were spent and why. Again, people were drawn into the decisions of government and the actions of the state, to an extent unparalleled in history. In the past, political elites were relatively isolated from the general population because the state lacked the means powerful enough to affect large numbers of people. A particular case in point was war. Before industrialization, it affected relatively few citizens, but the First and Second World Wars changed that forever. These mass wars involved almost everyone: as soldiers or military personnel, as workers in the military-industrial complex or the remainder of the economy, now directed towards supporting the war effort, or as volunteers providing a variety of essential services. No one was immune to enemy attacks and bombardments. These wars required massive preparations before, and rebuilding efforts after. Such life and death struggles could not be sustained unless everyone was convinced that this was 'their' war, with everything they stood for in the balance. Yet most citizens were not in a position to understand the many factors that had led to war or how such factors might have been reversed. They had to be informed and motivated to make the supreme effort, even unto death. Only propaganda could perform this feat. Without it, the strain of combat, bombardments, food shortages, extra work, loss of family members and friends, and enemy counter-propaganda would soon have broken the morale of citizens.[73]

Peacetime also was not without its tensions, dislocations, and ongoing adjustments. We have examined how the massive restructuring involved in industrialization continually uprooted everyone. To help people cope, governments and corporations had to create a variety of social and health services. These formed a safety net that dealt with the symptoms but not the root problems, and further tied government and corporate decision-making to daily life. Once again, the issues were very complex, and citizens and workers had to be informed. The techniques of choice for doing so were public relations and human relations.[74]

The above examples confirm a pattern described earlier: the social cohesion of a mass society depended on a massive flow of information from its large centralized institutions to the citizens. These developments were both cause and effect of the emergence of the mass media and their growing importance. They poured out a great deal of information that complemented almost every facet of the weakening role of culture in individual and collective human life. The newscast became increasingly important as a way of staying in touch with the world. Doing so in a traditional society was based on experience and culture, since events were mostly local and thus directly experienced by a number of members of the local group. Those who did not, experienced them indirectly via the intermediary of one or several persons. There were the occasional travellers who brought accounts from people, places, and events far away; but since these rarely had any effect on the local group in question, they might as well have been stories. Not so with the advent of mass society. Events from around the world could trigger a war or an economic recession, or exercise more subtle but not insignificant consequences for people's lives far and near. For this reason, people had to make sense of these events, and they were supported in this effort by the formation of public opinion.

Our present situation still reflects these developments. Presumably, people form their opinions from newscasts and discussions with others. This is surprising, given the fact that most news items last no more than a few minutes for each of the main events. The others are even shorter.

Why are people interviewed on the street not indignant about the presumption of their having an opinion about the news event? To add to the puzzle, imagine making an inventory of news events covered over an extended period of time. It becomes immediately evident that the news offers us the most incoherent, discontinuous, and almost random panorama of our world. For example, for three successive days we may be informed about a political event in the senate of the United States. We are told very little about the developments that led up to the event, and then for months we may hear nothing more about it and its effects. When we multiply this over thousands of news items, it becomes clear that being informed about an event has little or nothing in common with understanding it. The latter involves a knowledge of its origins, evolution, and decline, including the way it affects and is affected by its context. Although the human brain-mind has evolved to create metaconscious patterns from limited, divergent, and constantly changing experiences, making any sense of the world as portrayed by the newscasts would require additional external information.

Imagine attempting to play the role of one of several reporters sent by a television station to cover an event in a faraway capital. Before departing, we were handed a few books, with the suggestion that we might skim them on the plane to acquire some background information. What happens is that we interpret the event with the cultural resources we have internalized, that is, as if the event were happening in our own society. At best, this is somewhat reinterpreted by what little background information we may have been able to gather about the people, culture, and history of the place. With the best intentions in the world, we will not be able to avoid distorting the event in ways that may or may not be significant. Besides, whatever report we give will not 'work' on television. Someone thoroughly familiar with this mass medium will have to 'translate' our report into something that 'works.' Professional journalists will report the event in terms of the requirements of television as the primary context for the event. This is no different from the way we have learned to compensate for the fact that telephones mediate a conversation in a non-neutral way by stripping away much of the context of a face-to-face conversation, including facial expressions, gestures and everything but the sounds. In the same way, professional journalists have learned to translate 'event dots' into 'media patterns.' Instead of reporting an event in the context of experience and culture, it is first and foremost reported for the immediate context of the mass media.

It is estimated that over half the events reported by the mass media are created and managed by public relations firms.[75] Since these items are indistinguishable from those reported by journalists, further insight into this translation can be obtained by examining the explicit and well-documented techniques of public relations. These techniques create the possibility of using an event as a springboard for achieving a desired effect. This is done by placing the event in a suitable context to give it the desired 'spin' by the 'spin doctors,' who appear to attract little attention except during election campaigns.

Living in a mass society requires that people be more informed than they have ever had to be, and that this process be tailored to the mass media. Jacques Ellul summarizes the situation as follows:

> As a result, he finds himself in a kind of kaleidoscope in which thousands of unconnected images follow each other rapidly. His attention is continually diverted to new matters, new centers of interest, and is dissipated on a thousand things, which disappear from one day to the next. The world becomes remarkably changeable and uncertain ... Even with major events, an immense effort is required to get a proper broad view from the thousand little strokes, the variations of color, intensity, and dimension, which his paper gives him. The world thus looks like a *pointilliste* canvas – a thousand details make a thousand points. Moreover, blank spots on the canvas also prevent a coherent view.
>
> Our reader then would have to be able to stand back and get a panoramic view from a distance; but the law of news is that it is a daily affair. Man can never stand back to get a broad view because he immediately receives a new batch of news, which supersedes the old and demands a new point of focus, for which our reader has no time. To the average man who tries to keep informed, a world emerges that is astonishingly incoherent, absurd, and irrational, which changes rapidly and constantly for reasons he cannot understand. And as the most frequent news story is about an accident or a calamity, our reader takes a catastrophic view of the world around him. What he learns from the papers is inevitably the event that disturbs the order of things. He is not told about the ordinary – and uninteresting – course of events, but only of unusual disasters which disturb that course. He does not read about the thousands of trains that every day arrive normally at their destination, but he learns all the details of a train accident.
>
> In the world of politics and economics, the same holds true. The news is only about trouble, danger, and problems. This gives man the notion that

he lives in a terrible and frightening era, that he lives amid catastrophes in a world where everything threatens his safety. Man cannot stand this; he cannot live in an absurd and incoherent world (for this he would have to be heroic, and even Camus, who considered this the only honest posture, was not really able to stick to it); nor can he accept the idea that the problems, which sprout all around him, cannot be solved, or that he himself has no value as an individual and is subject to the turn of events.[76]

For people to be informed, this kaleidoscope of 'facts' must be translated into a world in which people can actually live. This involves affirming an order so that everything makes sense, finding places in that order by contributing to it, confirming the value of their persons by doing so, and by sharing with others the confident hope that problems can and will be resolved. Integration propaganda steps in to meet the deep needs of the citizens of democratic mass societies. It shows that we are jointly travelling in the direction of progress and history, that democracy can overcome tyranny, that the free market and global trade can cure all our economic woes, that medicine can save us from disease, that education keeps us in the global race, that consumer goods can make us happy, and that the goodness of human nature continues to break through. Integration propaganda includes every event in these patterns by associating them with prejudgments, stereotypes, commonplaces, ideologies, beliefs, and myths. The meaning and value of any event is thus no longer determined by means of a culture but by contextualizing it in the 'universe' created by integration propaganda, thereby creating 'myth-information.'[77] The process of 'understanding' it is thus separated from experience and culture, as is the universe into which it is integrated. As was the case with knowledge separated from experience, this universe continues to depend on culture for its 'myth-symbols' in the same way as advertising does for its symbols. Because of desymbolization, the universe of integration propaganda imposes itself on the symbolic universe created by means of culture. It is this universe that creates public opinion, which is polled on an ongoing basis by governments, corporations, and other organizations. These institutions have little interest in private opinions unless they are converted into 'myth-symbols,' enter the universe of integration propaganda, and shape public opinion. Everyone lives as though this universe is reality itself. It moulds public opinion in a mass society.

The following two events illustrate the situation. A colleague encountered the president of the university having his hair cut and reading a despicable excuse for a newspaper. She enquired why he bothered reading that trash, to which he patiently replied that he had to keep up with what the public was reading regardless of its merits. Most members of the general public would only encounter the university via these channels, and whatever was depicted there was their 'reality.' Another colleague had reluctantly agreed to participate in the production of a series of short television spots designed to help the public think about the complex issues created by high technology. For one of them, he was quoted on the effects of cellphones on human life, from a clip recorded when he was asked to 'just say something' so that the recording levels on the equipment could be adjusted. The quote said nothing that anyone would not already know. Seeing it on television for the first time, he telephoned the producer, informing him that he would no longer participate. The producer asked him to reconsider, suggesting that he telephone a few people to ask them what they thought he had said: 'You will see that the only thing people will remember is that you were on television. They will not remember what you said, except in the vaguest of terms.' This turned out to be true: the medium had created its own message.

In traditional societies, the daily-life world was that of experience symbolized on the basis of a culture. In modern societies, the mass media superimpose another reality that coexists with the former, in a manner analogous to the coexistence of knowledge separated from experience and knowledge embedded in experience. This development further extends the technology-based connectedness into human life and consciousness. Ellul sums up the situation as follows:

> The development of propaganda is no accident. The politician who uses it is not a monster; he fills a social demand. The propagandee is a close accomplice of the propagandist. Only with the propagandee's unconscious complicity can propaganda fulfill its function; and because propaganda satisfies him – even if he protests against propaganda *in abstracto*, or considers himself immune to it – he follows its route.
>
> We have demonstrated that propaganda, far from being an accident, performs an indispensable function in society. One always tries to present propaganda as something accidental, unusual, exceptional, connected with such abnormal conditions as wars. True, in such cases propaganda may become sharper and more crystalized, but the roots of propaganda go much deeper. Propaganda is the inevitable result of the various components of the technological society, and plays so central a role in the life of

that society that no economic or political development can take place without the influence of its great power. Human relations in social relationships, advertising or human engineering in the economy, propaganda in the strictest sense in the field of politics – the need for psychological influence to spur allegiance and action is everywhere the decisive factor, which progress demands and which the individual seeks in order to be delivered from his own self.[78]

According to Ellul, for the individual in a mass society, integration propaganda means the following:

Man, eager for self-justification, throws himself in the direction of a propaganda that justifies him and thus eliminates one of the sources of his anxiety. Propaganda dissolves contradictions and restores to man a unitary world in which the demands are in accord with the facts. It gives man a clear and simple call to action that takes precedence over all else. It permits him to participate in the world around him without being in conflict with it, because the action he has been called upon to perform will surely remove all obstacles from the path of realizing the proclaimed ideal.

Here, propaganda plays a completely idealistic role, by involving a man caught in the world of reality and making him live by anticipation in a world based on principle. From then on man no longer sees contradiction as a threat to himself or as a distortion of his personality: the contradiction, through propaganda, becomes an active source of conquest and combat. He is no longer alone when trying to solve his conflicts, but is plunged into a collective on the march, which is always 'at the point' of solving all conflicts and leading man and his world to a satisfying monism ...

... propaganda also eliminates anxieties stemming from irrational and disproportionate fears, for it gives man assurances equivalent to those formerly given him by religion. It offers him a simple and clear explanation of the world in which he lives – to be sure, a false explanation far removed from reality, but one that is obvious and satisfying. It hands him a key with which he can open all doors; there is no more mystery; everything can be explained, thanks to propaganda. It gives him special glasses through which he can look at present-day history and clearly understand what it means. It hands him a guide line with which he can recover the general line running through all incoherent events. Now the world ceases to be hostile and menacing. The propagandee experiences feelings of mastery over and lucidity toward this menacing and chaotic world, all the more because propaganda provides him with a solution for all threats and a posture to assume in the face of them.[79]

The conditions of human life in a mass society are such that we cannot help but contribute to the need for, and ready acceptance of, integration propaganda since we all need to make sense of and live in the world. We can no longer rely on culture, whose role in individual and collective human life has been decisively weakened by the developments of the past two centuries. Nothing in our modern world can be understood in isolation from this fact, just as nothing can be understood about the local group in prehistory and history without reference to culture. In politics, this situation creates the equivalent of the revised sequence in the economy. Government advertising informs citizens of the problems that must be confronted and the solutions that they wanted all along, because these affirm the order and direction of the world. To the extent that this political advertising is successful, democracy is undermined. The revised sequence operating in both the economy and in politics is part of a much broader and deeper phenomenon, namely, a comprehensive revised sequence in which the technology-based connectedness imposes itself on the culture-based connectedness of human life – a subject to which we will turn shortly.

In the decades following the Second World War, politics in democratic mass societies underwent a significant change that was some time in the making but was awaiting the necessary technical means for its completion. I have attempted to show that the chain-reaction-like process of industrialization imposed one necessity after another. The pattern was always the same: as long as happiness and freedom meant high standards of living, high levels of mobility, and numerous holidays, every possible use had to be made of the prime factors of production; and each and all accommodations required were seen as necessary to accomplish people's goals and aspirations. Politicians considered to be ineffective in helping society to pave the way for what was technically 'necessary' did not last long. Modern politics has become the domain of whatever is necessary to keep the chain-reaction-like process of growth going, but especially of the ephemeral, to assure ourselves that we are still the political masters of the system.[80]

If democratic politics is the art of guiding a way of life by means of a people's values and aspirations, the political adjustments accompanying the growing use of a universal science and technology ought to make us reassess the situation. Once more the deeper underlying development was that of a further weakening of the ability of cultures to sustain human life in the world under the pressure of an increasingly autonomous technology-based connectedness. Insofar as this threatened certain spheres of human life with chaos, the state had no choice but to step in. Very little of a society's way of life remained unaffected by the need of the state to coordinate and regulate activities no longer adequately sustained by a culture. Since such coordination and regulation was based on interventions determined by the application of scientific and technical knowledge by technostructures within government bureaucracies, this development was self-reinforcing, encountering few if any constraints. The end result was a vastly expanded political sphere more or less coextensive with collective human life. The growing autonomy of the state with respect to human life and society required new flows of information. These took the form of extensive political advertising, which sought to shape public opinion and convince the members of society that the state was serving their best interests. The possibility of a reasonably well-functioning socialism or democracy was all but eliminated, but this was made existentially acceptable by giving politics a secular religious character rooted in new cultural unities, to be examined in Part Three. Once more, necessity was turned into 'the good' by means of secular political religions based on a new secular sacred. In this secular sacred world so dependent on integration propaganda, there was no room for the *homo economicus* of the nineteenth century who controlled the economy through the Market, nor for the free citizen who directed the state through the ballot box. The other-directed person gradually became *homo informaticus.*

6.5 The Intellectual and Professional Division of Labour and the Poverty of Nations

The separation of knowing and doing from experience and culture rapidly spread beyond industry and government, eventually transforming almost every aspect of the industrially advanced nations. For example, Keynesian economic policy began to fail when the ratio of desired to undesired effects of technological and economic growth declined as their dependence on highly specialized scientific and technical knowledge grew. The role of culture was transformed even further. The consequences affected every trend I have described thus far. This began in the 1970s and led to many new phenomena, to be discussed in Part

Three. All are now fully rooted in a technical approach to life, which has grafted itself onto the cultural approach. Their fundamental difference resides in the extent and kind of context that is taken into account. The cultural approach to life makes the greatest possible use of context by symbolically giving every experience its place so as to reveal its meaning and value relative to all other experiences and thus to life as a whole in a world that includes the unknown. As such, it supports individual and collective human life. The technical approach, on the other hand, makes a minimal use of context by restricting it to a relatively narrow specialty that is separated from experience and culture. The context used frequently amounts to little more than the technology-based connectedness of a society. As a result, this approach ends up strengthening the technology-based connectedness of a society, whether it is intended or not. It owes a great deal to science, which, as we have seen, behaves as if reality is unmanageably complex and cannot be known as such. Instead, most of the connections between the object of study and reality are severed by means of a process of abstraction. The object of study is thus placed in the manageable intellectual context of a particular scientific discipline and sub-specialty, supplemented by the limited physical context of a laboratory experiment designed to examine a few variables, preferably one at a time. Parcelling out the task of knowing ourselves and the world in this way turned out to be so manageable and efficient that science outdid its culture-based competitors.

It is gradually beginning to dawn on us that in the absence of a science of the sciences capable of *scientifically* integrating their findings, each and every discipline separately contributes to an exponential growth of knowledge of things in a context of its own making, as opposed to their 'real world' context. When such knowledge becomes more highly valued than its traditional counterparts, this simultaneously creates an exponential growth of ignorance of how things fit into, contribute to, depend on, and are inseparable from everything else. Discipline-based knowledge treats the findings of all other disciplines as knowledge externalities, even though these might provide clues as to the significance of what was lost in the process of abstraction. In this way, each discipline establishes exclusive authority over a portion of the knowledge frontier, and the validation of any findings can be restricted to its practitioners. Transdisciplinary studies seek to reduce these knowledge externalities by placing a number of disciplines within a common context. However, the influence of such studies has been so small that disciplines continue to evolve within highly limited intellectual and physical contexts of their own making. Contrary to what appears to be the case on the level of an individual discipline, science as a whole has an anti-world and anti-life bias.

We have seen that a similar approach for acting on the world by making minimal use of context (separated from experience and culture) emerged primarily within technology but quickly diffused to every other sphere of modern life, all but replacing the roles of tradition and culture. Specialists of all kinds are now engaged in adapting and evolving contemporary ways of life. Once again, they all but sever the connections between what they are dealing with and the world by means of a triple process of abstraction. First, they know the world beyond their domains of competence only in terms of the requisite 'inputs' it delivers and the desired 'outputs' to be returned to it. A second abstraction is required because any domain of competence does not correspond to a 'chunk' of the world. For example, no specialist knows everything about hospitals in terms of the process that takes in sick people, treats them, and discharges them back into the world. Doctors, nurses, pharmacists, technicians, social workers, nutritionists, office personnel, cleaning staff, volunteers as well as relatives and friends all participate differently and partially in converting the inputs of sick people into desired outputs of patients on the mend. The more specialized the experts' knowledge, the more they deal with everything in terms of a 'technical shadow' of its former self. Under extreme pressure, patients may become 'the appendix' in Room 1, 'the gall bladder' in Room 2, 'the broken hip' in Room 3, and so on. What is coterminous with each specialist's domain of competence has been abstracted from the fabric of relationships associated with converting inputs into desired outputs.

Third, decision alternatives cannot be adjudicated in terms of what is best for human life, society, or the biosphere because this would require that the domain of specialization be situated in relation to everything else. In the absence of this knowledge, what is 'better' is operationalized in terms of how the technical shadow of a 'chunk' of the world contributes to obtaining the greatest possible desired outputs from the requisite inputs, or the same outputs from reduced inputs. All other implications have been eliminated by this third abstraction. Success is measured in terms of output/input ratios including efficiency, productivity, profitability, cost-benefit ratios, risk-benefit ratios, and the GDP (which is obtained from relatively fixed inputs). Such performance ratios provide specialists with no guidance as to whether any improvement is partly or wholly achieved at the expense of the integrality of what is made 'better,' of the compatibility between what is made 'better' and the broader context in which it operates, and of the ability of what is made 'better' to evolve and adjust on the basis of self-regulation using negative feedback. It is obvious that the resulting 'system' may accidentally get it right some of the time, but in most cases the 'better' is achieved at the expense of context compatibility. Experts must shift their attention from How can this improve our lives? to How can this be made to yield its greatest power through converting requisite inputs to desired outputs?

To sum up, the technical approach begins by abstracting whatever is to be made 'better,' representing the remainder of the world only in terms of the inputs it must provide and the desired outputs it will receive. Next, whatever has been abstracted in this way is studied by further abstracting those features directly relevant to the goal of transforming the inputs into outputs, as well as being coterminous with an expert's domain of competence. These are included in some kind of model, while the remaining features are excluded from it. The model is then manipulated to determine how whatever it represents can be made to function 'better' in terms of contextless output/input ratios. The previously-neglected contexts continue to be excluded. Finally, the results are used as a basis for reorganizing the portion of reality originally abstracted. It is a strategy for reorganizing human life and the world piece by piece, with minimal consideration being given to the contexts within which they occur, other than the local technology-based connectedness. The technical approach to life builds a new order. Insofar as it may be regarded as an order at all, it is an order of non-sense, because it is established and evolved outside of the domain of sense (i.e., the realm of culture). A distinction must be made here between nonsense and non-sense. Something is nonsense when it belongs to the domain of culture but violates its order of meanings and values. Something is non-sense, on the contrary, when it does not belong to the domain of culture even though it is a human creation. It creates an 'order of disorder' within a culture, representing the equivalent of pollution within the cultural order.

The story of the third megaproject of humanity involves a shift from an almost exclusive reliance on the cultural approach to life on the part of traditional societies, to a primary reliance on the technical approach to life (at the expense of the cultural approach) on the part of mass societies. The strengths and weaknesses of each of these two approaches provide us with a key to understanding our recent journey. Based on the above descriptions of the two approaches, it is possible to deduce some of these. The replacement of experience and culture by the technical approach to life involves dealing with people and the world in a manner that externalizes everything not immediately relevant to the goal at hand. Nevertheless, throughout the entire process, whatever is abstracted, studied, modelled, simulated, and reorganized remains connected to everything left out. Consequently, the technical approach is very effective at obtaining the greatest possible desired results from whatever is required to produce them, but also in straining, distorting, or breaking many relations in the fabric of reality that were neglected, diminished, or marginalized by it. It also excludes the possibility of being guided by any human values, since the context of human life and the world is represented exclusively in terms of the desired outputs and the requisite inputs. It can be guided only by what masquerades as values, but which are really nothing but output/input ratios. As a result, it disorders all that is shaped by the historical and natural processes throughout which everything in human life, society, and the biosphere evolves in relation to everything else. The technical approach to life is unmatched in getting results, but does so at the expense of the integrality of the human and natural worlds. It creates a diversity of reorganized elements that in some respects are 'carcinogenic' with respect to the cultural and natural orders.

The technical approach to life cannot be applied in a piecemeal fashion. Since the very beginning of industrialization, it has been an all-or-nothing affair. Bettering one aspect of human life or the world in one place, and another aspect somewhere else, does more than locally introduce an element of chaos. Each technical improvement must be supplied with the requisite inputs and be relieved of the desired outputs it produces. Consequently, such improvements must be made in such a way that the outputs of one technical improvement constitute the inputs into another, linking all of them in a technical order that emerges by creating chaos in the cultural and natural orders. There can be no intermediary cultural or natural processes because these would not match the performance of the technically improved areas. Hence, the technical order takes the form of a network of efficient transformations connected by exchanges of inputs and outputs. Where the technical order is connected to the cultural and natural orders, bottlenecks occur because the transformations within the network are more efficient than their predecessors and hence incompatible with the cultural and natural relations outside of it, which remain untouched by the technical approach. These bottlenecks cry out for further technical improvements, compelling an ever-expanding technical order. In sum, the technical order represents a domain of non-sense lying outside of the domain of sense (i.e., experience and culture based on meanings and values). The technical approach to life reduces relationships that participate in cultural and natural evolution to transformations of inputs into outputs, and externalizes everything else. The result is constant tension between the technical order on the one hand and the cultural and natural orders on the other. The technical order must expand until it is coextensive with the others because it feeds on them. All this simply restates in more general terms what we have already found, namely, that the technology-based connectedness of society expands with industrialization.

In contrast, the culture-based approach to human life in the world is much less successful at obtaining desired results, but it respects as much as possible the way everything interacts with, depends on, and evolves with everything else. The above analysis of some of the strengths and weaknesses of the technical approach to life and its cultural counterpart can illuminate the successes and failures of the ways of life that emerged during the last three decades of the twentieth century. We excel at improving the power and performance of cultural and natural relationships at the expense of the integrality and sustainability of human life, society, and the biosphere. Examples abound. The materials we create are extremely well-suited to the functions for which they are made but collectively undermine the health of ecosystems and, through them, human health.[81] We have surrounded ourselves with time-saving gadgets but have less and less time to ourselves.[82] The proliferation of communication technologies has been unable to overcome the 'lonely crowd' or compensate for the kind of support individual persons once experienced from the kinds of social relations more common in a traditional society than in a mass society.[83] Computers are well-suited to a variety of tasks, but spending too much time in front of these machines negatively affects our image of what it is to be human and hence our relations with others.[84] Our means of transportation allow us to travel anywhere, only to find increasingly the 'geography of nowhere.'[85] High technology promised to deliver us from the smokestack industries and thus help solve an environmental problem, but instead it simply changed the kinds and quantities of pollutants being released.[86] The green revolution did not succeed in increasing the biomass obtained from photosynthesis, but instead boosted the edible portion of plants at the expense of the functions of the other portions, thereby necessitating more pesticides and herbicides, with all the negative implications for soils and ecosystems.[87] Biotechnology, by not respecting the integrality of the DNA pool of the biosphere, is beginning to produce genetic pollution, with imponderable consequences. Our weapons systems are now so powerful that their all-out use can no longer defend anyone but only destroy everyone.[88] The information highway promised to deliver us from rush-hour traffic by enabling us to work at home, thereby reducing pollution levels in cities and making them more liveable. It would also provide more access to information, democratize the world, and a great deal else. Pedagogical and educational innovations would help children and young people better adapt to the new realities, but instead they found it increasingly difficult to make sense of the lives of their parents, while parents intuitively realized that the new connectedness of the world had passed them by in ways that they could not grasp.

This pattern of technologies that produce spectacular but specific results by undermining the cultural and natural orders goes back nearly a hundred years. Human expectations tend to be based on these results without taking into account their downside. The construction of the electrical grids was supposed to decentralize industrial production back to the home or small workshop, thereby eliminating many negative features of industrial society.[89] Nuclear power would feed these grids at a cost 'too cheap to meter.'[90] Workerless factories would shift the balance between work and leisure, to the point that we were supposed to worry about how we were to spend all this free time.[91] Educational television would bring the best teachers and professors within everyone's reach, thereby reducing the information gap between rich and poor, north and south to create a more just world order. The microprocessor would revitalize democracy and decentralize large institutions. For a while, all this was celebrated as a move towards the more rational and secular society, until the 'noise' caught up with the 'signal' of technological and economic growth.

Our present situation is a continuation of these patterns. Engineers, managers, and regulators are accustomed to respectable figures indicating our success in obtaining desired outputs from requisite inputs as measured in terms of output/input ratios. However, when we regard the industrial-economic system as a whole, things appear very different. Alternative indicators to the GDP – they subtract costs incurred in the production of wealth from the total value of goods and services produced by an economy to arrive at net wealth production – show that the latter has been declining for decades.[92] The American Academy of Engineering estimates that 93 per cent of materials extracted from the biosphere do not end up in saleable products.[93] Some time ago it was reported that Blue Cross was GM's primary supplier.[94] We can reconcile such contradictory impressions of how well modern ways of life serve us when we begin to interpret our successes in terms of 'signal-to-noise' ratios of desired to undesired effects of design and decision-making. The former derive directly from what is abstracted from reality and included in the model, and the latter from what is externalized in the process. The former contribute to the successes of our economies, and the latter to their failures to prevent or greatly minimize undesired consequences. To the extent that we pay for these failures, they directly undermine gross wealth production. In other words, the 'signal' of technological and economic growth appears to be threatened by the 'noise' of unwanted and unexpected effects. It would appear that the more we work on increasing the output of goods and services of modern economies, the more we are impoverishing ourselves. Economic development appears to be going in reverse.

The same kind of conclusion may be reached when we examine how industrially advanced societies deal with the unwanted effects of technological and economic growth. Contemporary societies are based on an intellectual and professional division of labour by which specialists of all kinds make decisions whose consequences fall mostly outside of their domains of competence, to be dealt with in an 'after the fact' manner by others in whose specialties these undesired effects fall. As a result, the 'system' first creates problems and then 'solves' them. It is next to impossible to get to the root of any problem in order to prevent or minimize it. The 'system' displaces rather than resolves the problems it creates, thereby feeding on its own mistakes and trapping us in a labyrinth of technology.[95] For example, we first produce pollutants, then install control devices to remove the most dangerous ones from waste streams, and then landfill them, which merely transfers these pollutants from one medium to another without solving the real problem. Also, we continue to feverishly restructure corporations to improve the productivity of labour with the result, as shown by socio-epidemiology, that human work has become one of the primary sources of physical and mental illness.[96] The situation has compelled us to add social and health services at great expense.[97] Since these do not prevent the problem either, their costs can only grow, to the detriment of corporations, employees, and society. All this raises the question as to whether the primary outputs of technological and economic growth include waste and unhealthy workers. As these and other difficulties steadily increase, societies react by doing almost anything except going to the root of the problem. For example, we debate whether to privatize some or all health services, how best to increase the productivity of medical personnel and facilities by improving information systems, or how best to apply operations research to the scheduling of operating rooms. When this increases the stress levels of personnel even further, we add stress management clinics in an ongoing chain of compensating additions. Doctors order blood tests and other examinations, but many of them have little or no understanding of how workplaces affect us psychosocially and physically. From socio-epidemiology, we know that human health cannot be 'produced' by means of disease care. Instead, it is sustained by meaningful and satisfying work, wholesome nutrition, adequate shelter, social support, and a fulfilled need to love and be loved. From this perspective, what we call the health care system is in effect an end-of-pipe disease care system, which has all but forgotten what health care is all about.[98]

Many other consequences are similarly compensated for, as opposed to being resolved. The addition of control devices and services to the 'system' and their undermining of gross wealth creation reduces the pool of net wealth. The result is a distortion of income distribution, manifested by a growing gap between rich and poor individuals and nations, since some can better protect themselves than others.[99] Along with this distortion comes a variety of social ills, placing growing demands on social safety nets and making health care increasingly unaffordable. The response has been to attempt to increase the productivity of social services in order to justify keeping public expenditures level or even reducing them. Hope for a brighter future has been commercialized by lotteries and casinos as another compensatory service. Many governments add to these problems by downloading services to create the illusion of tax and deficit reductions. Such downloading has to stop somewhere, but it does so with those institutions, communities, and individuals that can least defend themselves, thus introducing a further element of chaos into society.

In sum, decision-making in contemporary societies may be compared to individually and collectively driving a car by concentrating on its performance as indicated by the instruments on the dashboard as opposed to watching the road. The result is a great many collisions with human life, society, and the biosphere. The metaphor is an appropriate one because the consequences of the decisions made by specialists fall mostly outside of their domains of expertise where they cannot see them. This leaves them little choice but to concentrate on representing the contexts in terms of the inputs and outputs these yield and absorb. Success is measured in terms of performance ratios that, by means of cultural myths, are transformed into human values. It appears extremely rational to have experts tackle only those tasks for which they have the requisite competence, but this creates an irrational system because of the highly interconnected nature of our world.

What I will seek to show in Part Three is that what we have progressively invented piece by piece during the last two hundred years is a way of reorganizing our world and human life within it, without paying much attention to the fact that this simultaneously disorders it in other ways. The environmental crisis is the result of straining our dependence on the biosphere, and the appearance of a mass society the result of straining and dissolving our relations with one another. Our civilization has introduced chaos into the natural and social orders by selectively reintegrating a variety of elements into technical networks and structures. This new technical order feeds on a growing chaos in the human and natural spheres, which we deal with by creating a labyrinth of compensations. We are thus undoing the complex order inherited from natural evolution and human history. The weight of this labyrinth risks a disintegration of human life, society, and the biosphere, or, at least, their mutation in ways that no one can foresee. Our present path is leading nowhere, and we need to assert a 'somewhere' by means of human values and aspirations.

PART THREE

Our Third Megaproject?

chapter 7

Technique and Culture

7.1 The Disenchantment of the World Revisited

The full significance of the developments described in Parts One and Two may now be seen in the light of my earlier claim that humanity has begun its third megaproject: creating a global civilization primarily based on the technical approach to life and only secondarily on the cultural approach, thereby creating first *homo economicus,* and later *homo informaticus.* In Part One, we observed how human life broke its bonds with local nature and the gods, and in Part Two how it also began to break its bonds with local cultures. When this detachment spread throughout all of life, human beings and societies reattached themselves to everything via technology-based connectedness (which became technique-based connectedness when it enveloped culture-based connectedness). The profundity of these transformations was, as we will see later, clearly intuited by art and literature.

The umbilical cord with local nature was severed by the necessary restructuring of the technology-based connectedness of industrializing ways of life. The network of flows of matter was no longer contained within a larger network of flows of matter occurring in a local ecosystem. The situation was quantitatively and qualitatively different from the way empires had extended their ecological footprints by means of trade. Matter now began to participate in a newly emerging technical order. To do so, it had to be systemically 'denatured,' to be fashioned into relatively pure materials 'decontaminated' from much else that naturally occurred with them. Alloying these pure materials, or in other ways combining them, created materials with a technical as opposed to a natural structure. In turn, these materials helped to construct an entirely new physical habitat. Biotechnology is now extending this approach to nature by encompassing living entities. Agrobusinesses are connecting monocultures into a global food system that seeks to keep nature at bay as much as possible.

Energy underwent a similar transformation. Its delivery by nature through wind and water, and by society through domesticated animals and human toil, declined. Energy became separated from all this to become 'pure' energy obtained by extracting, processing, and transporting fossil fuels. Their conversion powered a multitude of devices that delivered services to a way of life, or did so indirectly by powering energy systems that distributed energy carriers, such as electricity.

In the social world, work had been separated as much as possible from the remainder of a person's life and extended family to become 'pure' labour. Capital had become detached from all cultural values to be temporarily attached to a material thing: gold. Flows of commoditized matter, energy, labour, and capital were initially coordinated by the Market. When technological knowing and doing separated themselves from experience and culture, they became dominated by technical information acquired from, and applied through, scientific disciplines and technical specialties. These studied and organized everything in terms of their own context, making them autonomous with respect to all others. Technical knowing and doing could now be assessed on their own terms, and this made them very different from culture-based knowing and doing. The organized application of this technical information created a factor of production in its own right that increasingly organized the flows of commoditized matter, energy, labour, and capital in a manner that became universal and hence non-cultural.

Severing the umbilical cord with local cultures began with the introduction of the technical division of *physical* labour into the culture-based connectedness of a society. People delivered commoditized labour: ideally, an imitation of how a machine would carry out a technical task. They thereby extended the technology-based connectedness of society. Later, the division of physical labour was complemented by a division of *intellectual* labour based on highly specialized knowledge separated from experience and culture. Within technostructures, specialists of all kinds participated in a manner radically different from their involvement in daily life. Since most of the consequences of their decisions now fell beyond their domains of competence, they could not exercise responsibility for them the way they did in daily life. This was delegated to other specialists in whose domains of expertise these consequences fell, thereby creating a labyrinth of technology. Nor did their decisions deal with a 'chunk' of the world. They were only able to deal with its 'technical shadow,' which was all that remained after several processes of abstraction. Dealing with things in this way in order to technically improve them took them out of the natural and cultural orders, to incorporate them (along with everything else that had been technically improved) into a new technical order. Technical knowing and doing had more in common with information than with traditional knowing and doing.

The technology-based connectedness of a society was now built up with flows of matter, energy, labour, capital, and technical information, with the latter becoming the prime factor of production. At the same time it underwent a qualitative change, gradually reversing the relationship with the culture-based connectedness of human life and society. With the establishment of the Market, the cultural approach first came under pressure from the economic approach. With the separation of knowing and doing from experience and culture, the technical approach to life gradually gained primacy over the cultural approach. In other words, at first the culture-based connectedness was merely strained by the technology-based connectedness. Gradually the latter began to envelop the former, turning it into what will be referred to as the technique-based connectedness of human life and society.

A new technical order was thus constituted as a global network of technically reconstituted elements from human life, society, and the biosphere. The technical approach abstracted elements from the natural and cultural orders, transformed them, and reinserted them into a technical order to which they were now perfectly adapted. Everything now took on a 'meaning' and 'value' according to its place and significance within the technical order. It appeared as if humanity was succeeding in escaping the natural and cultural orders. Natural evolution would now be brought into human history by 'managing planet Earth,' and since the cultural gods were dying or already dead, everything would now become possible. Life, language, and cultures became functions, structures, and systems, and the external world became a human construction. Any murmurs to the effect that these were the mere 'technical shadows' of natural and cultural life were dismissed, as one scientific and technical fashion succeeded another in a scramble to make everything conform to the new technical order or to attempt an escape. In other words, reification succeeded alienation. Universal science and technology replaced culturally appropriate sciences and technologies. Mass societies replaced traditional societies. Unsustainable ways of life replaced mostly sustainable ones. I will seek to show that these were symptoms of the new emerging megaproject, based primarily on a technical approach to life and secondarily on a desymbolized cultural approach.

Although dealing with everything in terms of its 'technical shadow' permits humanity to succeed beyond its wildest hopes and dreams in the domain of performance values, it simultaneously compels humanity to live in the torn fabric of relations that constitute individual life, society, and the biosphere. It means that our successes are directly linked to our failures: it is impossible to have the one without the other. It also means that as we live in this torn world, the cultural cycle links its evolution to that of human life. There is a powerful tendency to select those aspects of what it is to be human that make this torn world more liveable, which in turn makes further tearing possible, and so on. All of this is exemplified by the latest technical frontiers in information technology, nanotechnology, and biotechnology. In the process, the role of culture is transformed. We will need to explore what happens to it when more and more areas of a way of life become dominated by a knowing and doing separated from it. We will need to explore how in daily life people cope with a deluge of new situations having little or no precedent, given their reliance on prior experience. We will need to explore how specialists divide their being between the technology-based connectedness of their professional lives and the culture-based connectedness of their daily lives, and how on the collective level a society avoids a kind of dual history as its technology-based connectedness and its culture-based connectedness pull it in different directions. Is such a society headed for collapse? How can babies and children grow up into healthy adults under these conditions? We will encounter these and other issues as we pick up from where we left off in Part Two.

In the debate over the nature of modern societies that emerged in the 1960s and 1970s (whether post-capitalist, post-industrial, technological, or other), the declining role of culture was overlooked. Surely one thing that set those societies apart from all others was that a wide range of activities were no longer based on custom or tradition grounded in culture. These societies systematically began to research virtually every sphere of human activities in order to improve them. Success was now measured in terms of performance values. To make something 'better' in these terms means comparing it to other forms of the same activity, and to alternative ones accomplishing the same goal, in order to choose the best one. Little or no attention is paid to the cultural meaning and value of that activity in terms of the contribution it makes to human life, society, and the biosphere. This technical approach to life implies living as if the quality of individual and collective human life can be improved in a piecemeal fashion by making the means of human existence more efficient. It represents a sharp departure from social evolution based on culture. Here, everything was understood, dealt with, and valued relative to human life. Anything (including a technological artefact) was dealt with first and foremost in terms of its meaning and value relative to everything else in individual and collective human life. The universe was endowed with human significance. That same something has now become a cold fact in a reality structured by a new technology-based connectedness. The technical approach situates everything in relation to it, even though people continue to live as if the cultural approach has remained more or less intact. It is only after a fact has been established that the question of its significance for human life may arise, thereby 'disenchanting' the world, to use Weber's term.[1] An entirely new kind of civilization has emerged in which human life is increasingly out of context with itself, society, and the biosphere. The integrality of human life and the world, lived in cultural terms, is now dominated by the reality of isolated facts 'lived' via the technical approach. The result is the recreation of human life in the image of technique, for which technology is the model, just as physics was the model for science.

The above thumbnail sketch of where the developments described in Parts One and Two are heading will be elaborated in this part of our narrative. It is the story of the building of humanity's third megaproject and a civilization dominated by its most powerful creation.

7.2 The Invention of Universal Knowledge

The process of humanity breaking its bonds with local nature and local cultures must not be confused with the late Greek invention of universal knowledge and the attempts by some Greek philosophers to make this knowledge a foundation for Greek culture in order to arrest its decline. Although these efforts failed, the invention of universal knowledge so profoundly marked the journey of Western civilization that it is essential to briefly examine and contrast it with the events described in Parts One and Two.

From the perspective of the skill-acquisition model developed by

Stuart and Hubert Dreyfus,[2] the acquisition of a culture as described in chapter 4 may be interpreted as people becoming experts in living in the world by means of that culture. It enfolds all the individual skills involved in making sense of and living in the world. When such a culture is highly creative and dynamic in its evolution, coping with daily life in the world appears effortless, not unlike the performance of any master or virtuoso. People spontaneously respond to situations, and when such responses are later examined with the benefit of hindsight, they appear remarkably effective. By means of their culture, they seem to get it right most of the time. There is little need to stand back to think a situation through.

Culture-based expert living in the world is diametrically opposite to expertise based on universal knowledge. Its acquisition by rational detached observers excludes everything cultures rely on: perception, emotion, feeling, intuition, skill, reason, and tradition. In order to acquire it, people must leave much of themselves and their culture behind, as a philosophical absolutization of the acquisition of knowing and doing separated from experience and culture. Even more important, universal knowledge redefines what is ultimately real. It is no longer a matter of what is real for a particular time, place, and culture, but what is true everywhere. Rational knowledge thus represents an attachment to the universal and eternal, of which the finite and temporal are but a shadow; while cultures represent an attachment to the finite and temporal, and through these to the eternal by means of religions. The human body represents an obstacle to detachment from experience and culture, but it is the centre of cultural embodiment in the world. For example, to be 'reborn' as a Platonic philosopher requires a person to 'die' to their body. The contemporary version of this is well known. On the Internet, people 'die' to their bodies because what matters is their minds and 'lives' on the Net. Once we know our own 'program,' we will be able to make a backup and 'live' eternally on the Net. Universal rational knowledge acquired by a detached observer thus has an anti-cultural and anti-life orientation built into it.

From time to time, we who are 'experts' in living in the world by means of our culture encounter situations that are very different from anything that existed before, and which can therefore not be readily symbolized on the basis of prior experience. Every society faces these kinds of difficulties in its evolution, but usually on a modest scale. It regularly deals with discoveries and developments from within, or contacts with other cultures from without, that put its 'common sense' to the test. (By common sense, I mean the body of meanings and values shared by the members of a community by virtue of a common culture.) When this happens, societies usually turn to reason in an attempt to discover whatever meaning and value something of concern might have in terms of the way their culture symbolizes life in the world. If a particular interpretation makes sense to many people, the symbolization into their culture will follow. From this perspective, reason is a human creation of a particular time, place, and culture that a community relies on to facilitate the evolution of its way of life in the face of somewhat anomalous circumstances. Hence, this kind of reason is the opposite of rationality.

I am well aware that many Western philosophers would have argued that there is no such thing as common sense and that, in any case, nobody has been able to find it. From an ahistorical perspective, this is true. There is no common denominator underlying all cultures in history as a kind of absolute and rational foundation. The same kind of problem arises today when we search for universal rights that are to be the common denominator of all cultures in our world. Their affirmation at the expense of justice or compassion also creates barriers to cultural evolution, because it is always relative to a particular place and time as well as to the particular history of a people. As a consequence, foundational philosophy is unable to understand anything about human life unless it recognizes that experience and culture are a symbolic taking hold of the world to master the art of living in it. In this sense, foundationalism has essentially represented an agenda that is anti-life, anti-cultural, and anti-historical.

Western civilization was born in Greece. In the days of Homer, Greek culture was a highly influential and widely admired mastery of living in the world. However, Greek culture eventually went into a decline. Among the contributing factors were the consequences of deforestation and subsequent soil erosion, which limited what the Greeks could produce. Olive oil and wine had to be traded for other necessities. This brought Greek culture into contact with others, resulting in a degree of relativization. For example, in Athens women were uneducated and expected to stay at home. This remained entirely self-evident until contacts occurred with other cultures such as the Thracian, for which it was equally natural for women to accompany the men on overseas trade voyages, taking responsibility for keeping the accounts and doing the bartering in the markets. It was obvious to Greek observers that these women were very good at what they did, which raised the question as to why Athenians did not educate their women. In the same way, other elements of Greek culture were called into question through encounters with different traditions, some of whose customs appeared equally good or even better. Such relativizing influences led people to question their own traditions and customs.

Other factors that contributed towards a cultural decline resulted from an incompatibility between the Greek democratic process for decision-making and the military requirements to protect trade routes and small colonies from the influence of more powerful empires. Particularly during the Peloponnesian War, the Hellenic Greeks were not very effective in making good military decisions. Military decision-making was simply incompatible with their system of direct democracy. A sense of crisis was gradually heightened with the onset of a plague epidemic. Their confidence in their ability to more or less spontaneously make the right decisions was severely shaken. Socrates began to realize that the Athens he knew was coming to an end and that he had better do something about it.[3] His invention of universal knowledge was passed on by his students Plato and Aristotle. At the core of their 'solution' was the transformation of culture-based reasoning into rationality.

Socrates and Plato attempted to put all knowing and doing on a firm footing, drawing their inspiration from the endeavour in which the Greeks had excelled, namely, putting mathematics and science on a rational foundation. The resulting philosophical enterprise profoundly influenced the course of Western civilization and endures to the present, even though Nietzsche, Heidegger, the American pragmatists, and the postmodernists all broke with it.

In the domains other than mathematics and science, the Homeric Greeks relied on the epic poems the *Iliad* and the *Odyssey*. These poems of the deeds of their gods and heroes provided society with models from which people drew their own conclusions. They represented an attempt at articulating Greek myths and functioned in a manner not unlike religious texts, except that they contained no commandments.[4] The role of these poems was greatly weakened through contact with other cultures. Plato responded to the crumbling ethos of his city, Athens, by seeking a rational and universally valid foundation for the ethical practices of his culture so that they could be maintained and justified in the face of others. He expected to curb the threat of relativism and nihilism in this manner. Following his teacher, Socrates, he behaved much like a modern knowledge engineer seeking to build an expert system.[5] Recognized authorities are questioned to determine the rules that are presumed to underlie their knowledge. For Plato in particular, it was self-evident that such rules had to exist. His 'dialogues' were an enquiry into domains such as justice, virtue, friendship, and beauty and involved questioning members of the community recognized as being authoritative in a domain. Despite Plato's skill, all such people could provide him were examples that could not be converted into rules because there were always some unusual circumstances that called for their violation. Suppose Greek parents taught their children never to tell a lie. Also suppose that the family had attended a rather tedious birthday celebration of a friend. As they left, they overhead their parents say to the host that it was a marvellous party. If the children had interpreted the instruction of never telling a lie as a rule, they would have to invent a new rule to the effect that you may tell a lie if it avoids making others unhappy. Soon thereafter, they might have discovered another exception, when a drunken acquaintance threatened to kill their father over a gambling debt. To prevent a possible murder, their mother told the lie that her husband was not at home. There are two ways in which to interpret what the children are learning. One is that the instruction of never to tell a lie is a rule, and that learning to live in the world is a matter of acquiring virtues based on ever-better rules. The other holds that no such rules exist, and that instead it is a question of learning when to tell the truth and when not to, that is, acquiring the virtue of truthfulness. Socrates and Plato held the former interpretation, but Aristotle held the latter.

Aristotle recognized that apart from mathematics and the crafts, no such rules existed. True wisdom lay in the acknowledgment that there are limitations to one's knowledge and that this will always be so. Plato fundamentally disagreed with his student. Since people can teach others to be virtuous or to become an artist, for example, rules had to exist. According to him, human beings were born with all the rules required for the assimilation and organization of knowledge. The situation may be compared to that of a computer chip whose pathways await content to flow through them so that it can perform its functions. For Plato, experts had forgotten these 'pathways' through which the content of the world flows. His position of extreme rationalism distrusts experience, which, he argued, leads to opinions he called *doxae*, which mask the pristine rational knowledge embedded in the soul from birth. It was the task of the philosopher, who possessed wisdom, to reveal that knowledge.

Plato's *Republic* illustrates his position as well as that of his teacher, Socrates (seen of course through Plato's eyes). It is an attempt to lay down a rational foundation for the concepts of justice and a just society. Quite significantly, it opens not in Athens but with a return from Piraeus (the port of Athens), where a celebration had marked the introduction of a new and non-Athenian goddess. It contains many dialogues aimed at helping people to recall the pristine rational knowledge they had in their souls – a process Plato called *anamnesis*. In the absence of a concept of evil, it was important to do so correctly because unjust behaviour was the result of poor upbringing, which had obscured the issue with doxae. However, the dialogues aimed at bringing about anamnesis were always related to mathematics, which served as the model. One of the most celebrated of these renders an account of Socrates questioning a slave-boy who has had no education but who, through careful questioning illustrated with lines drawn in the sand, 'remembers' the Pythagorean theorem. For Plato this confirmed that the rules of mathematics existed in the soul of the boy, who, by means of being questioned, was able to remember them. Given this focus on mathematics, it is not surprising that, at the end of all the 'dialogues,' Socrates cannot come to any conclusion as to what constitutes virtue.

In order to understand Plato's blueprint for a rational republic, it is important to note that the Greeks had no concept of culture. Human life was conceived of as having three principal elements: the soul, the spirited part, and the passionate part. The soul contained reason, which, not being subject to time, place, or society, was eternal, and thus in the domain of the gods. Since the soul could not be contaminated by doxae but only be obscured by them, it constituted a potentially perfect foundation for a society. The virtue corresponding to reason was wisdom. The virtue corresponding to the spirited part was courage, while the virtue corresponding to the passionate part was moderation because it brought human beings closest to the earth through their needs for food, clothing, shelter, and sex. A person could be just and lead a good life only if reason ruled spirit and passion, or if wisdom guided courage and moderation. A just society had to be ruled by a philosopher-king who combined wisdom with the exercise of political power. He was assisted by other philosophers, who would help him to purge people of their doxae, to break through to the true and inherent wisdom they already had. The 'surplus' of wisdom at the top of the social hierarchy had to compensate for the 'deficit' of wisdom in other social strata, such as that of the merchants who had succumbed to their passionate parts.

A rational city or republic could thus be created by a kind of zero-sum game.

This concept of a rational city led Plato to a number of conclusions that were counter-intuitive and non-experiential. For example, according to this blueprint, the youth of the city should be educated (a process of anamnesis) by the philosophers rather than by their natural parents, who generally lacked wisdom. Aristotle disagreed with his teacher over these kinds of prescriptions. He accepted rationality, but only in the domain of mathematics. Rationality could not be applied in the domain of the natural sciences because they were totally dependent on human experience. In the domain of daily-life experience, not everything was considered mere doxae, every opinion as good as any other. For Aristotle, many such opinions were also reasons, although of a different kind from those found in mathematics. They were reasons that had been validated through the experiences and history of a society. Many different things were tried, and the best practices and habits retained. The reasons for a society's customs and traditions were therefore not rational and embodied no laws or rules. They could be justified only by having been passed on from generation to generation as the best of accumulated experiences. This led Aristotle to reject Plato's prescriptions for a rational city as counter-experiential. He would argue, for example, that if Plato's suggestion that children should be brought up by the wise was such a good idea, why have human beings always lived in natural families? Being a member of a natural family does not depend on rational rules or mathematics, but on experiential and biological factors. Aristotle was distrustful of abstract reason as much as Plato was distrustful of experience. Thus, for Aristotle, a declining Greek ethos could be addressed only through experience, and, for Plato, only by rational principles.

We do not know which side of the debate most Greek people identified with. There are grounds for believing that most Greeks were more Aristotelian than Platonic in their daily-life behaviour. Socrates was condemned to death by Athens, probably for the impact he had on the youth of the city. By questioning its youth about why they did things the way they did, and insisting that their usual answer of having been taught thus by their parents was inadequate, he intensified the decline of the Greek ethos by turning the youths into relativists and nihilists. The people did not take kindly to this, and it is supposed that this was the reason for his condemnation.[6] It would appear, therefore, that rationality did not come close to becoming a mainstream phenomenon in

Greek culture. However, the writings of Plato and Aristotle sowed the seeds for an unresolved tension between rationality and experience, with sometimes the one and sometimes the other gaining the upper hand in the development of Western philosophy.

From the perspective of Homeric Greek culture, the development of Hellenistic Greek philosophy and the search for universal knowledge was an unworldly endeavour not conducive to helping the Greeks to effectively deal with the situation they had to confront. In fact, it hastened the decline of Greek culture. As the centre of Western civilization shifted to Rome, we encounter a society that had made politics coupled to a unique legal system the centre of its way of life.[7] It was here that the three 'perfections' of Western civilization, namely, Greek philosophy, Roman law, and Christian revelation, first came together. Since all three were contradictory to one another, subsequent Western cultures distinguished themselves from most predecessors in terms of their dynamism, which was related to their internally contradictory character. For example, Jews and Christians had an entirely different view of what it was to be human than did the Hellenistic Greeks. Their emphasis was on being true to their God and fellow human beings. Truth was regarded in terms of being utterly committed to these relations rather than to a rational and universal knowledge. For the Jews and Christians, truth was revealed by a God who claimed to be wholly other than the religious gods. From the perspective of cultural anthropology, the First Commandment meant that the Jewish or Christian people could not develop cultures with a sacred and myths. In this sense, the revelation was anti-religious. There was no movement of the finite and temporal to the eternal. The reverse was the case, culminating in the Incarnation. Hubert Dreyfus sums up the difference between Greek truth on the one side, and Jewish and Christian truth on the other, as follows. For the Greeks, the gods loved something because it was the right thing to do, that is, it followed from rational and universal principles. For the Jewish and Christian people, something was the right thing to do because God loved it.[8] Initially, the Christians formed a kind of counterculture within the Roman Empire, until their faith was declared to be the official religion. At this point, its contradictions with Greek philosophy and Roman law had to be explained away, and this set Western civilization on a bizarre journey.

If we accept the premise that most great thinkers express what is thinkable in their cultures (i.e., within the reigning sacred and myths) in a non-iconoclastic way and that they do so well ahead of their contemporaries, then Augustine sought to fuse Christianity with Greek philosophy in general and Plato in particular. Thomas Aquinas instead concentrated on Aristotle. By the time Descartes appeared, it was possible to take it for granted that the god of the philosophers was one and the same as the Christian God. Pascal was the first to recognize that this could not possibly be the case. He systematically set out all the contradictions between Greek philosophy and Revelation. Hegel made the last attempt at providing a philosophical account of Christianity from a detached and objective point of view. The rise of Christendom and its subsequent decline cannot be understood without the way these contradictory elements in Western cultures worked themselves out. Kierkegaard was the first and the last existentialist philosopher to create room for critical reflection within a defining commitment to the Christian faith. Much later, Ellul went further by recognizing that this defining commitment to God and other people confronted the culture of the believer and his or her attachment to a sacred and myths, including the alienating religion built on them. In other words, the self must be understood in psychological as well as socio-cultural and historical terms.

Working out the implications of the contradictions between Greek philosophy and Christian revelation had a profound effect on the development of Western science and technology. Eventually rationality won out, paving the way for scientific and technical knowledge to separate themselves from experience and culture. As discussed previously, many scientific laws such as those of Newtonian physics, for example, are counter-intuitive and cannot be experienced directly. Special apparatus has to be constructed to show that any moving body maintains its velocity in the absence of forces acting on it. This is certainly not how we experience things in daily life. Similarly, a feather and a small lead weight do not fall at the same speed in our daily-life world. Galileo's rationalist hypothesis could not be confirmed until the appropriate technologies became available. The rationalistic view of technology gained the upper hand when technological knowledge separated from experience. It legitimated the separation of this knowledge from experience and culture. Today, science is reaching a point where the mathematical representations of reality are so abstract that it becomes increasingly difficult to verify and interpret their meaning and value.[9] In the secular religious context of contemporary cultures, this is regarded as a non-issue. If the gods are dying or already dead, then there is no creation and no nature. All we can expect 'out there' is what we have socially constructed. If this turns out to be useful and gives us greater power over our lives and the world, all the better. Nevertheless, scientific and technical knowing and doing are unthinkable without their practitioners behaving as if there is an underlying order to be discovered and manipulated. In other words, their behaviour implies the possibility and value of objective truth, and then we are right back to the Platonic Greek tradition. Some scientists continue to claim that science helps us understand the creator God and the manifestations of his wisdom and goodness. However, being separated from experience and culture, science is outside of the domain of meaning and in the domain of cold facts, to which only the weak are supposed to be in need of adding a metaphysics. The others are to conclude that there is only a chaos out there, from which we draw social constructs. This, as we will see, turns science into technical practices, but they too imply a myth (in the sense of cultural anthropology).

To sum up, I do not believe that the late Greek invention of universal knowledge marked the beginning of the creation of humanity's third megaproject. Human life in the world is an embodied one. Each and every experience is gained through the body, and it is through experience that the mind is developed. We need only speculate how experience would be different if one or more of the senses were alternatively embodied, in order to understand how much experience shapes the mind and hence our defining commitment to a time, place, and culture, and thus how much the body, mind, and spirit are enfolded into one another. Human life is profoundly historical.[10] It is therefore impossible to put life on a non-physical, asocial, and ahistorical foundation. Plato and Socrates, and their philosophical descendants, pushed human thought to its limits within the defining (spiritual) commitment of their time, place, and culture and then mistook it for a rational foundation, rather than introducing an iconoclastic element into the declining Greek culture, which might have permitted it to evolve.

Humanity's third megaproject did not come into being by placing human life on a rational, scientific, and technical foundation. Instead, it began with individual and collective responses to particular social, cultural, and historical situations, which gradually led to a detachment and reattachment to local nature and local cultures. This reattachment, as I will seek to demonstrate, includes but is no longer dominated by local nature and local culture. Individual and collective human life is now lived in the torn fabric of local relations and the global temporal technical order constituted by networks of commoditized natural, social, and cultural elements. It requires a technical approach to life – that of *homo informaticus*. Technical knowing and doing have been separated from experience and culture, only to be reattached to them in other ways. At no time was the separation of knowing and doing from experience and culture absolute. Nor can it become, or be based on, universal knowledge as understood by Socrates and Plato and their successors. Humanity's third megaproject is not a striving for their asocial, ahistorical ideals.

7.3 Rationality and Industrialization

The process of industrialization helped create conditions under which rationality became a way of life for the first time in human history. Until then, it had merely been a philosophical and scientific idea. I have shown how industrialization rapidly created new relationships and conditions in almost every aspect of a way of life. Making sense of these experiences was often difficult because traditional cultures gave people little to go by. Their cultures could not be paradigmatically extended on the basis of prior experience. Until then, people had evolved and adapted each aspect of their way of life as an integral part of a tradition, thanks to a knowing and doing embedded in experience and culture. When successful, the culture-based connectedness of a way of life was evolved in a manner that sustained human life. Industrialization changed all this because it required that ever-greater attention be paid to the technology-based connectedness of a society. As a result, adaptations primarily focused on this technology-based connectedness were less and less compatible with the culture-based connectedness, thereby introducing a growing chaos within it, which in turn necessitated a further strengthening of the technology-based connectedness. This was not a situation in which people were faced with too much change coming so quickly that it overwhelmed their cultural resources in a kind of 'future shock' scenario.[11] Instead, they were confronted with a quantitatively and qualitatively different kind of change related to a growing gap between the technology-based connectedness and the culture-based connectedness of their lives. For years, they may have responded to a particular situation in a manner that had become routine. When conditions changed, they suddenly recognized one day that this was no longer adequate, forcing them to think it through. When these were isolated events in the way of life of their society, such reliance on thought and 'reason' tended to be based on their cultural resources and thus cumulatively extended their culture and way of life. However, when such incidents were omnipresent and directly or indirectly related to disturbances in the technology-based connectedness of society, the usual cultural resources were inadequate. Reason became focused on evolving the technology-based connectedness of society, thereby producing a mutation that turned it into what Max Weber described as the phenomenon of rationality. It depended on technology as opposed to tradition and custom, which Weber very broadly defined as follows:

> The term 'technology' applied to an action refers to the totality of means employed as opposed to the meaning or end to which the action is, in the last analysis, oriented. Rational technique is a choice of means which is consciously and systematically oriented to the experience and reflection of the actor, which consists, at the highest level of rationality, in scientific knowledge. What is concretely to be treated as a 'technology' is thus variable. The ultimate significance of a concrete act may, seen in the context of the total system of action, be of a 'technical' order; that is, it may be significant only as a means in broader context. Then concretely the meaning of the particular act lies in its technical result; and conversely, the means which are applied in order to accomplish this are its 'techniques.' In this sense there are techniques of every conceivable type of action, techniques of prayer, of asceticism, of thought and research, of memorizing, of education, of exercising political or religious control, of administation, of making love, of making war, of musical performances, of sculpture and painting, of arriving at legal decisions.[12]

Weber went on to state that 'The presence of a 'technical question' always means that there is some doubt over the choice of the most efficient means to an end. Among others, the standard of efficiency for a technique may be the famous principle of 'least action,' the achievement of the *optimum* result with the least expenditure of resources, not the achievement of a result regardless of its quality, with the absolute minimum of expenditure.'[13]

This definition of technology, based on Weber's observations of a new development of his time, may be joined to the present analysis by asking how to make sense of this striving for 'efficiency' and 'optimum result.' Both concepts restrict human attention to deriving the maximum possible desired outputs from the requisite inputs, referring to 'the totality of means employed' as connected by their inputs and outputs. Technological activities contribute to what in the present analysis is referred to as the technology-based connectedness of a society and disregard their dependence on and contribution to the culture-based connectedness. In other words, such technological activities could proceed where alternative culture-based approaches failed, and they led to technology-based connectedness eventually becoming coextensive with culture-based connectedness.

Nearly half a century later, Jacques Ellul independently concluded that the phenomenon of technique had become the most decisive one in Western civilization at this point in its history. He defined technique as 'the totality of methods rationally arrived at and having absolute efficiency (for a given stage of development) in every field of human activity.'[14] This suggests a very different approach to human life and the world. The methods employed are no longer arrived at on the basis of experience, tradition, and culture handed down from generation to generation as accumulated experience turned into a collective wisdom: a design for living morally guaranteed by a metaconscious sacred and myths. The new methods are rationally arrived at, that is, by *ratio* rather than by context. Furthermore, these methods are not first and foremost concerned with human needs, desires, and aspirations or, for that matter, with human values of any kind. It is now a question of the greatest possible efficiency calculated as the ratio of desired outputs and requisite inputs, without any reference to their meaning and value for human life. Individually, these methods limit consideration of their surroundings and the world to the inputs delivered and the desired outputs they receive, that is, to the 'local' technique-based connectedness. Collectively, they restrict attention to the technique-based connectedness joining these inputs and outputs in every sphere of human life. Ellul's concept of technique describes, but is not limited to, what I have called the technology-based connectedness of a society, including as it does all methods rationally arrived at and excluding all those based on culture. These methods are characterized by their interdependent efficiency in transforming requisite inputs into desired outputs and the exclusion of any reference to their meaning and value for human life. The definition also recognizes that the development of technique will not work unless every field of human activity is involved, presumably for reasons analogous to the ones examined in Parts One and Two of this work.

Both Weber and Ellul recognize that one of the defining features of rationality and technique respectively is their reliance on what we have called performance values, of which efficiency is the paradigm. These decouple a rationalized or technicized activity from its human, social, and natural contexts, which runs counter to a culture-based approach that symbolizes everything by situating it in relation to everything else in its context. The new techniques that began to emerge around the turn of this century did not primarily rely on experience and culture the way traditional techniques had. They required that the knowing and doing on which they were based were separated from experience, in which case the standards for evaluation had to be self-referential and thus comparative in nature. What we are observing here is in essence the emergence of a technique-based connectedness in all spheres of life at the expense of the culture-based connectedness.

The emergence of the phenomenon of technique has many roots. One of them is worth noting. The birth of the mechanistic world view at the end of the Middle Ages is significant because a living world was now beginning to be thought of as adequately conceptualized in terms of dead machines (first the clock, and later the computer). Why did Western civilization begin to look for the living among the dead? This anti-life outlook bears the stamp of Greek philosophy and its influence, first on Christendom, and following its decline on the efforts of Western cultures to reinvent themselves. Once again, some of the greatest thinkers of that time sensed and put into words a shifting of the myths of Western cultures, eventually leading to technique-based connectedness gaining the upper hand over experience and culture. Their ideas gradually gained acceptance as 'great' because people recognized that they corresponded to something very deep and real within their lives and world. In turn, this thinking prepared for and reinforced the spread of knowing and doing separating from experience and culture in every domain.

7.4 Logic, Artificial Intelligence, and Culture[15]

Technique evolves within culture but it cannot separate itself entirely from it. Two developments can shed light on this relationship. First, the development of Western logic necessitated a complete and absolute separation from experience and culture, which implies the non-logical character of the latter. Second, artificial intelligence built up from logical rules, algorithms, and the like, which are completely and absolutely separated from experience and culture as well, has never succeeded in imitating human intelligence. Beginning with these two endeavours, I will show that the relationship between technique and culture is essentially parasitic because the former cannot survive without the latter, while, at the same time, it is far from clear whether culture can survive under the pressure of technique.

A distinction must be drawn between the logical, the illogical, and the non-logical. Something is illogical when it belongs to but violates the sphere of logic. Something is non-logical when it does not belong to that sphere to begin with. The development of logic in Western civilization increasingly implied that culture and language are non-logical. It gradually separated itself from language and culture, becoming the only body of human knowledge entirely separated from them. Aristotelian logic dominated from its inception in the fourth century BC until the rise of modern science and mathematics in the sixteenth century. It focused on sharpening human thought and debate, often comparing arguments to sequences of syllogisms and thus generally reinforcing language and culture. It was succeeded by modern science and mathematics, which Western civilization increasingly turned to as the only reliable way of knowing reality, thus devaluing culture to some extent. This eventually led to new forms of logic. One of the most important ones was that of Boolean algebra, developed in 1847. It was based on the recognition that algebra, until then used to manipulate numbers or quantities, could be used to represent ideas as well. One of its major interests was the extraction of information from a set of premises.

The complete separation of logic from daily-life thinking began with G. Frege, who founded modern logic. His argument was simple and decisive. If logic was an inborn engine of the brain, as Plato had argued, or if all human behaviour ultimately depended on rules and algorithms manipulated by the brain, as held for decades by artificial intelligence research, then logic should evolve just as the brain did during human evolution, and that would lead to a contradiction. The laws of logic cannot evolve because they embody 'eternal' mathematical truths. Therefore, reason based on culture had to be a mixture of the logical and the psychological. The object of research in logic was to make it pure by decontaminating it from its cultural elements. Logic had to become 'unnatural,' which meant a complete divorce from the world of culture. This meant that the world of logic could no longer be contradicted or verified by the world of experience and culture. It had to be built on the criterion of internal consistency. The connections between its elements must be consistent so that no paradoxes occurred. However, no consistency with any other world, including the one of experience and culture, was required. It was thus left with internal consistency as its only organizing principle, and it would therefore be the body of knowledge in which the separation of knowledge from experience and culture was, at least in principle, absolute.

Bertrand Russell also recognized the need for this radical separation when he encountered fundamental inconsistencies. The consistency of non-Euclidean spaces may be guaranteed if Euclidean space is consistent. This can only be ensured by the consistency of arithmetic, which in turn is based on modern set theory founded on a number of axioms that, at first sight, appeared to be entirely commonsensical. He discovered an inconsistency, however, which forced him to abandon common sense and to sever all connections with the real world. Initially, he deeply appreciated this development, until the shock of the First World War changed everything.

The world modern logic creates is very different from the world created by a society by means of its culture. Modern logic is not interested in the meaning or value of any statement that helps constitute its world. The only thing that matters is the quality of the connections between such statements. Each time a statement is consistent with the world of experience and culture but inconsistent with other statements in the world of logic, it has to be removed. It is apparent that the two worlds have entirely different natures, which makes them incommensurate. Modern logic is non-cultural and culture is non-logical.

Had internal consistency been the principal goal of culture, it would not have been able to support groups and societies in their efforts to make sense of and live in the world. Culture and language have to cope with a world of a hierarchy of wholes, many of which, particularly in the cultural sphere, are frequently enfolded into one another.[16] Each whole contributes to larger wholes on the next 'higher' level, whose characteristics cannot be derived from their constituent wholes. At the same time, each whole itself is composed of wholes on the next 'lower' level, and so on. For example, a person is a member of a group, which is a part of a community belonging to a society within a civilization evolving within a biosphere that envelops a planet within the universe. Consequently, the meaning and value of any whole is multidimensional, ambiguous, and, particularly where enfolding occurs, frequently internally contradictory or dialectical. Had culture been based on the principle of logical consistency, it would not have been able to cope with the world as we know it.

The human-social sphere of reality comprises many dialectical relationships that cannot be adequately represented through logic. Since it is common in modern societies to use the same words for machine processes and for human activities such as memory or communication, the differences have become all but invisible. However, the former can be approached only through logic, and the latter through culture. Recall what we said earlier about human communication. Human communication and social relations are based on a dialectical tension between individual differences and a shared cultural unity. For these and other reasons, no culture created by any society was ever built up on the principle of logical consistency. It is also clear that the word 'communication' should not be used for both human activities and machine processes.

There is also a practical reason why human language cannot be purely logical. It would become a very cumbersome means of communication. One need only consider a few examples couched in the form of syllogisms to appreciate this fact. A literal and logical frame of mind is ill-suited to cope with the pressures of daily life. The effectiveness of language is in part due to the fact that it leaves gaps in such chains of syllogisms, which a listener fills in as a matter of course. This is because the brain-mind helped to evolve precisely the kind of intelligence best suited for human life in a living world where everything grows and changes and nothing ever repeats itself in quite the same way. That same intelligence has difficulty coping with a world built around machines that are pure algorithmic repetition.

Indirect confirmation of the above argument has come from decades of research in artificial intelligence, based on the hypothesis that everything human beings do is the result of the application of rules and algorithms. It postulated what might be called *homo informaticus*, and sought to bring the world of culture and language into the world of logic. As is now recognized by almost everyone, this research failed to reach its original objectives. The theory of Hubert Dreyfus accurately predicts why artificial intelligence is unable to deliver on any of its original promises. The theory holds that because computers are not embodied in the world, they cannot 'experience' it, let alone 'live' intelligently in it.[17] I will limit myself to a few observations. Human embodiment in the world has both a physical and a symbolic dimension. The two come together in the brain-mind, which has co-evolved with the natural world and the cultural-historical world. When the search for a general artificial intelligence was all but abandoned in favour of creating a highly restricted intelligence by means of expert systems, Stuart and Hubert Dreyfus developed the previously discussed theory of the five stages of human skill acquisition to show that no machine would be able to go beyond the third stage, representing the limit of any rational scheme.[18] As I have noted previously, this is precisely where metaconscious knowledge, resulting from being physically and symbolically embodied in the world, begins to play a decisive role.

Artificial intelligence faces some additional limitations as well. Whenever 'pure' logic and mathematics are embodied into a machine or into a human brain, physical and temporal limitations are imposed that frequently cause algorithms to explode. Because of embodiment, it is impossible for any machine or the human brain to be entirely rational since they cannot deal with infinite series or infinite strings of digits, for example. It is in fact possible to use these limitations to have calculators produce utter nonsense.[19] Since all these systems are created, manipulated, and used by human beings, limits of embodiment always occur. For example, some proofs may be so long that it takes more than an entire lifetime to write them down or read them all. Imagine *homo economicus* deciding whether to get up at seven o'clock or seven-thirty. If she were to get up at seven, she would be able to make a big breakfast, call a friend, walk to work, and so on. If she were to get up at seven thirty, the extra half hour of sleep would leave her only enough time to get to work. Either of these possibilities would affect subsequent events, which in turn would affect others, and so on, which would all have to be calculated to ensure that the maximum utility is obtained from each moment of her life. An attempt to simulate this through conventional artificial intelligence approaches would quickly result in combinatorial explosions.

The problem of simulating common sense by means of computers runs into similar limitations of embodiment. Suppose that all the experiences of a person could be stored in a computer memory. The task, then, is to program this computer so that it can answer some simple daily-life questions or make an interpretation of a situation. When a teller screams in a bank, we do not encounter combinatorial explosions because each aspect of the situation has a meaning and value indicating its potential relevance to interpreting the situation.[20] We instantly know what is happening. Current computers, however, in the absence of meaning and values, would have to check for all possibilities: did the teller scream because a red car drove by, a mother and baby walked in, a client said good morning or was wearing a blue suit, and so on. The possibilities are endless. Even if some of these limitations could be removed by adopting computer architectures based on neural nets, there would still remain the problem of human awareness of oneself and the world.

Systems of pure logic and mathematics have their own limitations. In 1931, Gödel demonstrated that complex deductive systems are incomplete and that their internal consistency cannot be proven.[21] Turing later showed that some problems cannot be solved even if a Turing machine carried out an infinite number of steps, so that it could go on forever attempting to complete a proof.[22] Apparently, not everything can be expressed in terms of mathematics. Beyond it exists a non-logical domain, which is filtered out by modern ways of life whenever they rely on mathematics to improve something. It has also been argued that pure mathematics does not exist, and if this is the case, it will not be able to separate itself entirely from experience and culture.[23] The domain of experience and culture appears to be non-mathematical, that is, it cannot be contained within the domain of mathematics.

We may be coming to the end of a period during which the secular faith of rationalism was unchallenged in its heaping scorn on culture as being subjective, superstitious, and religious – in short, unworthy of an age of science and technique. Rationalism failed to recognize the accomplishments of culture as possibly humanity's most important and enduring creation. It goes without saying that culture, like any other human creation, is not omnipotent and does have its limitations. When these come to light, we must learn to recognize and live with them unless we are absolutely certain that we can create an alternative for making sense of and living in the world. I will seek to show that neither science nor technique provides such an alternative because humanity still relies on culture for primary socialization to give what little meaning, direction, and purpose desymbolized modern cultures are able to offer to children and young people.

Some observers speculate that soon we will have another basis for human existence. It is the continuation of the search for eternal life, but this time in secular dress. The arguments appear to amount to a belief that, with the arrival of *homo informaticus*, human life will soon be eternal. The genome project is supposed to reduce the code of life (DNA) to information that we will be able to store, manipulate, and resurrect into new life at our whim. A few researchers in artificial intelligence still expect to be able to reduce all human experience and culture to information. At that point, our bodies and souls will have been entirely reduced to the form of information, which will enable them to go on 'living' within computers, the Internet, or any other future information system.[24] These beliefs are based on a complete misunderstanding of both technique and culture and the influence these continue to have on human life. Simply put, since technique changes people more than they are able to change it, humanity is re-engineering itself in the image of technique. It is not a question of perfecting computers and robots to make them more like human beings, but of human beings becoming more like robots and computers.

The relationship between technique and culture has increasingly evident limitations. The environmental crisis brought a greater awareness of the importance of our natural commons. We began to miss entities and relations that we had made ineffective or destroyed altogether. It also brought us an awareness of the incredible resilience of the biosphere, which in a large measure must be attributed to the fact that everything has evolved in relation to everything else, with the result that each and every life form helps to support all others. In the same vein, traditional societies created social ecologies that were also highly resilient to things that spiritually endangered human life, by taking hold of an unknown reality in order to make it liveable. With the aid of technique, we are displacing this kind of connectedness with an entirely new kind. Each one has enormous strengths and weaknesses, but one thing is certain. Technique-based connectedness is not very conducive to the living of human life.[25] As a result, the role of culture can no longer be taken for granted the way it was, effectively doing its work mostly beyond human awareness in the metaconscious dimensions of the brain-mind.

It is no accident that a concept of culture developed in the social sciences only after the role of culture in human life began to falter, first under the pressure of industrialization and later under the pressure of technical rationality. It was this development that increasingly attracted scientific attention, leading to the present understanding of the role of culture in human life. Hence, in addition to destroying the natural commons, we are also destroying our symbolic commons, endangering our life support both from the earth and from the sky. I will continue to develop the claim that the relation between technique and culture is a parasitic one. If the non-logical and non-technical aspects of individual and collective human life are increasingly limited and restrained by present human evolution and adaptation, we may well be doing incalculable harm to ourselves. If living has little to do with quantification, rules, algorithms, logic, and neural nets, we are undermining what until now has made us human. The qualities of human life that are non-logical and non-technical, which are being trivialized in our modern world, may well turn out to be the most important, having so successfully facilitated the making of history. Sustainability no longer is merely a question of our relation with the biosphere but also a matter of relating to culture and symbolization.

7.5 On Creating a New Concept

In the process of making sense of our lives in the world, people have created concepts that symbolize 'locations' on their mental maps. When these concepts symbolize something new emerging in people's metalanguage, they may resonate with and become accepted by the members of a culture. Presumably, acceptance is based on a tacit recognition that such linguistic models help reduce the unintelligible complexity of reality to an accessible complexity endowed with meanings and values. It is virtually impossible to objectively test the validity of these concepts. Wittgenstein showed how the members of a linguistic category share not the same fixed list of characteristics, but a network of criss-crossing and overlapping family resemblances.[26] The acceptance of a new concept into a language depends on a collective discernment of its validity by the members of a culture. Such a discernment is based on their metalanguage as well as a great deal of discussion.

The validity of a concept is not merely a question of its factuality. This is obvious when we consider the description of a disease in a medical textbook. The medical community does not reject such descriptions, on the grounds that not all patients who have the disease exhibit all symptoms. Wittgenstein's insight accurately depicts the situation: the symptoms exhibited by patients will form a network of criss-crossing and overlapping family resemblances. A few patients may have most of the symptoms, other patients may have some of these symptoms as well as some others, and so on. It would appear that the validity of the description of the disease depends on its ability to help medical students become good diagnosticians and expert doctors.

Factuality can never be a sufficient criterion for judging the adequacy of a concept. A concept always simplifies reality for a particular purpose. As a result, whether a detail should be incorporated into the concept or whether it can be left out depends on its meaning and value with respect to the purpose at hand. In turn, this purpose presupposes a broader context and a vantage point. It is for this reason that all the major concepts in the social sciences are 'essentially contested.'[27] Any such concept has an objectivity relative to this context and vantage point, as determined by a school of thought within a discipline. At the same time, the practitioners of a discipline are also members of a society making sense of and living in the world by means of a culture. Whether a detail can be left out of a concept or should be included in it depends on these immediate and larger contexts and vantage points. Ultimately, experts in a field can only 'profess' how things look to them. There can be no fact without a theory, no theory without what Polanyi called tacit knowledge,[28] what Kuhn called a disciplinary matrix,[29] and what Foucault called an epistemé.[30] It may well be that these in turn cannot exist without the metalanguage established by a culture.

All this is complicated by the fact that no concept in the social sciences can be empirically validated the way it can be in the natural sciences. Since life never repeats itself, it is impossible to duplicate any experiment. For this reason, social scientists can only be engaged in praxis, which turns theory into practice and practice into theory. In other words, understanding a particular situation in the context of a school of thought in a discipline as well as a time, place, and culture will reveal some aspects that sustain human life and others that undermine it, some that are 'good' and some that are 'bad.' If, as a result of our understanding of the situation, an expert decides to intervene and recommends that this or that be done, the outcome should confirm that understanding. If it does not, presumably that understanding lacks some important details, or situates those that are included improperly in relation to each other in terms of their place and significance. It is in this way that theoretical understanding can be turned into practice, and practice turned into theoretical understanding, which is as close as we can get to the equivalent of an experiment in the natural sciences. With hindsight, it is possible to examine whether the formulation of a theory has anticipated subsequent major developments relevant to it. It is from this perspective that the creation of the concept of technique must be examined.[31]

Perhaps the following story may illustrate what this part of our analysis is about. Suppose two people who love one another turn to technique to enhance their relationship even further. One is interested in showing their love with more hugs and kisses. To accomplish this, she correlates these signs of affection with the kinds of circumstances under which they occur, for the purpose of engaging in behaviour that creates such circumstances more frequently as a trigger for the response. He is equally interested in discovering what makes her prepare a special surprise meal for him as a sign of how much she loves him. Such an approach to further improving the relationship is bound to have the opposite effect. Their experiences begin to build a metaconscious knowledge of the best intentions going astray. Instead of enhancing their love for each other, they begin to reify one another to improve the 'performance' of their love. This implies a subtle denial of the other person's freely giving of their love, and instead compels them to be bound by certain circumstances. To the extent that such circumstances elicit a response, they are not free; and if they are not free, they cannot love because the former is a prerequisite for the latter. The same kind of consequences result when two people attempt to improve the 'performance' of their friendship, or when a person seeks to improve the 'performance' of a neighbour. This pattern can also occur in relatively impersonal social relations between a shopkeeper and her customers, a teacher and the members of his class, an office manager and her employees, and so on. In a small and subtle way, the full subjectivity of the 'other' is not respected as they are turned into a 'producer of hugs and kisses' or some other form of desired behaviour. This illustrates how the spirit of technique permeates our daily lives.

chapter 8

Human Life Out of Context

8.1 The Technical Approach to Life

The technical approach to life has reached far beyond industry and the economy. We have examined how the culture-based connectedness of human life and society became torn as a result of the growing pressure from an ever more independent technology-based connectedness. The first two phases in the process of industrialization prepared the 'meta-experience' on which the phenomenon of technique was based. Initially, many new situations for which there was no precedent in experience or culture were thought through by means of reason, still deeply rooted in culture. The results were not satisfactory because they tended to be more compatible with the traditional way of life than with the new emerging one. The technical division of labour and the industrialization founded on it required the creation of their own 'world,' which, when interpreted through culture, led to the meaning and value of each machine and productive activity being related to their adjacent ones, and through them to the other entities of this new 'world.' These developments gradually created the meta-experience for performance values to emerge and for thought processes to become more and more rationally directed towards a goal. This tore the culture-based connectedness even further, making it still more difficult to involve everything in the context of everything else by means of the cultural approach, thereby necessitating a diffusion of the technical approach in a self-reinforcing process.

We have observed the emergence of a new intellectual and professional division of labour ideally suited to the development of technique-based connectedness. Each practitioner interprets a situation and acts on it by means of his or her substructure of experience, derived from secondary socialization and practice. In contrast with doing so by means of a traditional structure of experience, the context taken into account is restricted by a triple process of abstraction. Reality is reduced to the requisite inputs it delivers to a particular activity and the desired outputs this activity returns to it. The activity itself is reduced to those aspects that are coterminous with the practitioner's specialty. Improving the activity is reduced to varying these aspects, correlating them with variations in the performance of that activity, and retaining the 'best' one. In sum, the hallmark of technique is the endless improvement of every human activity based on comparing variations in its form and their correlation to performance, as opposed to doing so in terms of their meaning and value for human life and society.

I am not suggesting that in every society and activity things change exactly as described in this chapter. On the contrary. The socio-cultural context in which such transformations begin varies from society to society and from activity to activity. However, the results are delimited by the constraints on the transformation process imposed by rationality making minimal reference to context. We have become keenly aware of how a great deal of human life, society, and the biosphere became a market externality to the economic order. A parallel development occurs as a result of knowledge externalities created in building the technical order. In essence, the technical approach to life concentrates on the 'local' technique-based connectedness associated with what is to be made 'better.' The surrounding technique-based connectedness is represented by the inputs it delivers and the desired outputs it receives. The process that transforms these inputs into outputs is studied, and those aspects directly relevant to making the local technique-based connectedness 'better' are retained insofar as they are coterminous with the expertise of the specialists undertaking the study. The findings are used to construct some kind of model, which is then manipulated to determine under which conditions the process of converting inputs into outputs can be improved. Finally, the results are applied to reorganize something in order to make it 'better' by strengthening its local technique-based connectedness. The technical approach thus leads to techniques for organizing and reorganizing a particular human activity. What such techniques share is a network of criss-crossing and overlapping family resemblances and a set of consequences for human life, society, and the biosphere. These will be examined at the end of this chapter.

8.2 Sport

Except for its current form, sport has always been an expression of the culture that created it. When Greece was the centre of Western civilization, sport sought to develop the form and strength of the human body, which was idealized in statues.[1] When the centre of Western civilization shifted to Rome, sport aimed at preparing warriors for combat and battle. With the fall of the Roman Empire, sport ceased to be a mainstream event in Western civilization until the Industrial Revolution. The correlation between modern sport and industrialization is unmistakable. Jacques Ellul put it as follows:

> Sport has been conditioned by the organization of the great cities; apart from city life, its very invention is inconceivable. Country 'sport' is but a pale imitation of city sport and has none of the characteristics of what we know as sport.
>
> The sporting vocabulary is English; it was introduced to the continent when the continental nations came under the influence of English industrialization. After the industrial center of gravity passed to the United States, American sporting forms prevailed. The Soviet Union began to cultivate sport when it began to industrialize; the only country in central Europe which had organized sport, Czechoslovakia, was the only one which was industrialized.
>
> Sport is tied to industry because it represents a reaction against industrial life. In fact, the best athletes come from working-class environments. Peasants, woodsmen, and the like, may be more vigorous than the proletariat, but they are not as good athletes. In part, the reason for this is that machine work develops the musculature necessary for sport, which is very different from peasant musculature. Machine work also develops speed and precision of actions and reflexes.[2]

No modern athlete can achieve world-class status without the intervention of the technical approach. Every aspect of an athlete's life that has a significant impact on his or her performance in sport is carefully studied. Coaches, doctors, psychologists, engineers, nutritionists, managers, promoters, and public relations people can all play their part. For example, devices such as stopwatches and video cameras record the motions of the athlete, and the results are analysed by coaches to determine which strategies are better than others and how. Athletes are then trained accordingly. The results are once again recorded and analysed to prepare the next technical intervention. Thus, every aspect of the practice of a sport is timed, recorded, reviewed in slow motion, compared to the records of previous performances and to those of other athletes in a constant striving for optimal performance. The motions of modern athletes are no more the product of their experience, common sense, personality, way of life, and culture than are the movements of technicized work. The approach to coaching has taken the form of the technical approach to life. Knowing and doing become separated from experience to be guided by performance values. In the endless comparing and analysing of techniques, there is no room for context values. Nor can any questions be asked about the long-term health implications and their impact on the social lives of athletes, the effect of technicized sports on youth and their communities, and so on.

In the same way, every aspect of a sport evolves by means of the pattern I have called the technical approach to life. Sports doctors help to optimize every aspect of muscular and physiological performance. The contribution to performance of diet is optimized by nutritionists, that of equipment by engineers, and that of mental preparation by sports psychologists. Managers schedule training and decide on competitions for optimal performance. Promoters seek to increase sponsorships and endorsement contracts to maximize financial benefits. Public relations specialists try to obtain the maximum media exposure to raise the profiles of athletes, which in turn affects everything else. The technical approaches encountered in sport are more qualitative and intuitive than those used for the organization and reorganization of factory work, but they do not differ in kind. They involve series of comparisons of practices and matches, attempts to correlate details with variations in levels of performance, the development of metaconscious knowledge (separated from experience and culture) to yield intuitions and ideas for further perfecting the techniques used by the athletes, putting these ideas into practice, further comparison of the results to other performances, and so on.

Sport carries into the domain of leisure and recreation – precisely those qualities required for industrial production. Soccer became the most popular sport in Europe. It is a protest against industrial life, symbolized by the taboo against the use of hands, while at the same time it carries beyond the sphere of work those aspects of human life that are essential for industrial production.[3] The same orientation, attitudes, and morality are found: the predominance of reflex actions, the emphasis on speed and precision, and the desire for all-out competition to win. Particular social attitudes are encouraged. Sport is not played with friends, acquaintances, and neighbours, but with others who are good enough in a particular sport to make it to the team. There is strong encouragement to give up almost everything for the sake of competition and winning. A new moral outlook on sport develops, namely, that those who have the best-adapted physique and musculature for the sport, and train the hardest under the guidance of the best coaches, will win.

The development of sport based on the technical approach must be organized on a national basis, with the direct or indirect intervention of the state. It is necessary to identify those youngsters that have the muscular and physical characteristics crucial to a particular sport, coach them, and manage their progress to end up with enough world-class athletes to represent a nation and to maintain professional sport. The organization of sport ensures that spontaneity and being a person of your time, place, and culture is replaced with a strict regime of exercise, diet, and medication, in turn supported by a host of machines. Everything is dedicated to the pursuit of efficiency. The athlete is converted into an efficient piece of apparatus trained to fully exploit his or her body for the sake of winning, becoming a star, and being financially successful. The more the technical approach to sport is applied to each and every facet, the more performance is improved and the less anything can be left to chance and spontaneity.

It is not surprising, therefore, that the more totalitarian a regime, the more it can bring to bear totalitarian methods to the organization of sport. Success on the world stage is taken as a measure of the excellence of a regime. It is a fundamental component of the propaganda aimed at its own people and at other nations. The authority of such regimes facilitates a totalitarian approach to sport, which translates into a competitive advantage.

Sports events are an important source of entertainment as well as a public celebration of nationalism. They also help to diffuse the orientation and values of industrial production, and at the same time complement its effects on human life outside of the workplace. When people come home from work, they are now more nervously than physically fatigued. The suppression of the self in work is not easily reversed when people come home. There is a powerful tendency to fetch a drink, flop into a chair, and switch on the television set. Watching sport on television is essentially passive and involves only the most superficial contact with fellow viewers. Many children's toys prepare people for this passivity in leisure time: battery-driven toys encourage children to watch them do things and require only minimal intervention, as opposed to traditional toys that had to be played with. Whether as athletes or spectators, modern sport cannot be fully lived, and thus extends the patterns first established in the rationalization and mechanization of work.

8.3 Education

Modern systems of mass education organized by the state have been and continue to be shaped by two primary forces. We have encountered the first one in the discussion of the phenomenon of knowing and doing separating from experience. It created a need for mass education to produce ever-larger numbers of technical experts, first in the domain of industrial production and later in any sphere of the way of life of a society organized and reorganized on the basis of the technical approach to life. This system classifies the members of each new generation according to their ability to participate in the advancement and application of knowledge separated from experience, streams them, and educates them accordingly, with the primary focus being on the vital necessity of producing enough experts to evolve every aspect of society according to the technical approach to life. This gives modern education its technical dimension.

The second primary factor shaping educational systems arises directly from the influence the technical approach to life is having on society, to which its members have to adapt. From our study of the first two phases of industrialization, it is clear that this adaptation has three components. First, human beings had to adapt to the growing demands of work in industry or other large organizations. Cultural and historical developments did not prepare them for these demands, particularly once the role of culture itself became undermined by knowing and doing separating from experience and, later, by the technical approach to life. Second, the social milieu underwent profound changes, from one centred primarily on the extended family, and the communities they helped to constitute, to one largely centred on the individual receiving much less social support as a result of weakening relations and groups in a mass society. This too was exacerbated by the weakening role of culture. Third, the members of society had to adapt to an entirely new physical milieu, which changed from a largely rural and natural one, via an urban industrial one, to the present technical life-milieu (which we will discuss later). The required adaptations placed new burdens on children, parents, teachers, and educational bureaucracies, making the process of socialization both more difficult and complex.

Around the middle of the twentieth century, Jacques Ellul, in his landmark work *La technique*,[4] argued that these adaptations could not be made by individuals in a mass society receiving only a weakened social support and, I would add, a declining effectiveness of their cultural resources. A broad range of human techniques was required. These performed the essential function of helping the individual to adapt to and cope with the many changes engulfing life. Ellul's description of the need for educational techniques was further developed in his analysis of human existence in a mass society, set out in *Propaganda* and in *The Political Illusion*.[5] I will complement his arguments from my own perspective.

Imagine the situation in an industrializing society in which an experienced teacher moved from a school in the countryside to one in an urban industrial setting. In the former, the teacher knew her students, their family backgrounds, and the roles these families played in the community, and the community knew the teacher. There was a broad socio-cultural context in which the teacher-student relationship was embedded. In the emerging mass society, more and more of this was lost as people hardly knew one another. The teacher-student relationship must function with minimal socio-cultural context. Moreover, the educational process was less and less guided by a value framework of strongly shared vernacular values, and was plagued by a growing conflict between performance and context values. The teacher's prior experience could contribute little to this highly impersonal situation, and she turned to the school board for help. Initially, the school board could offer little help, but this situation began to change when society started to train educational specialists.

It is probably difficult for many of us to appreciate the difficulty the teacher faced because fewer and fewer of us have taught in and been a part of a traditional society. Of course, the problem did not stop at the relationship between teachers and students. The school board was hardly in a better situation than the teacher. It had to set policy without the benefit of knowing many members of the community and without a set of cohesive traditional values with which all parties involved in education had grown up. Their structures of experience no longer provided them with an effective mental map. There existed little relevant prior experience so that there were fewer cultural resources to fall back on. This was the dilemma posed by the transition from a traditional culture based on human (context) values to one based on technological (performance) values. The separation of knowing and doing from experience further aggravated this situation.

At first teachers did the best they could with little assistance. The result was the well-known traditional school. Ellul's description resonates with the experiences many of us have had:

> All of us who were adult in 1950 in France preserved a vivid memory of dismal schools where teachers were enemies and punishment was a constant menace; of narrow, barred windows, gloomy brown walls, and uncomfortable benches hollowed out by generations of bored students. The smell of sour milk, dirty smocks and snot-nosed kids made a unique impression that a young instructor would never forget. We well remember books without illustrations, incomprehensible lessons learned by rote, discipline, and boredom. We had a healthy fear of the masters, upon whom we played our tricks. We feared some of our fellow pupils too, especially those who sat behind us, against whom we were unarmed. The students were divided into the weak and the strong, much like an embryonic political structure where the weak quickly band together. There was pitiless competition in respect to studies, marks, and places. Categories were simple then: work was a curse, the school was a hostile world, and the greater society outside its walls seemed to be the same. All superiors were enemies. There were the snivellers who wanted only to get by, and the tough characters strong enough to dispense with the kind of success school life offered. All the rest were either cowed or rebellious, according to their natures. These were the ancient and familiar categories of school life which were suddenly overthrown by the extension of a series of techniques that we call *techniques de l'école nouvelle* – progressive education.[6]

This situation was clearly not to anyone's liking and invited the technical approach to life. How could education be made 'better'? How could teaching methods be improved? How could children best be classified in terms of the kind of education that would suit them best? How could school boards more effectively administer education? What role should parents play in schools? How could we best set national standards for education and ensure a well-educated workforce? These kinds of questions guided many studies, which quickly began to take on the form described as the technical approach to life. It was the obvious response to situations in which culture was no longer a guide. Besides, the cultural unity had become dominated by performance values, which exclude context and which do not extend the traditional role of culture but undermine it.

By way of illustration, take one of the problems referred to above, namely, how could teachers and school boards best cope in a situation where they did not know the students? It was not long before someone recognized that this situation had been dealt with elsewhere in society. The army provided a social setting in which there was greater freedom to do human and social experimentation, particularly when military necessity dictated it. Mass armies made possible by conscription created a unique setting because recruits were, to a great extent, separated from the connectedness of their civilian life to be reconnected into an entirely different order. They had to develop a new identity, personality, and social ties in a milieu that was simpler than that of society and could more easily be studied and modelled. It was in this setting that techniques were developed to classify recruits according to their suitability for various tasks and according to their intelligence.[7] These techniques were readily extended to the sphere of education, to test the intelligence and scholastic aptitude of children. They were soon joined by a whole arsenal of other techniques, including those of grading, vocational guidance, and group dynamics. They provided all parties with technical information that replaced, as best as possible, the knowledge embedded in experience that used to guide education. These and other techniques gradually transformed education into a process promoting conformity to the necessities imposed by a technical world. I know this goes contrary to our best intentions of ensuring the happiness of children and the unfolding of their full potential, but the structural economic difficulties of our time, growing global competition, and free trade are forcing education to face technical necessities. It has to help young people adapt to new circumstances, such as the growing competition for a dwindling number of jobs. In a modern society, this means that education must be technical.

Making decisions and shaping policy in the educational sphere depends on many techniques. Survey methodology is but one of these, and it will be briefly discussed because of its importance, not only in the sphere of education but in almost every other sphere of society, including business marketing, politics, health, and the mass media. It proceeds as follows. Some hypotheses about a particular problem or situation are formulated, to be explored in a set of questions used in a survey. Large self-administered surveys are usually designed in such a way that responses can be converted into variables so that correlations between these variables can be analysed by means of complex statistical multivariant procedures. The results are then used to guide decision and policy-making. This approach can be used to market a car, influence voters, maximize television audiences, or improve the diagnosis of learning disabilities. The method takes the form of what we have called the technical approach to life.

The process, once again, separates knowing and doing from experience. Many social scientists are trained in such a way that they appear to do little else but find a representative sample to administer a survey when confronted with a problem. It is a rare occasion, indeed, when they and the fieldworkers they employ have a direct experience or knowledge of the situation being studied. In the case of a contract, there is usually a minimal awareness of the organizational context of the agency paying for the study. The people being surveyed that are perceived to have a 'problem,' or whose behaviour is to be modified, are asked to respond to hypothetical situations embodied in the survey questions. It is assumed that the responses are indicative of how they would actually behave. Respondents who are familiar with the situation being surveyed often find that the standard responses do not describe what they know about it, the possibility of which is sometimes acknowledged by the category of 'other.'

Many social scientists appear to be losing a sense of when survey methodology is appropriate and when it is not. They are often divided into two camps: those who emphasize the conceptual approach, and those who do the number crunching. Rarely do they cooperate. Consequently, very few surveys, if any, are founded on a thoroughly researched understanding of what is to be surveyed. The preoccupation is with developing an adequate report by the deadline specified in the contract.[8] That the researchers are not experientially involved through their own lives and are thus minimally affected by whatever decisions are supported by the survey generally leads to their preoccupation with efficiently getting the task done.

Those who are directly involved in the situation and thus affected by the consequences may be frustrated by the survey. They wonder why certain questions are asked and not others, and suspect a lack of awareness of the situation by those who design and administer it. They may wonder who is administering the survey and why. Why is it being studied the way it is? What are they trying to find out? What will be the implications of responding to the survey?

Asking these kinds of questions could also help those who carry out the survey to understand its broad context and its implications. However, there is rarely time. An issue has come up, and an action needs to be taken. There is no time to conceptually study the situation. Survey questions have to be dreamt up quickly or borrowed from other surveys. Under these conditions, there is a strong temptation to second-guess the kinds of answers a sponsor is looking for. Yet the questions on the survey constitute an implicit model of the situation and its key aspects. Stage one in the technical approach is thus frequently dealt with in a minimal fashion, and this affects the next three stages. Respondents often have difficulty recognizing any of the choices provided as grounded in their experience. They react to relatively arbitrary questions in terms of preconceived forced choices as a technical form of the behaviourist's stimulus-response paradigm.

The methodology becomes even more problematic when it is aimed at some intimate aspect of people's lives. Respondents are assumed to answer truthfully because they are presumed to have a vested interest in helping to create a 'better' world. This assumption is seriously flawed. People may exaggerate their sex life, underplay deviant behaviour, or avoid painful memories. When minorities are surveyed, they may be suspicious of the whole process and modify their answers according to their perception of the situation and how it may potentially affect them. Coders are always tempted to modify a response if it appears inconsistent with some of the others or to fill in unanswered questions according to previously encountered patterns, thus contaminating the data.[9]

Underneath all these problems lies an even more fundamental one. Human life as lived and interpreted on the basis of a culture assigns everything a multidimensional meaning and value, some of which may be contradictory, in a manner that may be referred to as dialectical. Hence, what do the stimuli and responses really mean? Despite the well-recognized risks of 'garbage in, garbage out,' this methodology is a dominant one in the social sciences, converting many of their activities into social techniques. Here again we encounter the difference between culture and rationality. The assumption is that the data obtained by a survey must be logically consistent. There should be no contradictions. This more than anything else forces the whole procedure out of the domain of experience and culture into an abstract, logical world separated from human life and society. Some of these difficulties might be overcome if, instead of doing research *on* people, research is done *with* people.

To sum up, survey methodology used to explore particular situations for the purpose of making them 'better' exactly fits the technical approach to life described earlier. It begins with studying a particular area for a particular purpose that can usually be expressed in terms of one or more performance values. The responses to the survey are then compiled into a set of statistical correlations, which are next examined to formulate recommendations that, if implemented, are expected to yield the desired result. The development of systems of mass education owe a great deal to this technical approach. However, the sphere of education, like most other human endeavours, is not clear-cut and suited to 'yes,' 'no,' or 'no opinion' types of answers. Especially for an issue such as education, which is complex, controversial, and value-laden, this research approach falls far short of creating adequate decision-support models.

Similar difficulties occur when the comparative approach is guided by performance values. Suppose group learning is compared to the traditional approach to classroom teaching in order to determine which of the two is the most effective. Some indicators of educational 'efficiency' would have to be developed. These constitute an implicit model of classroom learning, which could then be analysed and compared in order to determine the better pedagogical approach. The pattern is a variation of the technical approach to life.

Very few of the educational strategies and policies devised and implemented by huge bureaucracies appear to be guided by clear thinking about the almost impossible task expected of modern schools, given the societal context in which they operate. If my analysis is on the right track, the implications for growing up in a modern society are profound. There is little social support in a mass society, with many weak and dislocated families. A profound contradiction is experienced between the human (i.e., context) values taught and the technological (i.e., performance) values experienced. The support derived from culture to help make sense of and live in the world has been considerably weakened, and the void is not being filled by the technical approach to life, as we shall see. Is it surprising, therefore, that teenagers attempt to set up their own 'culture' and create their own 'society' through a heavy reliance on their peers? At least it permits them to develop some fragile moral basis, which makes the normal behaviour of the members of the peer group normative. The encounter of knowledge separated from experience also poses a difficulty because it introduces 'worlds' that are frequently counter-intuitive and counter-experiential. We have already discussed the example of learning Newtonian physics, which runs counter to much of the metaconscious knowledge about the physical world developed in young people's structures of experience in the course of growing up. The adults they encounter are themselves members of a mass society with minimal social support. Together these conditions make growing up exceedingly difficult.[10] Part of the existential void is filled in by the integration propaganda of television, and the schools are left to deal with the remainder.[11]

The above two factors shaping the evolution of modern education systems combine in the influence various educational technologies are having. The computer is an obvious example. For many educators, the computer is the answer to many problems that plague modern education. Unfortunately, there is much more preoccupation with the answer than whatever questions might have led to it. Such questions should be founded on a comprehensive analysis of modern education and the socio-cultural setting in which it takes place.

The fervour for the computer in modern education is reminiscent of that for educational television some decades ago.[12] There were almost no limits to what this technology would permit us to do. All one had to do was to use one's imagination. For example, students and teachers across the world would soon have access to videotapes of the best lectures given by the most renowned experts. Educational television would thus diminish, if not eradicate, the gap between the quality of education in the northern and southern hemispheres, and tremendously improve the standards of the lesser universities of all nations and peoples. The same could be done on the elementary and secondary education levels. Since knowledge is power, it does not take a lot of imagination to understand how people felt that the world could be made a better place through educational television. Unfortunately, the real questions of education could not be answered by educational television. There is much more to education than the distribution of information. Educational television, because of the cultural unity of the time, was highly overvalued, and thus could not deliver on the widespread expectations. The debates over educational television were a first signpost that education was beginning to be rethought in the context of an emerging new 'human nature.' *Homo economicus* was being replaced by *homo informaticus* under the influence of technique. From this point on, educational strategies had an uncanny resemblance to adaptations to accommodate the latest information technologies that happened to come along, as opposed to being adaptations to the evolving needs of human life and society.[13]

The use of computers in education will almost certainly encounter the same fate. Far too little attention is being given to the fact that computers have a negative influence on the mental images we have of what it is to be human, which, in turn, affects social relations that, in turn, weave the fabric of a society.[14] Is education gradually mimicking the internal workings of a computer? What is called 'research' in high schools and a great deal of undergraduate education is beginning to resemble information retrieval and the organization of that information in a logical and coherent fashion. Questions about the meaning and value of that information, expressing its significance for human life and society, appear superfluous. Some students are clearly irritated by such questions. 'Did I not obtain all the latest information on the subject, and arrange it into a good paper? You obviously agreed, since I obtained a good grade. Why do you want to know how I felt about it and what I thought it meant? I really do not see what you're driving at. Please excuse me because I have another paper to write.' As a result, information manipulated by human beings increasingly resembles what happens within a machine. It is disconnected from the lives of those who retrieve it, organize it, and assess it in terms of its relevance to a goal related to performance values. This disconnectedness bars the way of information becoming knowledge and of knowledge contributing to wisdom.

It is essential to recognize that modern education takes place in the context of a mass society. The student-teacher relationship tends to become impersonal and disconnected from individual and collective life. We have tried to make the best of it by devising pedagogical techniques, techniques of human relations and group dynamics, techniques for measuring performance including grading, techniques of administration, and others. Experts determine the material to be covered in a particular subject; educational psychologists devise the best strategy for teaching it; the teachers implement these from the guidelines and manuals that accompany the texts. Rarely do the different parties know one another, so that genuine feedback about what is and is not achieved is almost absent. To compensate for this, they must further rely on performance values, thus reinforcing the whole process. The institutional aspects of education are similarly affected. Teachers receive instructions on how best to deal with concerned parents, principals on how to deal with the media, and so on. This is not to deny that there are genuine personal relations, and that they play an important role. However, they function in spite of an educational bureaucracy that operates like any other bureaucracy, whose relations are not structured in terms of experience and culture but in terms of the technical approach to life.

8.4 War

War has not always been in the grip of performance values. Throughout much of history, it was largely conducted on the basis of experience and culture. In many cultures, the conduct of war involved lining up two armies against each other to charge at the appropriate moment. Other cultures practised battle strategies. In Western civilization, techniques of warfare were first developed by the Spartans and later by the Romans.[15] Soldiers were trained to carry out a certain number of standard procedures upon orders. In all cases, the scale of warfare was very small by modern standards since the state had neither the means nor the resources.

The rise of the nation state and the Industrial Revolution dramatically altered the scale and nature of war.[16] The state had at its disposal a growing tax base for the financing of war. It acquired the powers to conscript all young men who could fight. Backed by industrial might, these men could be outfitted with uniforms and guns, be transported, fed, and trained. The result was mass warfare that could no longer be left to chance, experience, and the whims of generals. Techniques of all kinds had to be developed.

I will limit myself to one example. Consider the task of sinking enemy submarines in the Second World War. Imagine a British reconnaissance plane looking for German submarines in the Atlantic. The characteristics of the planes and submarines involved were well known. Assuming various kinds of weather conditions and their effects on visibility and other parameters, it is possible to construct some simple models. The moment a pilot spots a submarine, he alters the course of his plane toward it. It is reasonable to assume that the crew of the submarine spots the plane roughly at the same time and begins to dive. The depth it will reach can be calculated from the average time in which the plane can be expected to reach the submarine under various scenarios. From the characteristics of the submarine, obtained from espionage, the depth that can be reached during that time can be estimated. By setting the depth charges to go off at a pressure level that corresponds to that depth, their efficiency can be greatly increased. Once again we encounter the technical approach, where an attempt is made to make a particular operation more effective by constructing a model, analysing it, and modifying the activity accordingly. This was the beginning of operations research, which, after the war, became applied to a great many other activities. Today this has been taken many steps further. Huge computer systems are used to run complex models of battle scenarios to examine the effectiveness of strategies and counter-strategies. Such simulations guide all modern military activities in both peacetime and war. These, in turn, require other techniques for the organization of the flows of men, equipment, supplies, fuel, and information. The 'one best way' must be determined for every detail, since so much depends on it.

The application of science and modern engineering techniques to improve the power and efficiency of weapons systems has created a situation in which the traditional distinctions between defensive and offensive weapons become almost meaningless because their all-out use affects friend and foe alike. It has also facilitated the automation of war, where a few highly trained technicians can do the work that used to be performed by hundreds of thousands of soldiers. Conscription becomes less essential than the contributions citizens make to the resource base through their taxes. Military strategists can no longer rely on experience since many new weapons systems cannot be fully tested because of their awesome power. They can only gather 'experience' from computer simulations. After the collapse of the former Soviet Union, the development of weapons systems continued almost unabated, pointing to the deeper underlying structural problem of the military-industrial complex being an integral part of most industrially advanced societies.[17] In the United States in particular, defence policy appears to be inseparable from, and in some areas the only instance of, a science policy. There is no real technology policy.[18] In daily life, we find the reflection of this reality in the many video games that simulate warlike scenarios. In this way, war and violence become an integral part of life in contemporary mass societies.[19] How different might the world be if so-called defence expenditures were directed to the prevention of war by dealing with situations that, if left to deteriorate, almost inevitably lead to armed conflicts? Even the most casual inspection of some of the statistics on military expenditures provides a numbing picture of how deeply the preparation for technical warfare has woven itself into many spheres of contemporary life.[20] The negative effects of the military-industrial complex on the civilian economy exhibit the high price societies pay for their dependence on war. Much of the frontier of technique lies within the military-industrial complex.[21] This fact will take on a whole new importance with the growing implementation of free trade agreements.

8.5 Commercial and Political Advertising

The traditional craftsman knew his clients and the community to which they all belonged. The modern relationship between producer and consumer is of an entirely different character. During the first phase of industrialization, it was merely depersonalized. During the second phase, producers could no longer rely on the market, and consumer demand had to be managed through advertising and other techniques. There simply was no other way in which the system could be made to work, and even then it suffered from crises of overproduction and underconsumption. Once again, we encounter the technical approach. Consumer attitudes are monitored through surveys and market studies.[22] Concepts for advertisements and symbols to be bonded to a product are carefully tested, again through surveys and T-groups. Once completed, the advertisement is carefully targeted at an audience through the best available means. This is facilitated by the fact that newspapers, magazines, radio, and television stations also monitor their clientele. For example, television stations determine the socioeconomic and demographic profiles of the audiences watching the programs to help maximize viewer ratings and advertising revenues.[23]

In the previous chapter, I showed that the many changes required to make the fullest possible use of knowledge separated from experience undermine democracy by creating a revised sequence. Governments have little choice but to monitor public opinion and to intervene when required through political advertising. This fundamentally transforms the relationship between politicians and citizens via the mediation of various techniques. What is important is not the personality of the politician, his ability to campaign person-to-person or to have a good presence in front of an audience of citizens, but the appropriate public image in the media. Such images can be crafted, by means of a variety of techniques, to the point that most citizens will have little knowledge of the person behind the image.

It should be noted that television decontextualizes and recontextualizes what is seen and heard by the television audience. The process begins with the camera. Despite cognitive psychology, there is a rather important difference between the way a camera functions and how human visual perception takes place. The detail of concern here is the fact that human perception is not framed with an 'edge' that marks the perimeter of our field of vision. We know from daily-life experience that no such edge exists. All we have to do is shift our focus. Hence, such an edge could have no meaning, and visual experience is interpreted accordingly. A camera, on the contrary, imposes an edge related to the field of vision of the lens and the frame size. Since our visual perception, based on experience and culture, seeks to interpret everything in the broadest possible context, the way this edge is placed by a photographer or a cameraperson can significantly change the context of what is perceived and hence its meaning and value. Similarly, each frame is carefully contextualized by others, and further contextualized by sound.

Manipulating the process of decontextualization and recontextualization can create effects, impressions, and moods that can vary significantly, with a considerable impact on what is visually experienced. The spin doctors know this very well, and have perfected a variety of techniques to craft the right image of politicians, media stars, and others who depend significantly on the mass media.[24] What is true for people is also true for issues. When everything of importance in a modern society passes through television, it inevitably places the visual dimension of that issue in the foreground and relegates everything else to the background. In other words, there is a rearrangement of the senses that 'works' very well for some issues and not at all for others. Everything has to be translated into images that 'work' on television. We have become a society of the spectacle and the spectacular. Language, and thus experience and culture in the traditional sense, has become devalued.[25]

In a general sense, everything has to be 'marketed' through the media by processes of decontextualization and recontextualization guided by the desire to create the greatest possible effect. This is superimposed upon the technical structure of the media, which is clearly not neutral in the sense that some aspects of human experience and culture are easily transmitted while others must be 'repackaged' to suit the medium. We conduct our lives in a bath of images technically crafted to create a variety of effects. To pretend that these images represent reality to us, and permit us to remain in touch with our world, is an illusion that is easily shattered when a person has the opportunity to live through an actual event and then, later, watch it on television. The process of knowledge separating from experience has reached the level of much of our daily-life awareness of the world.

8.6 Organization

Modern organizations' dependence on knowing and doing separated from experience has reached the point where managers can no longer rely on their own experience and knowledge and that of their employees to get things done. They turn to organizational techniques that break the mission statement down into its functional components, to be 'wired' by means of an organization chart, much as electronic components are wired with a circuit diagram. Functions can then be translated into job descriptions that are connected through reporting lines. For these organizations to function, it is imperative that people do not relate to one another on the basis of experience and culture but on the basis of the technically determined structure. Informal networks do occur as a result of people knowing one another outside of this structure, and their limited function can be useful when formal lines of communication break down. However, such networks do not account for the daily-life running of modern large organizations.

Organizations that utilize a great deal of technology must also structure themselves in such a way as to take maximum advantage of the capabilities of the personnel and the machines. The use of the most expensive machines must be carefully scheduled for them to produce their highest yield. Nothing is exempt from this technical development. The shrinking health budgets and increasingly complex and expensive technological infrastructures of modern large hospitals necessitate that organizational techniques be applied to all their operations. Even dying does not escape. In a mass society, a modern hospital must cope with more patients who face death alone. What can be done? Technique provides the answer. The stages of dying are carefully studied. The results are examined in terms of how patients can best be supported as they move through these stages. The findings are used to train volunteers on how to best accompany these patients as they face their final hours.[26] For now, I simply wish to point out that this accompanying has little in common with the family gathering around a dying member. Our funeral homes, too, are structured on the basis of a variety of techniques that leave no detail to chance, from the selection of a casket to the burial.[27]

Everything must be organized. From the flow of telephone messages in a communications network, the arrangement of goods in a store, the flows of doctors, nurses, and patients through an operating theatre, the movement of passengers and luggage in an airport terminal, to the flow of traffic through an intersection – all these have been analysed and reanalysed, organized and reorganized on the basis of the technical approach, to produce the best possible results.

Even religion has not escaped. Imagine the following well-intentioned development. The members of a local parish find the pastor's Sunday morning sermons particularly helpful for their lives. Soon the building is too small to hold all those who wish to come, and several parishioners urge the preacher to go on television so that many more may be helped. The idea finds acceptance and new problems arise. Bills must be paid and scores of letters must be answered. To carry out these and other activities an organization must be put into place. How can the preacher possibly answer hundreds of letters per week? Predetermined paragraphs must be written that can be selected by volunteers or paid staff to help constitute the replies, and this can only be achieved by means of computers. Parishioners notice that their pastor does not quite make the same impression on television as she does in person, and advice is sought on how to deal with the situation. It is inevitable that things will go out of control. The means employed have fundamentally transformed the entire situation. There exists no common denominator between the parish gathered together on Sunday morning and the television audience. Will the same means used for commercial and political advertising now suddenly respect one's neighbour as oneself, or will the means employed inevitably undermine the original ends?

8.7 Agriculture

Traditional agricultural methods selected varieties of plant species that were particularly well suited to local soil and climatic conditions. Over time, these were further adapted to the local context. In this way plants thrived and the best possible crop was produced. Modern agricultural methods are the exact opposite in their approach. A particular variety of species is identified as having the greatest desired output per plant, and cultivated in large monocultures with a minimum regard for context. Such monocultures must be defended by pesticides and herbicides, and soils must be adapted by fertilizers. Many crops today are grown from only a handful of varieties across the world, while traditional agriculture used hundreds of varieties. Traditional methods worked *with* nature, while contemporary ones work largely *against* nature.

From the beginning, industrialization had an enormous influence on agriculture. An industrialized urban area makes many additional demands on local ecosystems over and above those made by the people living on the land. Such demands are incompatible with the ways agriculture was traditionally organized, based as these were on people mostly growing for their own needs and bartering modest surpluses for those things they could not make themselves. These surpluses now had to be vastly expanded to meet the needs of the people living in the industrial centres. In addition, local ecosystems had to supply all inputs required to construct and maintain these urban centres as well as the raw materials for production, and to absorb all outputs of waste. Consequently, pressures rapidly built to reorganize agriculture to suit these demands, thereby converting the areas surrounding industrial centres into their hinterlands.

The consequences were many, but we will focus on the need to increase the productivity of agriculture. As this happened, more food could be produced with fewer people, and the rest had no choice but to migrate to the industrial centres to seek work. Those who could not be employed there essentially lost their place in society, with many deciding to emigrate to the colonies. Estimates for the period between 1846 and 1932 are as high as 50 million Europeans, with nearly half of them going to the United States.[28] The pattern is well known: colonization globalized the ecological footprint of industrializing societies to provide all the necessary inputs that local ecosystems could not supply, and provided a new home for the people who could no longer find a place in their own society. Traditional agricultural systems abroad were forcibly redirected to help supply the needs of industrializing societies. Local peasants were in one way or another driven from the best land to make way for agricultural systems required to support the ecological footprints of industrializing societies. Revenues paid for imports of industrial goods desired by the new settlers. Large plantations that grew crops for export proliferated, compelling local peasants to survive on marginal land.

This pattern took on a unique form in the United States, which subsequently spread to the entire globe as the basis for modern agribusiness. The context was a severe shortage of farm labour but an abundance of land lived on mostly by hunting and food-gathering indigenous people. Capital was substituted for labour, and the invention of interchangeable parts made this practical even in remote areas since no highly skilled craft labour was required to repair farm machinery. As the size and efficiency of this equipment increased, ever-larger parcels of land had to be planted with the same crop to make the ongoing substitution of capital for labour not only possible but also profitable. The traditional highly diversified family farm could not survive under agricultural industrialization. Too many different machines were required without the possibility of using each one to the fullest extent possible. Farms became more and more specialized in one or more monocultures.

Inserting monocultures into local ecosystems created large holes in their food-webs. The food supply of some insects and animals increased dramatically while those of others diminished, destroying the natural mechanisms of population control. As the regulatory services of local ecosystems were undermined, large quantities of pesticides and herbicides had to be substituted. At the same time, monocultures systematically depleted soils of the same nutrients year after year, thereby disturbing many soil cycles. The resulting 'mining' of the soils for fixed sets of nutrients needed to be offset by the use of fertilizers. Hence, the substitution of capital for labour that began with machines had to be expanded to include inputs of fertilizers, pesticides, and herbicides, all requiring significant quantities of energy for their production and use, estimated at 1400 litres of oil per person per year in the United States.[29] This agricultural system also required an ever-greater concentration of land ownership: it is estimated that 4 per cent of large landowners control half the world's crop land today.[30] Such concentration of ownership creates a difficult problem for every economy that cannot absorb rural people into the industrial or service sectors of their economies. Especially for the Third World, which will bear the brunt of the expected doubling of the global population, this amounts to a recipe for economic and political turmoil.

Today, the substitution of capital for labour is estimated to have reached a point where it takes ten times more capital to create a job in agriculture than in industry.[31] The growing capital intensity required that the technical approach to life be used to make every aspect of agribusiness as efficient as possible. Certain crop species were selected to obtain the greatest possible yield. These hybrid high-yield species produced the Green Revolution, but it is important to understand exactly what this means. The efficiency of photosynthesis could not be increased, with the result that the overall biomass did not change. What this 'revolution' did was to shift biomass from the non-edible to the edible portion of plants, but this came with the price tag of greatly weakening the plant and reducing its resistance to pests and weeds.[32] As more pesticides and herbicides were required, the capital intensity and the negative environmental consequences increased further. Additional interventions were needed in the form of biotechnology to genetically engineer plants to become more resistant as part of a chain-reaction-like process, not unlike the one we examined in industry. Like any other human activity, these interventions will produce both desired and undesired effects; and with minimal consideration being given to context, we can expect genetic pollution to become a major issue in the present century.

The food system based on substituting capital for labour, largely pioneered in the United States, is ill-adapted to the modern world, where we face exactly the opposite situation: too many people and too little land. The problem is analogous to the one encountered with the Fordist-Taylorist system, which in its modern form is equally ill-adapted to the conditions of the contemporary world.

In agribusiness we find an ever-smaller number of super-farms producing an ever-larger share of most crops. This results in a food system with three primary components. The first is the industrial sector, which manufactures the inputs of machinery, seed, fertilizer, pesticides, herbicides, and oil. The second component is the super-farms. The third encompasses the global systems operated by large transnational companies that process, package, and distribute the food. All three components have been made as efficient, productive, and cost-effective as possible by means of the technical approach. This has fundamentally transformed the nature of food. Processing living material with machines requires a significant adaptation.[33] It must be suitable for packaging and global distribution, necessitating a relatively long shelf life and an attractive appearance. Food has become adapted to a highly technical system that selects those features of plants and animals that allow further increases in efficiency. As agriculture organized and reorganized on the basis of technique continues to grow, it threatens the livelihood of many Third World farmers,[34] undermines the health of local ecosystems, and negatively affects what food is all about, namely, nutrition.[35] The third component of the system relies heavily on advertising, as it persuades consumers to want what the system can efficiently produce. Any critique that much modern food is unhealthy is met by public relations techniques aimed at undermining the credibility of any expert having the courage to speak out.[36] In many nations, consumers are not permitted to know whether their food has been genetically modified.[37] Agribusiness is one of the components of modern technique having one of the worst 'signal-to-noise' ratios of desired to undesired effects. The present situation is further aggravated by the patenting of life forms and the resulting biopiracy, the development and sale of 'terminator seeds,' the effects of free trade on local food production, the poorly regulated growth of biotechnology, and the extreme vulnerability of the global food supply because of its dependence on a few species.[38]

8.8 Living with the Technical Approach to Life

The technical approach to life is not culturally neutral, because culture as symbolization is based on contextualizing everything in relation to everything else. This, as we have seen, became very difficult during the first two phases of industrialization because the chain-reaction-like process of the transformation of both technology and society made the context of everything more turbulent and drove a wedge between the technology-based connectedness and the culture-based connectedness of societies. This made it more difficult to assign a meaning and value to anything in relation to everything else. Under these circumstances, the obvious response was to make any human activity as good as possible in itself and to evaluate it on its own merit. This led to the rise of performance values, their dominance over cultural values, and the technical approach to life. As the technical approach permeated a society, pushing back the role of culture, the overall result was an individual and collective human life out of context. This will be illustrated by means of the following seven consequences, of which the first four operate on the level of an individual human activity evolved on the basis of the technical approach, and the others on the macro-level of individual and collective human life. Even when the consequences (for human life, society, or the biosphere) of applying the technical approach to an individual activity are small and would be neglected in any technology assessment or environmental-assessment process, their being repeated over and over again in the fabric of human lives and society creates, as I will demonstrate in the next chapter, all the new and significant phenomena in the industrially advanced nations in the second half of the twentieth century.

The first consequence of the technical approach is the separation of knowing and doing from experience and culture, thus transferring the evolution of whatever is made 'better' from the domain of culture to a growing technical domain. Technical specialists no longer deal with anything according to their culture (i.e., according to its meaning and value, symbolizing its place and significance in human life). Their technical intervention transfers everything from the domain of sense into another that lies entirely outside of it: the technical domain of non-sense. This is inherent in the structure of the technical approach. It begins by abstracting whatever is to be made 'better' from its context. Two implications follow immediately: externally, the context is re-placed by the desired outputs and the requisite inputs; while, inter-nally, whatever is not directly or indirectly relevant to the transformation of these inputs into outputs is not studied. As a result, the context is limited to the local technique-based connectedness. Everything else is thus excluded from the process of optimization and the application of the findings to achieve the desired results. Nowhere in this technical knowing and doing is there any room for considering the relevant aspects of the culture-based connectedness of what is to be made 'bet-ter' and how it is embedded in human life, society, or the biosphere. I am simply restating in more abstract terms what was happening in the many previously discussed examples.

The second consequence of the technical approach is a fundamental change in the way people are involved in any activity that destroys its self-regulating character to be compensated for by external regulation. When knowing and doing are embedded in experience and culture, the people who know what is going on also carry out the activity and ensure that the desired results are obtained. When knowing and doing separate from experience and culture, they involve different people. Each person can no longer fully live the related experiences. Once again, I am simply restating in more abstract terms what we found in the previous examples. The people who carry out the study, build the model, optimize it, and apply the findings are not the people who actually carry out the reorganized activity. When these people face a problem or encounter an unforeseen change, they will not be able to interpret it in terms of the technical approach, to come up with the appropriate technical response. Hence, a technically reorganized activ-ity now requires knowers, doers, and controllers (from supervisors to senior managers). The activities are thus externally regulated by means of longer negative feedback loops involving several persons. In con-trast, culture-based activities mostly involving a single person are self-regulating by means of very short negative feedback loops. In such activities, the knowing, doing, and controlling are carried out by a single person, permitting full engagement. No such engagement is possible once an activity has been reorganized on the basis of the technical approach, which limits participation either to the brain-mind or to the hand. It transforms the way knowing, doing, and controlling participate in people's lives and consequently the nature of this knowing, doing and controlling. If people are knowers, then they are spectators to the doing and controlling of an activity; if they are doers, they are spectators to the knowing and controlling; if they are controllers, they are spectators to the knowing and doing. Neither the knowers, the doers, nor the controllers behave as human subjects with regard to the technical order they are building. The knowers deal with everything in terms of 'technical shadows.' The doers can no longer make sense of a growing domain where sense does not apply. The controllers must mediate between the technical domain of non-sense and the cultural domain of sense, which cannot be done in terms of the one or the other. The distinction between the human subject and the growing technical order begins to blur.

The third consequence of the technical approach is a distortion of the integrality of whatever is made 'better.' It follows directly from the fact that only the technique-based connectedness of whatever is to be made 'better' is studied, incorporated into a model, and optimized. In other words, making something 'better' in terms of performance values creates a hierarchy among the many internal aspects as to whether they are directly, indirectly, or not at all relevant to obtaining the maximum possible desired outputs from the requisite inputs. Hence, the model will represent what is to be improved in a distorted form, and this is further exacerbated by the optimization process, guided by performance values. Applying the results to the reorganization of what was originally studied will therefore inevitably distort its integrality. The overall result of applying the technical approach is to select and strengthen those aspects of a human activity directly relevant to the broader technique-based connectedness within which it functions, and to weaken and even undermine those aspects that are important for the culture-based connectedness of the human life and community of which it is a part.

The fourth consequence directly follows from the third. There is a simultaneous improvement in the compatibility between whatever is made 'better' and the technique-based connectedness within which it functions, and a degradation of the compatibility with the culture-based connectedness of individual and collective human life, and through it, with the biosphere. This is a consequence of the limitations of performance values. These are essential to ensure the effective use of scarce resources (including human 'resources') but are entirely mute on whether any improvement has in part or as a whole been gained at the expense of degrading the context of whatever has been improved, namely, human life, society, and the biosphere. Performance values are a necessary but not sufficient measure to ensure genuine improvements. Without specific attention being paid to the context implications as an integral part of making things 'better,' it is possible to not degrade the context by accident, but the odds are against it. Genuine improvements can be ensured only by a process guided by both performance and context values. In the nineteenth century this was hidden from human experience and culture by myths (in particular, those of progress, work, and happiness) that extrapolated one kind of improvement in a specific area measured in terms of particular criteria (all related to the technology-based connectedness) to the whole of human life, society, and the biosphere. These myths extrapolated the growing technology-based connectedness of society to potentially include everything, thereby masking the growing divergence between it and the culture-based connectedness of human life and society. As will become apparent in chapter 10, our present myths do exactly the same: they extrapolate the technique-based connectedness of contemporary societies to include their culture-based connectedness, thereby obscuring the enormous tensions between the two.

When a great many activities in a society are made 'better' by means of the technical approach to life, macro-level consequences also occur. We have found that the technical approach organizes a human activity to contribute as much as possible to the technique-based connectedness and then continually reorganizes it to evolve with that connectedness. It thereby converts it into a 'foreign body' within the culture-based connectedness of human life and society and its historical development, as well as within the natural order of the biosphere and its evolution. Such 'foreign bodies' would have been quickly rejected had they not linked together into a technique-based connectedness coextensive with the culture-based connectedness of human life and society. At present, we see a further extension of this pattern into the biosphere by means of biotechnology, which transforms and patents species of plants and animals into 'foreign bodies' in order to integrate them into the technique-based connectedness. This process is the exact parallel, with respect to the biosphere, of what we have been describing with respect to human life and society. Making 'foreign bodies' out of aspects of human life, culture, and the biosphere involves transforming their internal integrality into extensions of technique-based connectedness, transforming their participation in human life, society, and the biosphere into the extension and strengthening of technique-based connectedness, and transforming people fully engaged in their own lives into *homo informaticus*. As we will see in the next chapter, this involves the conversion of experience into information, knowledge into 'knowing *that*,' expertise into technical virtuosity, and wisdom into power. The expanding technique-based connectedness introduces a growing chaos into human life, society, and the biosphere, producing a variety of difficulties that quickly grow and interfere with this technique-based connectedness, at which point they can no longer be ignored. Further interventions of the technical approach are required, with the result that technique feeds on the disorder it creates. The implications for human life, society, and the biosphere are summed up in the following three macro-level consequences of the technical approach to life.

The fifth consequence of the technical approach is the reification of human life by turning it into an object for endless technical improvement. Human life in a technicized society is made up of some activities in which a person is a 'knower' and a great many others in which he or she is a 'doer,' largely controlled by others. To some extent, this has always been the case, but the rise of the phenomenon of technique has massively tipped the balance against self-regulating activities. The result has been a fundamental insecurity about activities that human beings have successfully engaged in for tens of thousands of years. For the first time, a person is not to rely on his or her own experience but on expert knowers, to a degree detrimental to the human individual. There is an excessive dependence on 'how to' books for bringing up children, communicating with one another, making love, and a host of other human activities. This situation requires that a great many activities are externally regulated and interconnected by the complex policies of large institutions – policies that are neither intelligible nor accessible to a great many people. The result is an erosion of meanings and values as well as a reinforcement of previously examined developments: the 'other-directed' personality, public opinion, and a statistical morality. It also reinforces a sense of no one being responsible. When people face problems such as a complex set of allergies, an inability to find work, a sense of being depressed most of the time, a drug problem, or a tendency to resort to violence, they often have difficulty finding the source of the problem and the responsible party. A common result is a sense of helplessness, anxiety, frustration, and discontent. The consequences of this reification and fragmentation of human life are immense, undermining, as we will see, the very possibility of a genuinely human civilization.

This reification of human life is superimposed on the ongoing problem of alienation. Since the interaction between individual and collective life on the one hand and the phenomenon of technique on the other is a reciprocal one, the question of whether human beings can influence technique more than they are influenced by it is one that should be posed on an ongoing basis. The best a person can hope for in a society with a technique-based way of life is to be an expert knower and have some control in at least one domain, and be on the receiving end of the technical approach to life in all other spheres. It would appear, therefore, that even in this most favourable situation, a person ends up serving the 'system' more than it serves the person. Once again, we are faced with a situation in which the lives of most members of society are possessed by a 'system,' much like a master possessing a slave. It also means that the interconnectedness of people's lives is greatly weakened by being reduced to a bundle of loosely connected social roles, technically structured and restructured on an ongoing basis. Books such as *The Organization Man*, *The Pyramid Climbers*, *The Minimal Self*, *The Lonely Crowd*, *Growing Up Absurd*, *The Broken Connection*, *Propaganda*, *The Culture of Narcissism*, *The Culture of Cynicism*, *The Second Self*, *Technostress*, *Experts in the Age of Systems*, *The Corporation*, and many others have touched on some of the symptoms of these developments.[39] It is as if technique-based connectedness selects those human characteristics that assure successful participation in it, suppresses those that interfere with it, and atrophies those that are irrelevant to it. It all adds up to a fundamental mutation of what it is to be human, creating a new 'human nature' that I have referred to as *homo informaticus*.

The sixth consequence of the technical approach to life is the creation of a mass society in which many activities are structured, managed, and externally regulated through large bureaucratic organizations. The technical approach to life is decisive in completing the transformation of a traditional society into a mass society. The creation of a mass society has its roots in the enclosure movements and the accompanying migration to the emerging industrial centres. These all but destroyed rural peasant life and put little in its place in the industrial towns, where people worked such long hours that there was barely enough time left for eating and sleeping. The resulting weak social fabric did strengthen somewhat as working hours were progressively reduced, but it came under new pressures as a result of knowing and doing separating from experience, the accompanying rise of mass education, and the growing phenomenon of rationality that grew into the technical approach to life. This situation negatively affected the internal integrality and external compatibility of all social wholes, including individuals, families, neighbourhoods, parishes, and unions and, through them, the entire social fabric. The result was a mass society characterized by the individual being the fundamental building block of a social fabric made up from weak relationships and groups.

These trends are greatly exacerbated by the destruction of the self-regulatory character of human activities in a society dominated by the technical approach to life. We have seen that this is the inevitable consequence of knowing and doing separating from experience, dividing any social role into the three distinct technical roles of knowing, doing, and controlling. Individuals and groups representing these three different roles usually find it very difficult to understand one another. Relations frequently turn antagonistic, making negative feedback highly ineffective.

Many human activities now require external regulation and control, and the problem does not stop there. Because these activities depend on adjacent ones, which in turn depend on others to form networks of interdependent activities, it is impossible to supervise them in isolation from each other. Supervisors must report to higher-level supervisors all the way up to middle and top management. Control requires supervision, and supervision requires management. It is the direct consequence of the technical approach and the interconnectedness of activities. Control has to be centralized somewhere in a network of technicized activities. At first, this gave rise to the traditional bureaucracies. These continued to evolve under the pressures of the technical approach. In some cases, they took on the new form of techno-structures, which continued to undergo significant changes as the result of computer and information techniques. The extreme case is that of the modern state, which has grown to an unprecedented scale regardless of political orientation and ideology. In a large measure, this was the result of the state being compelled to deal with the disordering of human life, society, and the biosphere produced by the spread of the technical ap-

proach. The only feasible solution in many cases was to further centralize the regulation and control of ever-larger networks of technicized
activities. The development of the transnational corporation and international agencies showed similar trends. This cannot simply be explained by a drive for profits, the will to dominate markets, or the
expansion of institutional power. The structural forces unleashed by
knowing and doing separating from experience and the development
of the technical approach were not foreseen and certainly not planned
by any social group for its own interest. Having said this, there is no
doubt that their impact on the cultural unity of society greatly increased
the will to power,[40] since technique is essentially power, as manifested
in the predominance of performance values over context values.

The emergence of a mass society goes hand in hand with the erosion
of the symbolic and cultural commons, and both spring from the proliferation of the technical approach in the way of life of a society. When
the integrity and external compatibility of the social entities that make
up a society are weakened under the pressure of the technical approach
to life, symbolization becomes much more difficult. As a result, meanings and values become much more shallow and may even appear
somewhat arbitrary. This affects the structure of a culture all the way
down to its deepest metaconscious cultural unity. This process has been
referred to as desymbolization. Culture-based connectedness is no longer
able to adequately support individual and collective human life and
must be supplemented by integration propaganda. However, this integration propaganda does not reinforce the culture-based connectedness
but instils a new social conformity to the technique-based connectedness of society. In this way, the dependence of human life on culture-
based connectedness is minimized. Culture becomes divided against
itself. On the one hand, it remains the accumulated experience of generations that have been members of a community, absolutized to eliminate its relative character. On the other hand, the new cultural unity
devalues this past experience and points to the future, where 'new is
better.' The youth symbolize that future, and the elders a bygone era.
Traditional values and institutions are undermined, and some disappear altogether. The other-directed person is not only fully open to
external influences but cannot live without them. Private opinions are
replaced by public opinions. Traditions are all but replaced by integration propaganda. Traditional morality is replaced with a technical (or
statistical) morality in which the normal is made normative. As a consequence of the growing technique-based connectedness, cultures are

now in the process of becoming the accumulated experiences of generations busy with technical knowing, doing, or controlling (but rarely all three). A mass society is characterized by a growing technique-based connectedness producing 'social pollution' within the culture-based connectedness of human life; at the same time, this culture-based connectedness is a source of disorder within the technical order. To survive, people desperately cling to the technique-based connectedness without any possibility of shedding their culture-based connectedness, even though the former tears apart the latter.

To sum up, there is a complementarity between three significant developments in contemporary societies, resulting directly from the proliferation of the technical approach in their ways of life. First, their social structures become those of mass societies. Second, there is a growing technical centralization of supervisory, managerial, control, and regulatory functions in very large, highly technicized structures. Third, the symbolic commons provided by their cultures is being eroded. Each of these three developments reinforces and complements the others. On the level of daily life, they manifest themselves as the disappearance of common sense and a growing dependence on lawyers, where the role of culture and community is too weak or has broken down altogether.

The seventh consequence of the technical approach to life is a further deterioration of the relationship between a society and the ecosystems on which it depends as a source for materials and energy, a sink for its wastes, a life support, and a habitat. When this fabric of interdependencies is organized and reorganized with minimal concern for context implications, the results are analogous to the ones we discussed for society. The integrality of ecosystems is undermined, fundamental cycles are forced out of their dynamic equilibrium, species are threatened to the point of extinction, and individual organisms are less healthy. Within society, the traditional fabric was replaced with that of a mass society and its coherence achieved by technical means, with profound consequences for human life. It is highly unlikely that biotechnology and nanotechnology can achieve the same thing for ecosystems and the biosphere. Some of the technical interventions that have already taken place out of necessity should serve as a warning that the consequences can be vast and unpredictable. Today, almost every living whole is in decline under the pressure of the technical approach to life. Environmental pollution is the exact parallel of the 'social pollution' created in society. Another Earth Summit, this time in South Africa, showed once again that humanity is not getting the message: the environmental crisis is as much a product of contemporary ways of life as are computers, cars, sick workers, and homeless people.

What is left after the technical approach to life diminishes the integrality and context compatibility of everything it touches? As a result, the ability of living wholes to regulate themselves with respect to their contexts is also weakened or destroyed. The fabric woven from reciprocally interacting and frequently enfolded living wholes, which constitutes the tissue of individual lives, societies, ecosystems, and the biosphere, is similarly affected. Life in general, and human life in particular, is pushed out of its natural and cultural context to be incorporated into a technical order that has none of the characteristics of what it replaces. As a civilization, we appear to be too mesmerized by our accomplishments in the domain of performance values to notice that there is no free lunch. A hefty price is being paid in the domain of context values. It may be objected that the new emerging context of the technical order is helping to create the kind of world humanity has always aspired to. Before admiring the emperor's new clothes, however, we do well to remember that human life and all other life forms have helped to create and have evolved within each other's contexts for a very long time, and it is precisely when this changes that the question of sustainability imposes itself on humanity. The consequences of pushing life out of its contexts are becoming increasingly visible and will no doubt take centre stage in the preoccupations of humanity during this century.

The ever-growing presence of technique functions as a kind of new selective pressure on cultural history and natural evolution. It concentrates on those threads of the fabric of lived relationships that make up human life, society, and the biosphere, which contain some fibres that can be strengthened by means of the technical approach to life in order to have them participate in a new fabric within the old. Performance values select these fibres according to their ability to contribute to power relations. As this new fabric of efficient and powerful relations continues to grow, it further distorts the fabric of living relations, from which it cannot detach itself. In a self-reinforcing process, things happen as though technique selects those distortions that interfere with the efficiency and power of technique-based connectedness, leading to further transformations and so on. As a result, human history and natural evolution are radically altered by the presence of technique-based connectedness. Contemporary civilization is now caught in a race. Can the growing technique-based connectedness continue to dominate and compensate for the growing chaos it is creating within human life, society, and the biosphere, or will this chaos cause their collapse, bringing technique down with them? Technique raises the question of human sustainability.

Our present situation may also be interpreted by returning to another metaphor. Before industrialization, each civilization and society created the 'one true way' of living in the world, providing the 'organizational principles' for making sense of and living in the world in terms of its unique 'cultural DNA.' There was no common denominator except that human life was symbolically constituted.

With the onset of industrialization, all this changed. There were two constraints, namely, the inability to create or destroy matter and energy, and the impossibility of allowing human life to fragment into pieces. The result was a growing technique-based connectedness, which was made liveable by a cultural mutation that subjected the culture-based connectedness to it. The cultural differences between industrializing societies rapidly diminished as they took on a common set of characteristics. Capital ruled over the new and non-cultural 'organizational principle' of the Market. However, all that this Market could organize were the many transactions between buyers and sellers within specific markets for particular goods and services. This regulation was piecemeal, involving one transaction at a time. All the consequences for these relations were external to the Market. In effect, the Market essentially regulated the technology-based connectedness and made everything else into market externalities.[41]

The decisions of buyers and sellers have numerous consequences for third parties that are not organized by the Market, which nevertheless pile up in and affect human lives, communities, societies, and ecosystems to constitute what have been called Market forces.[42] The ones that have a negative influence on human life, society, and the biosphere operate outside of the economic order, presumably without any regulation or control. This would include much of what has traditionally been 'organized' by culture. Once these cultures were integrated into the economic order by means of cultural unities based on the secular sacred of capital, this became impossible. Hence, we encounter a fundamental contradiction in the usual economic view of things. Surely it is difficult, if not impossible, to accept that much of individual and collective human life, now treated by the Market as externalities, is abandoned to chaos. A mass society is not a traditional society in the process of decline and collapse, nor is the environmental crisis without purpose in the economic order, since it has created one of its fastest-growing sectors.[43] What has gone largely unnoticed is that the invisible hand (and the invisible elbow) is now only indirectly related to the economic order, which is now subsumed under a new technical order. The technical approach to life has constituted an entirely new phenomenon, first described as rationality by Max Weber and later as technique by Jacques Ellul.[44] Technique, which rules over the technical order, produces its own positive and negative forces, which include but are not limited to those generated by what remains of the Market.

I have shown that when knowing and doing separate from experience and culture, one of the primary loci of technique, namely, the large corporation, can no longer rely on the Market to procure most of its inputs and sell its outputs at the quantities and prices required. Another locus of technique, the democratic state, can no longer rely on its citizens for direction and must resort to public relations and political advertising to manage public opinion. When these developments spread, technique becomes an 'organizational principle' for organizing and reorganizing every activity of individual and collective human life, as well as the relations with the biosphere. Technique now all but replaces the Market.[45] Whatever technique externalizes unleashes a series of tensions in human life, society, and the biosphere. The previously discussed consequences of applying the technical approach to human life constitute seven forces (analogous to Market forces), and there are many more, as I will show in the last chapter.

One particularly important force has been referred to as commoditization.[46] It represents the effect of reifying human life and society by technique, which creates privileged goods. These represent elements of human life, society, and the biosphere readily dealt with by means of the technical approach to life, thus to be turned into goods owned, produced, and globally distributed by transnational corporations. These act as a primary locus of technique in this instance. Whatever can be reified and added to technique-based connectedness is selected over everything else, thereby revealing the full meaning of 'development' and 'underdevelopment.' Whatever can be reified by technique (turned into a commodity) is 'developed,' and whatever cannot be reified remains 'underdeveloped.' For example, a mass society is a fertile soil for commoditization. Many children and adults are lonely, and it is impossible to 'produce' friends for them; but it is relatively easy to produce Barbie dolls, contacts for prospective dates, escort services, or pornography on the Net. It is much easier to sell pollution abatement equipment than to have engineers, managers, and accountants engage in pollution prevention. It is much easier to sell weapons systems than to preventively resolve potential conflicts. It is much easier to create a large organization to put a person on the moon than it is to deal with the poverty of inner cities. It is much easier to create privately owned toll roads than to build effective public transportation. It is much easier to sell pharmaceuticals than to teach people how to live well and to make sure schools, companies, and governments cooperate with such efforts. It is much easier to build new power plants than to elicit the cooperation of a community for energy-efficiency projects. It is much easier to produce goods than to make sure people can have their broken ones repaired. It is much easier to commoditize public education, health care, social security, and the like than to remedy the many difficult problems that plague these systems (many of which are the consequence of their becoming end-of-pipe solutions for problems created by technique-based connectedness). It is much easier to produce and distribute Prozac or Viagra than to deal with the root problems of contemporary life. The examples can be multiplied almost without limit as this pattern of events repeats itself everywhere under the influence of the reifying effects of technique.

Commoditization is closely linked to globalization. Commodities differ from traditional products by being produced not for a particular time, place, and culture, but for a particular technique-based connectedness. They become universal and can be produced in a few places, to be consumed all across the globe. This is true for both private and public goods. The result is a growing technical homogeneity all across the world, and a corresponding decline in cultural diversity.

Current integration propaganda plunges us into a universe of economic and political illusions. The economic world view still reigns supreme. After Keynesian economics failed in the 1970s, monetarism took over because people did not ask the following kinds of questions: Did the rule of the Market and expanding global trade not dominate the nineteenth century's industrial world? Did this not lead to the Great Depression, and was this not a monumental Market failure? Did Keynes not grapple with this situation to propose an alternative to leaving everything to the Market to find its 'natural' equilibrium? What can be 'natural' about the Market, since it is a human creation? How is it possible, then, that the events of the 1970s created an economic stampede to the 'good old days' of the rule of the Market and free trade?

Does this not represent a kind of secular economic fundamentalism every bit as dangerous as any other kind of fundamentalism? It was not long before economic fundamentalism swept Western politics. With hindsight, this is not surprising since the prevailing economic world view had become completely out of touch with reality. The theories and beliefs at its core can no longer make sense of our world. They are incapable of recognizing that new factors of production have emerged, that the conditions that justified the many assumptions on which economics was based have passed on with their time, and that almost everything that matters to human life has become external to this world view.

Politically, we have collectively wandered into the labyrinth of technique, and we now hope that somehow removing all barriers to the Market by means of globalization, free trade, and investors' rights can show us the way out of our dilemmas.[47] Surely this is a case of the wrong diagnosis. What free trade can deliver is the removal of the last barriers to technique-based connectedness and its accompanying commoditization. Whoever owns, produces, packages, and delivers the most effective techniques will have the right to impose them, regardless of the wishes and aspirations of local communities and nations. They will no longer be able to elect governments that can do very much to protect them or to maintain a public sector that delivers public education, health care, social security, occupational health and safety protection, environmental protection, standards for safe and healthy food, and so on. If we are to have these at all, it will be in commoditized form through privatization. What is new today is neither the Market (what little is left of it) nor free trade (expanding for some two hundred years), but a fundamental inability to understand what is happening in the world and to respond to it so as to ensure our ability to live meaningful and responsible lives. What we are liberating and globalizing are the very forces that make it next to impossible to develop anything except what can be commoditized by technique.

The commoditization of public-sector services by transnational corporations could be particularly problematic, since they are even less accountable to the people of democratic nations than their governments are. Democracy risks becoming an even greater shadow of its former self. Local ecosystems will be fully opened up to the expansion of the ecological footprints of transnational corporations and the societies they provide with commoditized goods and services. These are some of the effects of the invisible elbow of technique, as its invisible hand reorganizes the planet. I do not know of any political, moral, or religious tradition that would condone, let alone desire, such developments. All this so lacks a sense of the public good and our humanity that I can come to no other conclusion than that we continue to be possessed by a 'system.' What is new is that this time it is the result of the influence technique has on our awareness of ourselves and our world. Our civilization's coming of age, growing up, or having become rational are nothing but wishful thinking that denies our ongoing alienation. It is even debatable whether so-called free trade is really in the long-term interest of transnational corporations, given that they ultimately depend on stable, viable societies and healthy ecosystems. It is equally debatable whether this is in the interests of their shareholders, especially those whose pension funds support the invisible hand and elbow of technique. I assume, therefore, that the concept of technique, by offering us a different way of making sense of our lives and the world, may lead us to reinvent what it is to live a good and responsible life.

Here also we will encounter the effects of the invisible elbow on our minds, leading to secular religious commitments that dismiss all of this out of hand. Such a dismissal amounts to a denial that as people of our time, place, and culture we cannot help but be profoundly possessed by those things that really matter to our lives. A failure to critically counteract this influence frequently distorts our thinking.[48] If it turns out that the concept of technique has merit by helping us understand who we are and where we are going with modern science and technology, we will have to face the question of whether we wish to entrust our future to the invisible hand and elbow of technique by removing the few remaining barriers to them, such as by adopting free-trade agreements like the one proposed for the Americas. Does this not represent a replacement of the cult of the traditional gods with the cult of efficiency? Like our predecessors, we must not surrender our humanity to these new secular idols. A first step on the road to regaining a measure of freedom and control is to symbolize our lives and our world as best we can to discover the meaning and value of what we have created. In so doing, we distance ourselves from the meanings and values derived from a desymbolized culture under pressure from technique.

chapter 9

From Experience to Information

9.1 The Roots of the Information Explosion

The Second World War and its aftermath spurred industrial societies into making an ever-greater use of the technical approach for organizing and reorganizing their ways of life. The result was the emergence of a number of new phenomena that many observers regarded as over-shadowing industry. A debate ensued as to what was really happening to human life and society. Were societies becoming advanced industrial, post-industrial, or post-capitalist societies? Were they consumer societies, mass societies, neighbourhoods of an emerging global village, mega-machine societies, spectator societies, new industrial states, or technetronic societies?[1] The only consensus appeared to be that something radically new was emerging. With a great deal of hindsight, I will seek to demonstrate that there was no discontinuity or mutation in the evolution of industrial societies because the many new and important events were symptoms of an ongoing and much more fundamental change, the beginnings of which were described in Part Two. Technique was the outcome of knowing and doing separated from experience and culture diffusing into all spheres of life. The results were not only new kinds of corporations with different relations to the economy, government, and society at large, but also the many new and important phenomena described by these observers. These were all symptoms of the separation of knowing and doing from experience and culture. In other words, I will seek to show that technique is the root phenomenon from which all these developments grew.

By far the most important symptom of society's increasing reliance on the technical approach to life was the information explosion. No such information was required to evolve a traditional way of life on the basis of experience and culture, relying as it did on metaconscious knowledge acting as a mental map. Very little of this needed to be made explicit, and what did had no common denominator with information. Hence, a comprehensive separation of knowing and doing from experience and culture, leading to the phenomenon of technique, created an explosion of information necessary for evolving modern industrial societies. The threat of a potential bottleneck in their evolution was quickly eliminated by the appearance and rapid diffusion of an entirely new group of machines capable of processing this information by means of rapidly growing information techniques. Their widespread application generated even more information, creating a self-reinforcing cycle that affected everything else. The so-called computer and information revolution was the manifestation of a deeper and more fundamental development, namely, the growth of technique and its encroachment on culture.

There was no smooth transition. The spectacular increase in the use of information was initially anarchical with respect to technique-based connectedness. It resulted in a great many bottlenecks caused by information overload, thereby impeding the evolution of many areas of a way of life and straining their interdependence, both in relation to the technique-based connectedness and the culture-based connectedness of a society.

Consider the technical approach to life once again in terms of the quantity and quality of information it generates and utilizes. In order to study something to make it 'better,' it is abstracted from its context and represented in a model by those features that matter for a technique-based connectedness. The form of this model is next manipulated to improve it in terms of performance values. What was originally studied is then reorganized to accomplish this technically optimal connectedness. In other words, no knowledge is obtained of what is being studied and reorganized in any traditional sense of that concept. Knowledge has always been related to understanding things in the context of human life, society, and the universe through experience and culture. What we have instead is information that is decontextualized from culture-based connectedness and recontextualized in relation to technique-based connectedness. Information results from conceptualizing something in terms of technique-based connectedness. The technical approach to life is permeated by this disconnecting and reconnecting. Through it, everything becomes 'digitized' as a collection of disembodied bits and pieces, selected and manipulated according to their technical (as opposed to human) value. Experience is transformed into information, and knowledge into technically patterned information.

It is worth noting that Daniel Bell saw information purely in terms of machine functions: 'restoring, retrieval and processing of data as the basis for all economic and social exchanges.'[2] Claude Shannon founded information theory on the assumption that information can be treated like any other physical quantity such as mass and energy, thus opening it up to thermodynamics in general and the concept of entropy in particular.[3] This theory may be regarded as a technical extension of the thermodynamic constraints that dominated 'people changing technology,' as described in the first chapter. Hence, information theory contributed to the transformation of technology-based connectedness into technique-based connectedness.

The emergence of the information and computer revolution, and the many other new phenomena as consequences of knowing and doing separating from experience and culture, keep the focus of our study on 'people changing technique.' The reverse interaction leads to technique 'biting back,'[4] technique 'wounding,'[5] technique imposing a 'reverse adaptation,'[6] and a 'technical bluff.'[7]

9.2 *Homo Informaticus* and the Information Society

Many institutions required ever-larger quantities of information as they made increasing use of technical knowing and doing separated from experience. Without them, governments would have been unable to enlarge and manage the public sector in the economy, the military would have been much less effective in deploying mass armies and global weapons systems, the large corporation could not have grown to the size of the transnational with operations all across the globe, public education would have been swamped by academic and financial information, banks would have had difficulty launching credit cards and other services on the present scale, hotel chains would have been unable to keep track of reservations, and so on. The explosion of information thus became both cause and effect of the rapid development and diffusion of information machines of all kinds. The computer became the machine that gradually changed all other machines,[8] playing an instrumental role in the further evolution of the 'hardware' and the 'software' of technology, including the larger networks of techniques into which these were embedded. Bottlenecks occurred throughout these networks, creative attention was focused on these problems, and an extraordinary diversity of developments was set in motion.

In other words, another chain-reaction-like process of transformation was set in motion when a growing number of activities were organized and reorganized by means of the technical approach, thereby pushing back the role of culture. It is remarkable that almost without exception observers interpreted the situation as characterized by the many new phenomena that sprung up, their importance rivalling that of industry. Theories of what was happening made these into defining characteristics of the emerging societies. With a great deal of hindsight it is now quite evident that these attempts failed to grasp and foresee what was really happening. The only exceptions were Weber's theory of rationality and Ellul's theory of technique. According to a recent analysis, Jacques Ellul foresaw almost all the important developments of the latter half of the twentieth century.[9] I will briefly indicate how these phenomena, although genuinely new and important, were nevertheless symptoms of the growth of technique. What, after all, can be more fundamental to human life and society than a change in culture-based symbolization?

Daniel Bell recognized the growing importance of theoretical knowledge accompanied by new intellectual technologies, a large service sector in the economy, an expanded role for science in the economy, a knowledge class swelling and transforming the middle class, and a transformation in the character of human work.[10] He regarded such developments as primarily affecting the economy, social stratification, and the occupational system. The transformation of the other two 'dimensions' of society, namely, polity and culture, were interpreted by means of other 'axial principles.'[11] As a result, Bell was unable to see the more fundamental pattern that gave rise to these undeniably new and important phenomena that now, along with industry, influenced human life and society. This undoubtedly contributed to the invention of the rather meaningless concept of a post-industrial society. This concept did little more than acknowledge that something new had occurred but failed to come to terms with its true nature, other than that it came after industrial society. It overlooked the extraordinary continuity with prior developments. Everything described by Bell was rooted in, and derived from, deeper underlying patterns.

By way of example, consider the growth of the service sector in the economy. When knowledge separated from experience in the domain of industrial technology, and manufacturing techniques separated knowing, doing, and controlling, the latter two activities remained in the secondary sector, but the first could be transferred to the service sector when that was deemed to be advantageous in terms of performance values. The result was a proliferation of consulting services of all kinds. At the same time, the transformation of the extended family into the nuclear family and the progressive weakening of the latter necessitated that many of the services extended families had once rendered to society now had to be provided externally: day care, basic education, elementary health care, care for those who were ill at home, provision for people who (because of accidents) were no longer able to work, support for elderly parents, support for families who had lost one of the breadwinners, and so on. This partly explains the proliferation of private and public insurance services to deal with unemployment, long-term disability, sickness, accidental death, and old age. As more married women were admitted to the workforce, a variety of domestic activities were partly or wholly taken over by external agencies providing such services as fast food, housekeeping, nanny care, help-phone services for children, security response to burglar alarms, dog-walking, counselling services, and meals on wheels. New businesses and non-profit agencies emerged that organized and managed such services by means of a variety of new techniques. This corresponded to a component in economic growth that merely reflected a shift from unpaid work in the family setting to externally provided services, managed by the state, afforded by a second income, or donated by a voluntary agency. Many women now had an unpaid as well as a paid job. For the growing number of single parents (mostly women), essential services needed to be subsidized if serious social problems were to be avoided.

The supposed shift from products to services (and the corresponding dematerialization of the economy) is an oversimplification of an expanding drive for performance, leading to a variety of problems that were not addressed at their roots but were compensated for by adding services. As far as the rapidly expanding financial services are concerned, I have noted that in 1995 an estimated 95 per cent of the international flow of capital was speculative in nature, having no connection to traditional investments.[12] This growing speculative bubble has become a major component of many contemporary economies. Obviously, there was also the creation and rapid expansion of services related to the computer and information techniques.[13] A growing service sector must not be confused with a move to a service-based economy. In the former, profits still largely increase with a greater throughput of matter and energy, while in the latter, profits depend on delivering services with the least throughput of matter and energy.[14] In the same vein, the growing use of technostructure-like organizations for applying highly specialized scientific and technical knowledge to every aspect of large organizations involved an enormous expansion of the number of knowledge workers, a growth in the middle class, and an unprecedented role for science in the economy.

Other observers emphasized a greater continuity between the present and the past by designating what was new by the equally meaningless concept of advanced industrialism.[15] It provided no understanding of what was new, instead emphasizing that industry continued to advance, which it had done ever since the beginning of the Industrial Revolution. Others declared that society was becoming post-capitalist,[16] but generally failed to recognize that science and technique were rapidly replacing capital as the prime factor of production.[17] There were indeed many changes in the roles of capital and the Market. In Part Two, we examined the central role of knowledge separated from experience within the technostructure of large corporations, their growing ability to renew and accumulate much of their operating capital, the arm's-length relationship such corporations have with the market, and their dependence on a large public sector, particularly in the military-industrial complex.

As far as the celebrated economic take-off and self-sustaining economic growth are concerned,[18] two important factors derived directly from the spread of technique. First, the information revolution greatly strengthened the internal relations between the many branches of technique as each area of specialization increasingly acted as a transmitter and receiver of information. It also greatly strengthened the internal mechanism of technical development, where each invention and innovation triggered additional advances as a result of the flow of information. As information about millions and millions of smaller and larger advances circulated through an increasingly global system, a take-off in technical growth was indeed assured.

Translating this into an economic take-off was more problematic. Marxists were fond of pointing out that capitalism inhibited technological growth. It was certainly true that economic growth had been much less dependable, particularly in some sectors. Nevertheless, technical growth had gained a considerable measure of autonomy with respect to economic constraints as a result of the revised sequence operating between governments and their military-industrial complexes on the one hand, and the ability of the transnationals to plan the internal supply of operating capital on the other hand. A great deal of the technical frontier lay within the public sector of the economy, where it could remain uneconomic for a considerable time. Obvious examples are nuclear power, defence, internal security, and space. The recent rapid development of industry-university partnerships added to this trend. As a result, the most rapidly advancing technical frontiers in micro-electronics and information techniques, as well as biotechnology, were little affected by economic conditions during much of the second half of the twentieth century. It would appear that technical advances required to remove bottlenecks in the evolution of technique-based connectedness could remain uneconomic for a very long time, while those related to human life and society had to become economic in a very short time. In the nineteenth century, the thermodynamic constraints on industrialization were shaped by capitalism into the need for flows of materials and energy to contribute to the renewal and accumulation of capital. In the latter half of the twentieth century, that constraint was increasingly translated into the need to maintain and grow technique-based connectedness by information flows. From the point of view of human life and society, this meant that we were lost at the frontier.[19]

A second factor ensuring a link between technical and economic growth, not primarily structured by capitalism, derived from technique feeding on the problems it created for human life, society, and the biosphere, as discussed in the concluding section of chapter 6. Recall that the intellectual and professional division of labour institutionalized an end-of-pipe approach to its problems. It fed on its own mistakes, thereby trapping us in a labyrinth of technique, largely made up of compensatory technologies and services such as smokestacks, catalytic converters, health and social services for sick workers, and so on. As a result, the ratio of desired to undesired effects of technical and economic growth began to drop sharply in the late 1970s.[20] Paying for the growing costs incurred in the production of wealth contributed to burgeoning national deficits, which in turn led to the downloading of these costs from the federal to the state or provincial, and then to the municipal, level. Corporations cut back on health benefits for employees, and a growing number of institutions required people to do part of their information processing. Serving the 'gods of the market' did as little for us as the creation of false gods did in the past.[21] The concept of high technology was appropriate: it was indeed too high to benefit most people. It may be too high for nations as well. Investing heavily in the technological frontier in both the private and public sectors of the economy is no guarantee of economic growth. The increasingly important role of the global flow of information in technical growth, the resulting unpredictability and non-linearity of that growth, and free trade without social and environmental standards leave no assurances for anyone, although the most powerful developed nations have a better chance.[22]

Some authors sought to explain what was new as resulting from the decisive influence of a single phenomenon, such as a consumer society[23] or a global village,[24] thanks to the mass media. New and important as such phenomena were, they too derived directly from the spread of technique and the diminishing role of culture. For example, human beings are taken away from their world of experience and culture by the mass media. Everything is abstracted from a sequence of events in a time, place, and culture. Such parcelled bits of reality are then reconstructed and integrated into a new context of other such bits in a temporal series of images, to which sound is added in a manner that 'works' for a television show. This reconstruction is taken as more important than reality itself, but one in which we can be engaged only as spectators, given the one-directional flow of information. When toddlers and children spend a great deal of time in front of a television set, this bath of images helps build their structures of experience, which function as a metalanguage as they learn to master their mother tongue. It is difficult to know the precise influence this has on the role of language in experience and culture. In conjunction with other developments, to be examined shortly, it will undoubtedly contribute to a mutation of the role of experience and culture in human life.[25]

The managerial revolution,[26] the control revolution,[27] the lonely crowd,[28] and the organization man[29] – these were also phenomena that derived directly from the spread of technique in society. Guy Debord's claim that we live in a society in which we are spectators of much of our lives is a profound one, provided that we do not attribute this exclusively to capitalism.[30] Wherever a revised sequence operates, we are in danger of becoming, at least partially, spectators of our own lives. For example, to the extent that our choices as consumers are influenced by advertising, to that extent we have become spectators of these decisions, and the same holds for the political decisions we make as citizens. Hence, a society of the spectacle results directly from the spread of technique. Marcuse's 'one-dimensional man' is not the consequence of capitalism but of relating everything to technique-based connectedness and only secondarily to culture-based connectedness.[31] Also profound is the observation that what is emerging is a society of risk.[32] These risks are the direct consequence of technique, which improves performance by undermining the cultural and natural orders and their ability to sustain all life. We are now in a race between the rate at which technique creates chaos from the perspective of human life, society, and the biosphere, and its ability to feed on this chaos to expand and reinforce itself. How far can the role of culture be weakened and the remaining meaning and purpose of human life be desymbolized until modern civilization collapses the way others have?[33] How far can we undermine the integrality of the biosphere before it will be unable to provide adequate life support for the planet or it collapses altogether? From this point of view, technique continues to be the wager of our time.[34] The evolution of technique may well decide whether or not human life as we know it can be sustained and whether or not even more severe impairments of the functioning of the biosphere will occur. Technique's undermining of the integrality, external compatibility, and self-regulating character of everything it touches creates a multiplicity of 'local' risks.

Since the cultures of the 'developed' nations bestowed a very high value on technology, its further development was commonly interpreted as the key to a new promised land. High technology was replacing the smokestack industries. Microprocessors would bring information and thus power to citizens to help them deal with large bureaucracies, and would also revitalize democracy and education. The Japanese fifth-generation project and artificial intelligence promised new companions in our own image. There was a much-celebrated explosion of new freedoms. For example, people could now dress and wear their hair the way they wanted. Society no longer imposed a strict moral code and a set of religious beliefs; sexuality was opened up to exploration. A counterculture was burgeoning, and people started new communities.[35] The celebrations were short-lived as a new conformity imposed itself. The last vestiges of tradition, including its moral and religious forms, were disappearing simply because they were irrelevant to a society increasingly dominated by a technique-based connectedness. The old conformities based on culture were being replaced by new ones required to organize and reorganize everything to achieve the highest possible performance. As long as people conformed to this, it mattered little how they wore their hair, how they dressed, what their sexual habits were, and which religion of the day they espoused. As long as they were technically proficient in their work, and the influence of technique sufficiently oriented the remainder of their lives, the rest was irrelevant. Multiculturalism could be made to work as long as cultural unities were permeated by technique, and integration propaganda ensured the viability of a mass society. The new freedoms would have been genuine if society had continued to be based on a traditional culture but that was not the case, as technique changed people. As far as the many liberation movements were concerned, the real questions were, Can women change technique or will technique change women? Can environmentalists change technique or will technique co-opt or assimilate environmental consciousness? Will the anti-nuclear movement change energy technology or will integration propaganda marginalize it? Can the peace movement convince the world that high-performance weapons systems can no longer defend anyone but only destroy everyone, or are its members co-opted by paying their taxes? To answer such questions requires the best possible understanding of what is happening, and this is rarely the case in movements centred on a particular issue, no matter how noble and desirable their goals may be.

We are not moving towards a megamachine civilization, as Mumford feared.[36] Technique does not progressively absorb the cultural and natural into a mechanism. Our analysis of the technical approach to life makes this abundantly clear. Its four steps progressively externalize everything that is not relevant to the improvement of performance, but that does not mean that these externalities disappear. The technicization of work is typical. Workers do not become cogs in the production mechanism. Although their hands are integrated, the rest of their lives are not, as is evident from the problem of nervous fatigue and its many consequences. The technicization of work creates a tension between what in a worker's life is absorbed and what is excluded from technique-based connectedness. New techniques are then created to compensate for the problems that arise from these tensions. As a result, technique-based connectedness creates problems on which it feeds to grow and expand. The 'cells' of the 'body social' are only partly transformed, thereby threatening that body social even though technique-based connectedness cannot exist outside of it. The technique-based connectedness thus has a parasitic relationship to the cultural order. Its relationship with the natural order is equally parasitic.

Despite widespread belief to the contrary, we have never been, are not now, and will not in the future be heading towards a megamachine society. The technique-based connectedness is not an alternative to the cultural order of society, because it is not an order in any human or social sense. Nor is it an order from the perspective of natural evolution. It can establish itself only by creating chaos in the cultural and natural orders, on which it then feeds. It imposes an 'order' based on performance and power that is foreign to all life.[37] This is evident, for example, from what this technical 'order' produces in abundance and the scarcities it simultaneously imposes. There is an abundance of technical objects and techniques for improving performance, and a scarcity of everything that has thus far been essential for human life: enough free time to revitalize, social support from deep and enduring relationships, a way of life that makes sense and in which one can fully participate, and last but not least, spiritual roots. It may be objected that conditions in the past were also far from ideal. This misses the point. We as modern rational and secular people, with a vast storehouse of means at our disposal, appear to be cutting the ground out from underneath our feet. Just look at our children, many of whom are convinced that they will never have a regular job on which they can support a family, that their relationships with significant others will probably be temporary, that their health is increasingly fragile as the incidences of immunological disorders, depression, and mental illness are rising, that there may not be a social safety net much longer, that public health care may be a casualty of free trade, and that when they retire there may not be any social security. If people were becoming mere cogs in the machine, such concerns would be unthinkable.

The possibility that robots and computers would take over the world was a popular subject in the science fiction of the 1950s and '60s. For many, this was an exciting prospect because such computers would be based on an interface between information on the one hand and experience and culture on the other, so that we would be able to talk to them and they could respond to us in 'natural' language. None of this happened, of course. Instead, we adapted more and more to the computer, viewing everything from the perspective of technique-based connectedness and its 'lifeblood,' information. What the clock had been for the first generation of mechanistic world views, the computer became for a second generation, which regarded everything first and foremost in terms of the production and consumption of information. *Homo informaticus* inhabits a world built up from the ultimate contextless unit, namely, the bit.

Some of the debates of the time shed further light on the many attempts to make sense of what was happening to human life, society, and the biosphere. In one such exchange of views, Daniel Bell began his contribution by quoting T.S. Eliot: 'Where is the life we have lost in living, where is the wisdom we have lost in knowledge, where is the knowledge we have lost in information?'[38] Joseph Weizenbaum made this problem more explicit by arguing that the new approach to information excluded daily-life knowledge anchored in and framed by language.[39] He asserted that the computer has led to a naive scientism and positivism expressed in many ways, particularly that everything known can be clearly stated by language and captured by information. Although Bell acknowledged that all databases used by a society for making economic and social decisions are built up with assumptions, he nevertheless believed that societies can be governed with the use of such databanks for decision-making. Weizenbaum was concerned that the computer programs making use of all that information are built up with assumptions that may be transparent only to those who wrote the original code, less so to those who subsequently modify or extend it, and certainly much less to users. When such programs begin to involve large numbers of lines of code, they must be tested against a theory or fundamental understanding to ensure that they accomplish exactly what was intended. However, in most cases no such theory exists, making testing impossible. When information and programs are applied to the making of social and political decisions, there is the additional problem that we have no consensus of what this human reality is, so there is neither a theory nor a possibility of empirical verification. Weizenbaum asked, How are we to know when such complex manipulation of information systems goes wrong? Does this represent an evasion of human responsibility and a corresponding enslavement to data and algorithms? For example, should we entrust our security and possibly the survival of human life itself to a system like the Strategic Defense Initiative, whose programs are fundamentally untestable? Hence, Weizenbaum did not share Bell's confidence in the new information technologies to help us deal with the information explosion. If we cannot responsibly dominate this new technology, we need to deal instead with the sources of the information explosion. Unfortunately, Weizenbaum was not clear on how this might be accomplished. For him, the bottom line was the question of what kind of world these new technologies were helping to create, and whether it would be a livable one for ourselves and our children.

Bell's reply to Weizenbaum was that he went too far.[40] However, Bell does not give an adequate account of the integrality of individual and collective human life, permitting him to divide society into socio-economic, political, and 'cultural' spheres. As a result, the enfolded character of human life and society disappears as an issue, creating the possibility for a kind of intellectual 'divide and conquer' strategy so common in our age.

In related debates, some authors have pointed to another fundamental assumption: that the processes underlying human thinking and the functioning of computers are quantitatively but not qualitatively different. After examining Western philosophy and several decades of work in artificial intelligence, Hubert Dreyfus has argued that such an assumption is untenable in the face of a great deal of evidence.[41] Together with Stuart Dreyfus, he studied the extent to which human expertise can be captured by knowledge engineering and encoded in expert systems.[42] They concluded that, out of the five stages of skill acquisition that can be distinguished in becoming a human expert, machines appear to be unable to go beyond the equivalent of the third stage. This raises the practical and moral concern about what society can and should leave to machines and what to human beings. In the above exchange of views, Herbert Simon raised the issue of how computers may affect our sense of what it is to be human, which he tied to the notion of being unique in the universe.[43] He pointed out that this has progressively come under fire, beginning with Copernicus, who showed that humanity was not at the spatial centre of the universe; Darwin, who argued that humanity was just another branch on the evolutionary tree rather than a group of creatures endowed with reason and spirit by God; and Freud, who claimed that human beings were non-rational because of the presence of the unconscious. If we now take the next step, and assume that machines can learn and think much like human beings, what remains of being uniquely human, and what does this imply for our sense of identity? Weizenbaum was pleased that Simon could at least recognize that on this point the debate over artificial intelligence had something to do with ethics and fundamental philosophical questions.[44] In retrospect, what is rather surprising about these exchanges is the lack of collegiality and respect exhibited by the artificial-intelligence community of that time, whose members dogmatically asserted their assumptions and presuppositions. For decades, their secular faith in these could not be shaken.

There were also some fascinating ideological debates. I will restrict myself to a particularly significant secular religious quarrel between the Chinese Communist Party and its Soviet 'mother church' that began shortly after the Second World War. The latter regarded the former as heretical with respect to the great 'prophet,' namely, Karl Marx, who taught that a society had to go through the capitalist stage of human history before it could enter into the socialist stage. A Czechoslovakian heresy argued that neither the Soviet Union nor China was on the way to the 'promised land' as long as they did not recognize that capital had been supplanted by science and technology as the prime factor of production.[45] The liberation theology movements, as a synthesis between Marxism and Christianity, had still different interpretations of what needed to happen.[46] In the non-communist world, there was a similar debate as to whether the Third World could directly enter the information age and bypass the smokestacks industry phase of industrialization.[47] This debate had an impact on the foreign policy of a number of nations, including their strategies for Third World aid and technology transfer.

The above is a highly incomplete overview of the many theories about, and ideological debates over, the information revolution and the coming information society. I will add two observations of my own. During the first phase of industrialization, when the production and consumption of goods took centre stage in society, 'technology changing people' transformed our sense of identity into *homo economicus*, as long as the standard of living remained relatively modest. As the information explosion burst on the scene, the production and consumption of information began to take precedence, with the result that 'technology changing people' transformed that identity once again into *homo informaticus*. From this perspective, what characterizes human beings above all else is their ability to process information, and human beings and information machines are simply regarded as two different species of the same genus of information and signal-processing devices. Our use of language speaks for itself. There is a growing vocabulary of words that are used both for human activities and the functions of information machines, including memory, cognition, communication, knowledge, and expertise. Since our structures of experience function as a metalanguage, this presumably means that metaconsciously we no longer distinguish between the two. Similarly, we no longer distinguish between neighbourhood groups and communities whose relationships are mediated by culture, and 'chat rooms' mediated by machines. Special 'dialects' are springing up in these new 'neighbourhoods' and 'communities.'[48] Yet this confusion over our identity as human beings is creating many problems in human relationships.[49] Research continues to provide evidence that communication via the Internet is not the same as traditional communication, and that the substitution of the former for the latter creates many problems.[50] The shift from the culture-based approach to life to the technical approach implies a shift from experience to information. As people change technique, technique changes people into *homo informaticus*.

My second observation is that in most theories and discussions the information explosion is treated as an independent variable. The question as to why this explosion has occurred is rarely addressed. Much of Parts One and Two of this work can be interpreted as furnishing a partial explanation. What I am suggesting is that the information explosion is a signpost of the extent to which technique-based connectedness has come to dominate the culture-based connectedness of human life and society. *Homo informaticus* and the associated information society thus derive from the evolution of technique built on the rationalization efforts that began well before the Industrial Revolution. In sum, it would appear that the concept of technique is a model that can explain many of the new and significant phenomena that emerged during the second half of the twentieth century, including the relationships between them, thus providing a comprehensive understanding of what was happening to human life and society.

I have noted that no model, including the concept of technique, can ever be objective because the simplification of the unmanageable complexity of reality to a more meaningful and intelligible complexity must be guided by values and a vantage point, without which all details would be equally relevant or irrelevant, hence no model could be built. Technique is not a description intended to be objectively valid for all places and cultures in the second half of the twentieth century. It seeks to provide us with the best possible understanding of what is happening to human life and society during that time. The picture may not be reassuring, and some may reject it as utterly pessimistic. Although this is a common reaction, it is a rather strange position to take. Would a cancer patient accuse his or her doctor of pessimism? Surely the patient wishes to have the best possible diagnosis and treatment. Especially during times of rapid change, humanity requires the best possible diagnosis of what is happening in its historical journey. If technique is the nature of our modern world, we must learn to live with it since it is impossible to sustain the current population without it. Once again, humanity must symbolize this new 'nature' of our world to grasp its meaning and value in order to relate to it, not on its terms but on ours. A measure of freedom will then be created with respect to the influence this new 'nature' has on us. Only then is there a chance of reversing the process of desymbolization.

9.3 Technique and Industry

Industry itself underwent a fundamental transformation over and above what has already been described in Part Two. The Fordist-Taylorist production system officially operated exclusively in terms of knowledge separated from experience. There was an almost complete lack of awareness of the role of knowledge embedded in experience, and in any case, it was not officially acknowledged. It made the production process highly repeatable, predictable, controllable, and measurable, opening it up to information. Later on, the principles of scientific management were also applied to mental work, breaking down complex operations into their constituent elements and recombining them into logical rules, algorithms, and programs. Thus the mental operations required to plan, execute, control, and coordinate the horizontal division of labour were complemented by a vertical division of labour, tending towards a kind of vertical intellectual assembly line. It would 'assemble' the information that top management required for the formulation and execution of their orders. The emergence of the Fordist-Taylorist production system also marked the beginning of the shift from production based on experience and culture to that based on information. The horizontal division of labour atomized work into rational steps by stripping away all context that interfered with efficiency, creating either mindless human work separated from experience and culture or opening the door to information-based mechanization, automation, and computerization by machines. The vertical division of labour atomized mental work into rational steps from which all context was also eliminated as much as possible, for the same purpose. These developments opened up a growing need for integrating and coordinating these systems by means of flows of information.

Between the two world wars, the Fordist-Taylorist system represented the highest form of rationality possible in a unique U.S. setting: the making of complex products with a low-skilled workforce of immigrants, many of whom could hardly understand each other or their bosses. This situation stemmed from the lack of knowledge of a common working language in a culture evolving from a frontier society to a mass society situated in resource-rich ecosystems. The Japanese lean production system adopted the assembly line but incorporated it into a very different production system that made a good deal of use of experience and culture, reflecting very different socio-economic and natural conditions. Following the Second World War, Japan was still a highly traditional society in which 'technique changing people' was just beginning to make its impact felt. It led to a more highly rationalized form of production, with substantial advantages in terms of performance values, which would throw the industrial world into a tailspin.

The lean production system was developed by the Toyota company in Japan.[51] Although founded in 1937, the development of this system did not really get under way until after the Second World War. Much inspiration was drawn from visits to Ford in 1929 and 1950 by Kiichiro Toyoda and Eiji Toyota respectively. The latter studied Ford's Rouge complex at Dearborn, Michigan, in every detail as the largest and most efficient manufacturing facility in the world. Back home, he and Taiichi Ohno came to the conclusion that mass production would not work in Japan. Not only was the domestic market extremely small compared to that of the United States, it required a wide range of vehicles including heavy and light trucks as well as small and luxury passenger vehicles. The workforce was no longer willing to be treated as a variable cost and was in a better bargaining position, thanks to the labour laws imposed by the American occupation. Company unions negotiated for everyone, thereby eliminating the distinction between blue- and white-collar workers and securing a share in the profits through bonuses. It is worth emphasizing that the workers were Japanese citizens and not immigrants or guest workers willing to put up with substandard working conditions for high pay. Following the war, the Japanese economy was short of capital and foreign exchange, prohibiting the acquisition of the best Western production technologies. At the same time, the Japanese government had blocked all foreign investment in the Japanese car industry. All this and more contributed to Taiichi Ohno's decision that an entirely new approach was necessary.

This search produced some counter-intuitive and unexpected results. Relatively modest production runs made dedicated die presses uneconomic. Hence, Ohno developed a way in which the time to change dies was reduced from one day's work by highly skilled people to three minutes by much less skilled workers. This was accomplished by putting these dies on tracks so that they could be rolled in and out of the die presses as required. It led to a surprising result: making small batches cost less because it eliminated the huge inventories, while permitting the almost immediate discovery of any mistakes since parts did not sit around very long before being assembled. The resulting shortening of negative feedback loops brought with it a greater awareness of quality and the possibility of continuous improvement. The result was a much tighter coupling of the production system: with only two hours or less of inventory, any serious problem could bring the whole assembly process to a halt. What was required, therefore, was a highly skilled and motivated workforce capable of anticipating and solving problems before they occurred. This is in sharp contrast with the behaviour of the workforce in a mass production system, where the withholding of such knowledge and effort is an everyday occurrence. One reason that this difference could be sustained in Japan was a genuine reciprocal commitment, in which the company offered lifelong employment and access to company facilities in return for workers using their heads to promote the interests of the company and doing the work that was required. It made the workers into the most significant fixed cost of a company, since machinery could be written off and scrapped while workers could be expected to remain for some forty years.

Incorporating the commitment and brains of the workers into the system also permitted the elimination of many indirect workers required in mass production, because lean production workers did more than just one or two assembly operations. This reduced or eliminated altogether those who cleaned the work area, repaired tools, filled in for absentee workers, studied and improved work methods, and supervised workers to ensure compliance with these methods, substantially improving the ratio of direct to indirect workers. In addition, each lean production worker belonged to and was backed by a team with a coordinator as opposed to a supervisor. These quality circles met regularly to discuss how their part of the production process might be improved on an ongoing basis. Any problem was to be dealt with proactively, and any worker could stop the line if this was not readily possible. Once an error was discovered, it was traced back to and dealt with at its root, so that it hopefully would never occur again. All this was so successful that today, Toyota lines almost never stop. In contrast, mass production systems essentially encourage workers to park their brains at the plant gate and leave the thinking to industrial engineers. By reincorporating the brains of workers into the system, lean production seeks to reintegrate knowing, doing, and controlling as much as possible, thereby shortening negative feedback loops and pushing decisions down to the lowest possible level of the organization. Rationalization and technicization are internalized as much as possible by each worker, resulting in a spectacular improvement in performance.

The use of statistical process control techniques illustrates this crucial difference between lean production and Fordist-Taylorist mass production. In the latter case, data are collected on the shop floor to be sent up the vertical intellectual assembly line for interpretation and the formulation of a response by middle management, which is to be transmitted back down to the shop floor. The result is a long and ineffective negative feedback loop. In a lean production system, workers are taught to interpret the data by plotting it on charts and to take any required corrective action, resulting in very short and highly effective negative feedback loops. Lean production workers are encouraged to be their own industrial engineers. They apply time-and-motion study techniques to their ideas of how work methods might be improved. If an idea looks promising, it is discussed with other team members, to be adopted if there is a consensus that it constitutes an improvement. In this and other ways, the brains of the workers are reintegrated into the system as much as possible, thereby making the fullest use of knowledge embedded in experience and complementing this with knowledge separated from experience. This creates a much better synergy between the two than is common in Fordist-Taylorist mass production.[52] There are many spin-off benefits, including a greatly reduced need for space because little inventory needs to be stored thanks to the just-in-time approach and few, if any, vehicles need to have defects corrected after assembly.

In these and other ways, many of the inefficiencies inherent in the Fordist-Taylorist production system were reduced or eliminated. The extreme horizontal division of labour had created machinery that was so specialized and dedicated that making any changes was time-consuming and expensive. The extreme vertical division of labour created a blue-collar and white-collar workforce that was highly specialized, forming a huge bureaucracy of mostly indirect workers. It vastly complicated the process of coordination to ensure that everything came together at the right time with the highest possible quality and lowest cost. The same was true for the design of new products and their dedicated production process. The design of a new piston, for example, involved coordinating the efforts of junior and senior piston design engineers, manufacturing engineers, and industrial engineers, and all this for just one component for a new vehicle. All this could be rendered economic only by very long production runs, requiring large and stable markets in order to realize the greatest possible economies of scale. This stability minimized the dependence of the system on effective negative feedback loops for its regulation. Instabilities inevitably exposed a potentially lethal flaw, namely, a lack of control, which necessitated short negative feedback loops. There was another problem as well. The demand-control model, which successfully maps the findings of socio-epidemiology, rates jobs characterized by high demand and low control as high strain and therefore unhealthy. The Fordist-Taylorist system had become so suboptimal that producing unhealthy workers was one of its costliest outputs.[53]

As is typical for any technical frontier, there was no shortage of ideological claims. Even the name 'lean production system' implies that all alternatives are 'fat.' It reveals the value judgment that lies at its origin and that guides its development. The landmark study entitled *The Machine That Changed the World* puts it this way:[54]

> The striking difference between mass production and lean production lies in the ultimate objectives. Mass producers set a limited goal for themselves, good enough, which translates into an acceptable number of defects, a maximum acceptable level of inventory, and a narrow range of standardized products. To do better, they argue, costs too much or exceeds inherent human capabilities. Lean producers, on the other hand, set their sights explicitly on perfection, continually declining costs, zero defects, zero inventories and endless product variety. Of course, no lean producer has ever reached this promised land and perhaps none ever will but the endless quest for perfection continues to generate surprising twists.[55]

No level of performance was ever good enough, so that everything had to be sacrificed to performance values. It amounted to fully bringing technique into design, production, distribution, and sales. It began with motor vehicles and was later applied to other industrial products. As people changed production techniques, little attention was paid to how these in turn changed people and society. As a result, as production became lean, the humanity of the people involved became anorexic. In Japan, surveys show that over 70 per cent of workers are mentally exhausted when they come home, and job insecurity is on the rise.[56]

Socio-epidemiology is painting a grim picture of how people's lives are being devastated by work.[57] In the country of lean production, *karoshi* (death from overwork) has become recognized as a major problem.[58]

By the early 1960s, Toyota's lean production system, in comparison with the mass production system, claimed to use 'half the human effort in the factory, half the manufacturing space, half the investment in tools, half the engineering hours to develop a new product in half the time. Also it requires keeping far less than half the needed inventory on site, results in many fewer defects, and introduces a greater and ever-growing variety of products.'[59] All this led to an ideological zeal expressed by K. Matsushita as follows: 'we will win and you will lose. You cannot do anything about it because your failure is an internal disease. Your companies are based on Taylor's principles. Worse, your heads are Taylorized too. You firmly believe that sound management means executives on the one side and workers on the other. On the one side men who think, on the other side men who can only work. For you management is the art of smoothly transferring the executive's idea to the worker's hands.[60]

Such a diagnosis of the situation was only partially correct. When everything is said and done, Fordist-Taylorist mass production and lean production share the assembly line as the technological infrastructure. The just-in-time supply of materials, parts, and sub-assemblies tightly couples this technological infrastructure to those of the first- and second-tier suppliers, making short negative feedback loops possible. The reciprocal commitment between management and labour allows the vertical assembly line to be substantially shortened by making the fullest possible use of the experience of workers and their commitment. It permits lean production systems to encourage and more fully utilize the development of metaconscious knowledge in relation to both knowledge embedded in experience and knowledge separated from experience. It is as if these systems had learned a lesson from the failure of artificial intelligence. From this perspective, it is a small step in the direction of socio-technical approaches that argue for the need to jointly optimize technology and the organization making use of it.

The lean production system may be regarded as the next step in rationalizing the Fordist-Taylorist mass production system, as long as we recognize that this is only in the domain of people changing technique. Both kinds of production systems fail to take into account the fact that technique changes people at the same time. Ultimately, both systems are incompatible with human life and society. The greater control workers have in lean production is afforded not by human freedom, but by what serves technique-based connectedness. Lean production does not go to the root of the problem of assembly line work and the fundamental inefficiencies associated with it. Only Volvo's Uddevalla plant has gone to the root of the problem.[61] Introducing this assembly process into lean production, thereby replacing the assembly-line, might create a more genuine hybrid of craft and mass production by combining the strengths of each to offset the weaknesses of the other. It would inscribe technique-based connectedness into culture-based connectedness, with the latter possibly able to keep the upper hand. In the meantime, the influence technique has on people in the workplace in particular and in society in general weakens culture and the strength of any social relations, including the strong sense of loyalty traditional Japanese culture inculcated in the group. As the traditional sense of duty and commitment evaporates under the pressure of technique, the lean production system becomes vulnerable in terms of a declining commitment on the part of the workers. The problem can be deferred for a while by carefully screening potential workers in order to select those who still have a strong sense of traditional duty and commitment, but in a mass society the work ethic is under a great deal of pressure. Of course, 'technique changing people' is not the only threat to the effectiveness of 'people changing technique,' since the organization of lean production based on a reciprocal commitment between management and labour continues to be tested by economic turbulence and spiralling health costs.

I will not examine how Japanese lean production further rationalized the relationships with first- and second-tier suppliers and with customers.[62] It should be evident by now that lean production is a more highly technicized form of production than the Fordist-Taylorist one. Its clear superiority in the domain of performance values has threatened economies all across the globe, but there is little question that it, too, will exact a heavy price in the domain of context values. Apart from health issues, continuous improvement means that human capacities are stretched to the very limit, quickly moving from 'lean' to 'anorexic.' The tightly coupled system puts enormous pressures on workers since any mistakes can bring the entire system to a halt. There may also be a considerable impact on the local ecosystem as the just-in-time supply lines require an increase in the use of trucks.

We can now shed light on two bottlenecks in the development of the Fordist-Taylorist system that were avoided by the lean production system: an explosion of information for top management, and a limited ability to compete in terms of quality. As the transformation of the corporation described in Part Two advanced, management was inundated with information, particularly at the highest levels. At the same time, the distance between management and the shop floor increased also in terms of remuneration, further aggravating the 'us-them' polarization.

Management could rely less and less on direct hands-on experience and instead had to rely increasingly on information. In other words, the balance between knowledge embedded in experience and knowledge separated from experience in management activities tipped in favour of the latter, with the result that the production system increasingly disappeared behind the numbers. For example, employees became little more than entries on a budget line. When managers relied primarily on hands-on experience, their thoughts and actions took on a meaning and value in the context of all their experiences of the company and its interactions with the economy and society. As the role of management experience was undermined by an explosion of information, the technical approach once again produced a new technique. Strategic planning was to formalize the processes, taking all this data and putting it into patterns and models in order to provide a clear view of the position of the company and the direction in which it was heading.[63] These models could then be analysed in terms of the company's strengths and weaknesses in order to develop a new strategic vision of the one best way of strengthening the company's position. Strategic planners were to be the new 'knowers' that would help managers to do their work better.

Strategic planning emerged during the 1960s and was rapidly adopted by corporate executives for the purpose of developing and implementing strategies for improving competitiveness. It failed to deliver, and was replaced by other techniques. Many of us can readily identify with the reasons for the failure of strategic planning. We have sat through too many planning retreats where mission statements are invented in the morning, the strengths and weaknesses of a business unit or academic department are assessed in the middle of the day, and strategies are developed and finalized before dinner. The report then goes on the shelf, since everyone knows that life never works out that way. Instead, managers have to think strategically, which relies much more on intuition and creativity than on processes that can be formalized. Managers depend extensively on metaconscious knowledge related both to knowledge embedded in experience and knowledge separated from experience. Their hunches may have many sources, being based on that extraordinary capability that the human brain-mind has evolved to deal with a world that is ever changing and where patterns are always incomplete, fragmented, and even composed of contradictory elements.[64]

Strategic planning failed for the same reasons that artificial intelligence could not achieve its goals, why expert systems cannot match human experts, and why operations research does not work in the socio-cultural and organizational domains.[65] Life never repeats itself, and that greatest of human inventions, namely, culture, evolved to make the best possible use of this situation. Scores of many imperceptibly small changes that occur each day in people's professional and private lives are enfolded into everything else, with the result that gifted people can often detect embryonic beginnings of something new long before any clear patterns can be captured by information. Artificial intelligence, expert systems, and operations research are based on principles and methods that excel in the world of the dead (machines), where no constituent is enfolded into any other and where that world is not enfolded in some way into any constituent element.

There is an additional reason why strategic planning was bound to fail. Behind all the numbers, models, and plans lies an organization that critically depends on the morale and commitment of people. The rationalization of how people change technique provides no insight into how this simultaneously affects people. Without paying special attention to the latter, it is highly probable that the technique-based connectedness of the organization improves its performance at the expense of the people working there. The use of the technical approach to life, without regard to its limitations, inevitably leads to undermining all human functions in an organization. How can people be expected to show enthusiasm, and ignore their anxiety, for processes that they know (intuitively) are treating them as just another resource (the term 'human resources' speaks for itself)? How can corporations expect to maintain the good morale of their workforce? To be successful, an organization must synergistically combine the effectiveness of people with that of technique. For people, this means that the organization stimulates creativity, builds and utilizes skills, nurtures motivation and morale, and respects the person.

As we saw in Part Two, blue-collar workers constantly have to bridge the gap between a design for a product and its manufacture, established primarily on the basis of knowledge separated from experience, and how all this works out on the factory floor. This is equally true for white-collar work. The boundary between that domain of a work organization where knowledge separated from experience is dominant, and the domain where knowledge embedded in experience rules, can become a fault line where problems pile up unless there is mutual understanding and determination to make it work. More than ever it is essential for organizations to learn, and for this learning they still rely exclusively on people, despite all the rhetoric about artificial intelligence and expert systems. Much of this becomes more problematic when external consultants are brought in, those young MBAs who complete their tasks without any real knowledge of the ethos of the company, including its practices, values, and leadership style. In fixing some problems, they almost inevitably create others that may not manifest themselves until they have departed with the cheque. Other consultants may then be brought in, causing corporations to become hooked on consultants that propagate an endless cycle of management fashions that take a legitimate issue or concern and blow it up out of all proportion. It becomes almost impossible for a business executive not to join the latest stampede towards the one and only best way of doing business.

We have shown that work in the Fordist-Taylorist system is monitored and controlled by long, fragmented, and ineffective negative feedback loops. This makes it almost impossible for anyone to understand the effect they have on the quality of products and processes, thereby placing limits on what can be accomplished in this domain. As competition with lean production systems intensified, it became increasingly evident that product quality was inadequate. W.E. Deming,[66] as one of the pioneers of the quality movement, trained workers in statistical process control and used socio-technical approaches to give them greater responsibility over decision-making. Everyone was expected to take the initiative and experiment with how best to do their work, and to discuss their findings with others. Management's part of the bargain was to place a greater trust in workers as responsible beings and reward proactive work patterns with job security. Although regarded as idealistic by many, the quality movement was seen as necessary in order to compete with lean production systems. This management fashion also fell as quickly as it rose, being swept away by another in the stampede to re-engineer the corporation.

Business process re-engineering was launched by the publication in 1992 of *Process Innovation*, written by Tom Davenport,[67] and *Re-engineering the Corporation*, by Michael Hammer and James Champy, the following year.[68] The rapidity with which these ideas spread suggests that they struck a chord with many senior executives.[69] The issue being addressed by the rapidly developing next management fashion was called the technology paradox.[70] It referred to the widely recognized problem that the rising level of investment in information technology by U.S. organizations was at best translating into modest gains in labour productivity. The trend was most pronounced in the service sector of the economy.[71] The response to the growing gap between investment levels in information technology and increases in labour productivity was to redesign American business organizations to take greater advantage of information technology. It shifted the focus back to technology and away from people, where it had been during the heyday of the quality movement and socio-technical approaches. Business process re-engineering became the new secular gospel of salvation: do it and you will be saved, or leave it and you will disappear because of global competition. There was little or no middle ground until the disastrous consequences became evident. Although this management fashion has largely gone out of style, redesigning the organization in the image of information technology has not. The trend continues to gain momentum and may well constitute the triumph of what Brödner calls the technocentric alternative.[72] More fundamentally, it may be regarded as the triumph of information over experience and of technique over culture in the economy.

At first glance, this transformation of work may be regarded as the repeat of an earlier situation. Initially, productivity gains achieved by investments in word processors were modest. It is well known that the exceptions were the companies that regarded this technology not as more advanced typewriters with many new features, but as something that required a rethinking of office organization. In other words, the computer represented something that was discontinuous with what went before, and this was true wherever the computer replaced other business machines such as data-processing devices based on punched cards. In all such cases, the computer remained a tool in the hands of workers in bureaucracies. Not so with the new transformation. The separation of knowing and doing from experience had created an information explosion that necessitated more and more computers to deal with it. The flow of information within computers had been carefully optimized, so the problem to be faced had to do with the flow of information outside the computer in the organization. The solution was evident: the organization needed to be re-engineered (the term was well chosen) to ensure that the flow of information to and from the computer would be just as rational and optimal as the flow inside it. This shifted the focus away from the *functions* performed by human beings within the *structure* of a bureaucracy to the *flows* of information involved in any *process*. Computer engineering, in essence, furnished the model for reorganization. The fundamental principle was that data should be captured only once, which meant that all business processes should share their information, thus integrating them into what at that time were called enterprise systems or enterprise-wide systems. These systems would assemble all the information required to deliver the products or services, with human beings accomplishing the steps that were not yet automated.

In terms of Fordist-Taylorist thinking, information technology helped create a vertical intellectual assembly line 'producing' the added value desired by an internal or external customer. Outside the technostructures of the Fordist-Taylorist systems, low-level white-collar workers became the tenders of information machines to which they added information in predetermined ways, just as their blue-collar colleagues added physical components on the assembly line. There was one crucial difference: to ensure that information was captured only once and shared throughout the system, the intellectual assembly line had to be both horizontal and vertical. In this way, parallel lines were integrated, coordinating flows of information across functional boundaries. The vertical division of labour and its organization were now primarily executed by a large integrated database tended to by information workers. Technique-based connectedness became a network of flows of information within a large socio-technical information system that could no longer be acted on by means of experience and culture.

This picture fits but is not limited to business process re-engineering, defined as 'the fundamental rethinking and radical redesign of business processes to achieve dramatic improvements in critical contemporary measures of performance such as cost, quality, service and speed.'[73] A business process is defined as 'a collection of activities that takes one or more kinds of inputs and creates an output that is of value to a customer.'[74] Several examples of business process re-engineering are documented in the literature.[75]

At IBM Credit, two senior managers decided to literally walk a representative's financing request for a potential sale from person to person involved in the five-stage approval process.[76] It required ninety minutes, instead of the usual six days to two weeks, for a quote to be issued. Normally, the work was split among several departments, each of which completed a functional step, beginning with an intake person who logged the request and then forwarded it to the credit department, which checked the customer's creditworthiness using its own information system. The results were then forwarded to the business practice department, which set the terms, again using its own information system. Another worker then determined the appropriate interest rate, and, finally, an administrator turned all this information into a quotation that was couriered to the sales representative. The turnaround time was unsatisfactory from the perspective of the potential customer; the representative, who could easily lose the sale in the meantime; and the design and organization of the work, because of the fragmentation of information and the segregation of computer systems between functional units of the organization. It is easy to understand how a traditional bureaucracy, to which information technology had been added in a piecemeal and fragmented fashion, could end up in this situation. It is equally obvious that sooner or later someone would recognize that, given the potential of the new technology, this situation made no sense. The organization had to be redesigned to fit the new technology. This required the creation of business processes that could cross functional boundaries and integrate business functions.

The two executives had, in effect, created an intellectual assembly line by moving the documents past each person to add a piece of information in the process of 'assembling' a quote. Conceptually, the next step was straightforward: a model of the process had to be created in the form of an algorithm that would take a new multifunctional worker through the required steps and provide access to the relevant databases. Once optimized, the algorithm and databases were installed in an information system. Given the limits of expertise that can be captured by an algorithm, this worker would have access to an expert if a request was anomalous with respect to the algorithm. This aspect of 'people changing technique' had to be accompanied by a redesign of a portion of the organization in the image of the new 'technology,' which was usually referred to as the management of 'technological' change. It was carried out by organizational psychologists and human relations experts using a variety of techniques. This redesign of jobs and the organization constituted one component of 'technique changing people' insofar as people's working lives in the corporation were concerned. People's expectations needed to be redesigned. They had to become multi-skilled members of teams serving flexible organizations operating in a turbulent global market. Their interpersonal skills were now very important, as was their openness to regular retraining, placing a premium on endless flexibility.

The kinds of expectations that came with work in a Fordist-Taylorist bureaucracy, with its career ladders that people would ascend at a fairly predictable pace of a rung or so every several years, were incompatible with the new flexible organization that moulded and remoulded itself in the image of the technologies and techniques it used. In passing, I will note that this is one of the reasons why many business process re-engineering endeavours failed. The new skills, values, and expectations pointed to a new organizational culture that was incompatible with those of a traditional Fordist-Taylorist organization, with the result that, ultimately, corporate re-engineering was an 'all or nothing' proposition. People and organizations had to adapt themselves to the new 'niches' of the new 'ecology of technique.' The overall result was a dramatic improvement in performance of the technique-based connectedness. In the case of IBM Credit, business process re-engineering reduced average turnaround time from seven days to four hours, and the number of quotes each person could deal with increased a hundredfold![77]

Other business process re-engineering efforts followed much the same pattern. One of the pioneers in the insurance industry, Mutual Benefit Life, at first processed applications much as its competitors did.[78] This involved thirty steps performed by nineteen people belonging to five different departments, and required a turnaround time of five to twenty-five days. It was estimated that only seventeen minutes of this represented time actually spent on the application, with the remainder accounted for by transit from one department to another. The entire process was redesigned to permit one caseworker to complete the entire process using an information system. It included all the necessary databases as well as an expert system, with the result that even caseworkers with little experience could do the work. Gains in the domain of performance values were impressive – turnaround time was reduced to two to five days – but more than one hundred field office positions were eliminated because caseworkers could now handle double the volume of applications.

Before business process re-engineering, paying a bill at Ford involved three departments: Purchasing, Receiving, and Accounts Payable.[79] The latter department employed more than five hundred people, while a lean producer like Mazda had only five such workers. It turned out that most of the time was spent not on paying bills, but on investigating cases in which the purchase order, receiving document, and invoice disagreed. Business process re-engineering led to invoiceless purchasing using a central database. Purchasing would enter any new order into it, and Receiving would likewise record the arrival of the goods. The system automatically matched the two and issued the cheque, making the invoice of the supplier redundant. Accounts Payable could now do its work with only one hundred and twenty-five people.

The best documented case is that of CIGNA, a major U.S. insurance company that employed over fifty thousand people in nearly seventy countries.[80] Senior management attributed an 11 per cent drop in income to mostly unproductive uses of information technology, and decided to change this using business process re-engineering. Over a five-year period beginning in 1989, twenty re-engineering projects were undertaken. What was referred to as the first wave of these projects concentrated on reducing costs and improving service delivery, but half of them failed. A second wave took a 'softer' approach by paying careful attention to the management of technological change. In terms of performance values, the overall results were impressive: operating expenses were reduced by 42 per cent, turnaround time was significantly lowered, customer satisfaction was increased, and quality jumped.

Additional gains in the domain of performance values can be realized if information is captured only once throughout an entire organization. This can be achieved by what is referred to as enterprise integration: combining all business processes into an enterprise-wide system (EWS),[81] now commonly referred to as enterprise resource planning (ERP). The following example illustrates what can be accomplished:

A Paris-based sales representative for a US computer manufacturer prepares a quote for a customer using an EWS. The salesperson enters some basic information about the customer's requirements into his laptop computer and the EWS automatically produces a formal contract in French, specifying the product's configuration, price and delivery date. When the customer accepts the quote, the sales rep hits a key. The system, after verifying the customer's credit limit, directs the order. The system schedules the shipment, identifies best routing and then, working backward from the delivery date, reserves the necessary parts from suppliers and schedules assembly in the company's factory in Taiwan. The sales and production forecasts are immediately updated and the material requirements, planning list and bill of materials are created. The sales rep's payroll account is credited with the correct commission in French francs and his travel account is credited with the expense of the sales call. The actual product cost and profitability are calculated in US dollars and divisional and corporate balance sheets, the Accounts Payable and Accounts Receivable ledgers, the cost-centered account and the corporate cash levels are all automatically updated. The system performs every information transaction resulting from the sale.[82]

As this example illustrates, an enterprise-wide system integrates information modules belonging to particular business processes into a central integrated database that captures all information relevant to the operations of a corporation.[83] It is evident that, if such a system is designed properly and is installed with adequate support and commitment from personnel, dramatic gains in labour productivity and other performance values can be achieved. However, technical and organizational difficulties abound. Re-engineering the corporation was a failed management fashion, but enterprise integration by means of information systems that address much deeper structural issues is here to stay. It is steadily resolving the technology paradox.

Enterprise-wide systems (or, more briefly, enterprise systems) will not remain for long as islands of technique-based connectedness in oceans of experience and cultures. There will be growing pressures on suppliers, wholesalers, and retailers to instal enterprise systems that can be linked to those of manufacturing organizations for the purpose of improving performance. Similar developments are occurring in other large organizations including universities, banks, and hospitals. The necessary complement is that of individuals using their microprocessors to do their banking and shopping, make travel arrangements, file tax returns, obtain information, and procure entertainment. There is a striking parallel with what was happening near the end of the nineteenth century when utilities were building power stations and distribution grids. Some observers thought that this meant the end of the growing centralization that had taken place since the beginning of industrialization, because it would now be possible for families to manufacture a great deal of what they needed in their own workshops. Manufacturing could be decentralized and a better society ushered in. We are currently building the information infrastructure for the new technique-based connectedness of contemporary societies. Enterprise systems currently function as the 'information utilities' of large national and transnational organizations. These information utilities are connected by the 'grid' of the Internet. As the volume of information continues to grow, 'information highways' in the form of fibre-optic cables are added to increase grid capacity. Everything within this technique-based connectedness evolves in relation to everything else for the purpose of increasing individual and collective performance. As a result, the system-like properties of this emerging technique-based connectedness are reinforced.

As large organizations help to evolve this new technique-based connectedness, they will have to adapt to it in several important ways. There will be an unprecedented reliance on information at the expense of experience. The limits of this reliance are coterminous with the scope of the enterprise systems. Beyond it lies what cannot be, or will never be, a part of the technique-based connectedness of organizations. All this requires significant organizational changes. including management-labour relations.

The tension between information and experience is much more than an abstract theoretical issue. It is lived by an ever-larger number of people whose work depends on computer-based systems. Zuboff provides us with many excellent accounts.[84] For example, the operators of a recently computerized pulp mill experienced a great deal of anxiety since the numbers on their screens could not readily be connected to their prior experiences, such as handling bits of pulp. They might eventually establish some correlations between these experiences and the numbers on their screens, but the next generation of operators will have to get along without them. The research shows the deep and widespread frustration of people faced with this kind of digital toil. Here is how some of them have expressed it:

> It's just different getting this information in the control room. The man in here can't see. Out there you can look around until you find something.
>
> The chlorine has overflowed, and it's all over the third floor. You see, this is what I mean ... it's all over the floor, but you can't see it. You have to remember how to get into the system to do something about it. Before you could see it and you knew what was happening – you just knew.
>
> The hardest thing for us operators is not to have the physical part. I can chew pulp and tell you its physical properties. We knew things from experience. Now we have to try and figure out what is happening. The hardest part is to give up that physical control.[85]
>
> With computerization I am further away from my job than I have ever been before. I used to listen to the sounds the boiler makes and know just

how it was running. I could look at the fire in the furnace and tell by its color how it was burning. I knew what kinds of adjustments were needed by the shades of color I saw. A lot of the men also said that there were smells that told you different things about how it was running. I feel uncomfortable being away from these sights and smells. Now I only have numbers to go by. I am scared of that boiler, and I feel that I should be closer to it in order to control it.[86]

With the change to the computer it's like driving down the highway with your lights out and someone else pushing the accelerator.[87]

Office workers expressed much the same thing:

You can't justify anything now; you can't be sure of it or prove it because you have nothing down in writing. Without writing, you can't remember things, you can't keep track of things, there's no reasoning without writing. What we have now – you don't know where it comes from. It just comes at you.[88]

Now, once you hit that ENTER button, there is no way to check it, no way to stop it. It's gone and that's scary. Sometimes you hit the buttons, and then it stares you in the face for ten seconds and you suddenly say, 'Oh no, what did I do?' but it's too late.[89]

A project manager expressed concern over the kinds of jobs he had helped to create:

It's reached a point where the benefits analysts can't move their fingers any faster. There is nowhere to go anymore if you don't want to sit in front of a terminal. The quantity of pressure may be the only difference. Labor will sooner or later get smart and see that unions are their only answer.[90]

Other workers complained about job degradation:

The computer system is supposed to know all the limitations, which is great because I no longer know them. I used to, but now I don't know half the things I used to. I feel that I have lost it – the computer knows more. I am pushing buttons. I'm not on top of things as I used to be.[91]

You don't have to remember things, because the system does. You could get a monkey to do this job. You just follow the keys.[92]

Deskilling has gone to a point where even the knowledge required to recognize mistakes is being lost:

> The computer system is giving you a message. It's saying, 'You don't have to be on top of it anymore.' The thing is, we make a lot of mistakes, and the less you encourage people to know things, the less anybody is ever likely to notice all these errors.[93]

Human factors engineering is hoping to alleviate this situation somewhat by designing more effective interfaces between experience and information. For example, for the pulp mill a pictorial representation of the system could be associated with information regarding local conditions. In this way, a technical context can be provided for the information as a clue to its meaning and value. These would vary substantially, depending on whether someone was interpreting this information about the system in terms of knowledge embedded in experience or in terms of knowledge separated from experience. The one cannot be translated into the other, since they are determined either in the culture-based connectedness of human life or in the technique-based connectedness of the technical order. The latter is at best related to the former by a triple abstraction (discussed earlier) carried out by each member of a team responsible for making improvements on the basis of the technical approach to life. Thus far, all efforts to translate experience and culture into logic, artificial intelligence, and expert systems have failed.

All this is closely related to another important issue, namely, what we should delegate to machines and what to human beings. For example, which skills for flying a plane should be converted into the monitoring of the flight management system and which, if any, should be left to conventional flying? If we delegate most or all of them to machines, where will the test pilots of the future come from? Is the experience of flying essential for the design of flight simulators and flight management systems? This is but one instance of a widespread problem: how do we deal with the interfaces between knowledge embedded in experience and knowledge separated from experience, and between the technique-based connectedness and the culture-based connectedness of contemporary societies? No organization can avoid these interfaces and the problems associated with them. At present, technical specialists are individually and collectively in a state of denial of the validity of any kind of knowledge other than their own.

These kinds of issues will become even more pressing with the rapidly growing use of enterprise systems and the future networks of such systems. Their technical sophistication is also increasing as enterprise integration has become a well-established area of teaching and research. People working at the interface between the 'world' of clients and the 'world' of these systems face the above problems on a daily basis. For example, what does one do when a client's situation does not fit any of the categories on which the system bases its operations? How can one decide between available categories when it is impossible to know what the consequences will be for the client or for oneself, given that most of these people have little or no knowledge of how the system actually works, including the models and assumptions on which it is based? These people assemble information with little or no knowledge of its technical meaning and value, which are determined by the technique-based connectedness of the system. Even if the categories were initially carefully designed by people who have a great deal of knowledge about the business before it was operated by an enterprise system (and this is almost never the case), their applicability can only deteriorate as the living world around the enterprise system evolves while the system does not. The situation may well worsen with future systems, because the kind of direct experience that may have been available to the designers of the original system will no longer exist.

The people working on these new 'intellectual assembly lines' will not be able to build up any kind of job skills, either on the basis of a knowing and doing embedded in experience or on the basis of a knowing and doing separated from experience. The growing alienation and frustration of these people, induced by the commoditization of their work by technique, could lead to a deterioration of the relationships between management and labour to levels not seen since the early days of the Industrial Revolution. A proliferation of destructive coping mechanisms in and beyond the workplace is likely to occur. Adapting these systems to their evolving surroundings by adding 'patches' will be challenging. If there is no direct experience, system programmers will have no alternative but to decompose all the functions of an organization by means of the technical division of labour and to push its horizontal and vertical dimensions to their limits. The organization will begin to mimic a computer chip, but the discrepancies between the system and the 'real world' will not disappear. Further patches will be required, setting off a futile cycle – a microcosm of what is happening with technique-based connectedness establishing itself within culture-based connectedness without being able to free itself from it.[94]

The more complex and extensive the technique-based connectedness of an organization and that of its surroundings becomes, the more difficult it will be for people within it to learn about the 'system,' either in terms of knowledge embedded in experience or knowledge separated from experience. To really learn something, people have to get involved, but this becomes very difficult as the work context becomes more opaque. Hence, it becomes more difficult to live our work-related experiences, even in a minimal fashion. Will work be turned into symbolic toil, as Zuboff argues?[95] Will the very possibility of what she calls informating, as opposed to automating, be squeezed out?[96] Without living our work-related experiences, at least to some extent, there can be no emotional involvement and hence no commitment. Given what we currently know about the role emotions play in learning and living, this may well turn into a very serious problem. Are we creating conditions under which the process of skill acquisition will be greatly undermined? How much can we learn from the manipulation of information with very little context? To what extent can human workers become uninvolved information processors to match the technique-based connectedness of their organizations? How well will our 'technical shadows' serve us? In sum, the transition from a reliance on experience to a reliance on information, which now appears to be entering into its final stage, brings with it a range of issues that will confront the organizations of today and tomorrow. If many people are unwilling to get onto a plane exclusively entrusted to a flight management system, why should we leave the evolution of our lives and our world to information-manipulating systems? We have now come back to the kinds of questions and debates referred to earlier.

Organizational changes necessarily accompany the shift from a reliance on experience to a reliance on information. Successive generations of enterprise systems become the organizational backbone. This paves the way for entirely new kinds of organizations that, in terms of their technique-based connectedness, are centralized to unprecedented levels, but, in terms of their culture-based connectedness, can be highly decentralized. All this is well known. Companies can move their call centres to any place they choose, or permit people to work from their homes. Since the flows of information are structured by the enterprise system, it does not really matter where a particular department, unit, or individual works. People no longer have to work together in an office in order to get things done. At the same time, there is an extraordinary centralization because everything anyone does can now be monitored, supervised, and controlled to an extent that was impossible with face-to-face human supervision. It is very easy to count the number of keystrokes a word processor performs or the number of files manipulated by a caseworker. Many middle-management functions can now be automated; hence the reliance of these managers on face-to-face contact in order to tap into the metaconscious knowledge of their colleagues has been largely eliminated. An entirely new kind of centralized-decentralized organization is emerging that is much more significant than the phenomenon of downsizing (or right-sizing). These developments have continued to reshape management-labour relations. The strategy of substituting capital for labour takes on a whole new dimension with information-based enterprise integration, and along with it a growing mistrust of management by labour and vice versa. Workers fear growing underemployment and unemployment, which undermines any commitment to an organization; while management interprets this lack of commitment as a sign that their strategy is the only correct one.

9.4 The Price to Be Paid

There are very few, if any, studies of how work in a lean production or re-engineered setting affects human health and the quality of life.[97] This is not surprising since our present cultural unity (to be described in the next chapter) deflects critical attention away from technique. The best we can do, therefore, is to extrapolate the extremely robust findings of socio-epidemiology related to other forms of work to the new work situations created by organizations designed to make the most efficient use of information technology. This extrapolation points to the following hypothesis: if the precautionary principle had been in effect and if it had been extended to protect society as well as the biosphere, this development might have been postponed until better ways of preventing or greatly reducing its harmful effects on human life and society could be found.[98] If only things were that simple!

The relationships we have with our workplaces are a microcosm of the reciprocal interdependence between us and our surroundings, involving the previously discussed physical, social, and spiritual dimensions. Hence, workplaces can be alienating if their effects spill over into the remainder of our lives and thereby exercise a decisive influence. Similarly, workplaces can lead to both physical and mental illness. If the spillover effects onto the remainder of people's lives are substantial, hiring someone for a wage may lead to a condition of significantly 'possessing' his or her life, which raises moral and ethical concerns.

As a consequence, our relationships with our work and work settings may be classified along the following spectrum. On the one extreme, we find relationships in which the demands our work puts on us far outstrip our resources. Such resources include education, experience, creativity, motivation, imagination, and commitment. On the other extreme, we encounter relationships in which our resources far outstrip the demands. Either extreme can crush our humanity, be it through unbearable stress or destructive boredom. In the middle of this spectrum we encounter work that slightly stretches our resources, thus leading to creative responses through which we learn new things, expand our resources, and grow. In addition, these jobs provide a great deal of satisfaction derived from having successfully met new challenges.

Superimposed on this spectrum is the issue of control over our resources. These may be of little value if the work is so organized as to depend on the most minimal subset of these resources, as exemplified by the traditional assembly line and the emerging intellectual assembly line. Hence, a lack of control over our resources can further narrow the spectrum of healthy relations between ourselves and our work.

All this must be further assessed over time. Work experiences are the uninvited guests to the remainder of our lives. However, if after work we can replenish our resources and recover our energy by relaxing, eating, and sleeping, our work will not 'mine' our resources and our humanity. If, on the other hand, we cannot recover our resources, then our work may gradually mine them. How well we can recover our resources is affected by many factors, including how supportive and socially healthy our families and communities are, the quality of our nutrition and lifestyle, and the many stressors (or lack thereof) generated by our urban habitat.

Karasek and Theorell have synthesized the findings of socio-epidemiology related to work by means of their demand-control model,[99] which divides jobs into four categories according to the demands made by a work setting and the control people have over their resources in order to meet these demands (see figure 1). They have identified five independent variables (not of equal importance) by which work set-

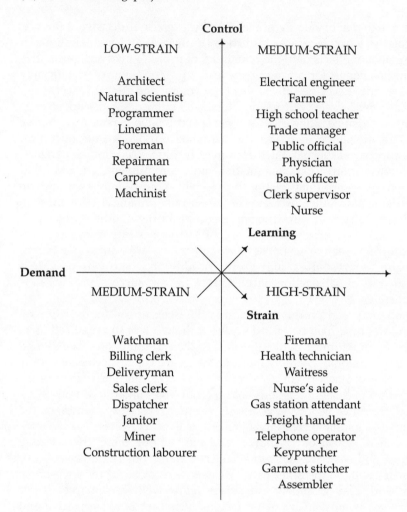

Figure 1: The Demand/Control Model

tings may be characterized: decision latitude, psychological demand, physical demand, social support, and job security. Not all of the demands imposed by work organizations produce stress and lead to disease. For example, they cite the findings of a study showing that engineers experience qualitative challenges as satisfying, but quantitative challenges (such as time pressures) as stressful. The difference is not difficult to comprehend. With sufficient control, the qualitative challenges permit engineers to creatively apply their resources, expand them by learning from these situations, derive satisfaction from success, and reduce initial stress levels. Quantitative challenges, on the other hand, permit learning but provide little satisfaction, and they don't reduce stress because there is always more to do. The two must therefore be distinguished in a model that attempts to predict unhealthy work situations that produce disease. Thus decision latitude (which includes skill discretion and decision authority) permits learning, the expansion of one's skill set, and the reduction of stress. Decision latitude was labelled as control and plotted on the y-axis, while psychological demand was labelled as demand and plotted on the x-axis. Their orthogonality was rooted in the fact that demand may represent a risk to human health while control does not. The remaining independent variables, namely, physical demand, social support, and job security, were found to be less important for creating healthier workplaces. This left Karasek and Theorell with a two-dimensional demand-control model. The four kinds of work experiences they identified were labelled as active jobs (high demand, high control), low-strain jobs (low demand, high control), passive jobs (low demand, low control), and high-strain jobs (high demand, low control). The active and passive jobs had medium strain. The model predicted that these four categories of work experiences would have different effects on human health, nonwork activities, and community participation. This is well supported by the evidence of socio-epidemiology.

The upper half of the demand-control chart reflects situations in which the influence people have over their work is more decisive than the influence this work has on them. The lower half represents the reverse situation. Mechanization, automation, and (later) computerization have pushed a growing number of jobs into this lower half.

The demand-control model shows that work can be designed to place relatively high demands on the skills and resources of workers but produce only modest negative effects on their health and lives, provided that they are permitted to exercise high levels of control. If, on the other hand, this control is minimal, serious negative effects occur.[100] Varying the level of social support has a much smaller influence. Further research has shown that control has two primary components: timing control and method control. The former refers to the ability of a person to pace and schedule his or her work, and the latter to a person's ability to select and use methods and resources he or she deems most appropriate.[101] Demand also has two primary components: monitoring demand and problem-solving demand. The former refers to the extent that passive monitoring is required, and the latter refers to the cognitive demands needed to prevent and recover from deviations.[102]

It is now possible to predict the effects of re-engineering work on human health with a high degree of probability. Organizing a business process in terms of an algorithm (including so-called expert systems) and databases operated by a machine, with an occasional step or decision assigned to workers, inevitably produces a substantial reduction in timing and method control. Variations in the work will be a matter of degree, not kind. In terms of Zuboff's important distinction between automating and informating, the former clearly dominates the scene.[103] This situation will lead to an increase in passive monitoring demand, since the worker is no longer the bottleneck in the information system handling a greater volume of work and thus information. The spectacular increases in performance provide a clear indication of how much more information must now be monitored. Stringent enforcement of this passive monitoring demand results from the system simultaneously monitoring the operator. The system keeps track of the number of finalized deals or transactions, and the quality of the service rendered can be ascertained from customer surveys. Problem-solving demands are bound to rise sharply, since a significant number of cases may be somewhat anomalous with respect to the categories provided by the algorithm. In the past, a clerk might make an annotation in the margin of a form, indicating the type of discretion that was applied, but most information systems leave no room for any kind of discretion. Yet the gap between the model of the world implied in the algorithm and databases of an information system on the one hand, and reality on the other, does not disappear.

Here we encounter the same problem as the one discussed for blue-collar work in Part Two. Brödner suggests that the more a production process is automated, the more these kinds of contradictions multiply.[104] At the same time, social support is all but eliminated as workers depend very little on others to get the job done, and when they do, communication is frequently mediated by the information system. The consequence is no more jokes or other interpersonal exchanges that provide some release from the intensive work pressures. Finally, to add insult to injury, job security is all but eliminated in the process of re-engineering. Significant reductions in skill levels are common, making workers more easily replaceable. Just-in-time labour becomes a reality as agencies supplying temporary workers thrive. It is not surprising that business process re-engineering as a management fashion undermined good management-labour relations almost everywhere.[105]

Interpreting these trends in terms of the demand-control model leads to the hypothesis that the restructuring of work to make a more efficient use of information systems increases demand and decreases control, thereby making work less healthy. Much of low-level white-collar work becomes what the demand-control model labels as high-strain jobs, which are the least healthy and most destructive of physical and mental health, social relations, and community involvement. In sum, it is probably difficult to exaggerate the enormous destruction done to human life and society that accompanies the further shift from experience to information and from a culture-based connectedness to a technique-based connectedness of human life.

Some statistics provide further clues. Between 1985 and 1995, more than 3 million white-collar jobs were eliminated in the United States.[106] Almost 8 per cent of the secretarial jobs (3.6 million jobs) disappeared in the United States between 1983 and 1993.[107] During the three years beginning in 1989, the wholesale sector lost 250,000 jobs.[108] It is estimated that in 1987 between 20 per cent and 35 per cent of all clerical workers in the United States were computer-monitored.[109] According to the American Management Association, more than two-thirds of U.S. firms now engage in electronic monitoring.[110] The trend will undoubtedly intensify as re-engineering opens up more jobs to electronic monitoring and as tightly coupled systems are more vulnerable to sabotage.[111] Manpower has become the largest U.S. employer. It is a temporary job agency whose employees have no job security and receive very low wages with minimal or no benefits.[112] Such just-in-time labour constitutes another important step in making human beings just another resource. Because of the strong downward pressure on wages in the United States, many people are compelled to work longer hours or obtain a second job in order to survive.[113] As more jobs move into the high-strain domain characterized by high demand and low control, the incidence of stress and mental illness increases. This was shown to be the case for white-collar workers operating a new computer-based billing system for a telephone company.[114] Over 60 per cent of Toyota workers have suffered chronic fatigue.[115] Based on research carried out at the Tavistock Institute, a growth in high-strain jobs is likely to increase smoking, alcohol and drug abuse, aggressive behaviour, and the like, which in turn undermine marriages, families and other groups, and thus the social fabric of any socio-economic strata affected.[116]

The trend towards deskilling, underemployment, and a decrease in control over one's work is steadily reaching up the organization to now affect middle managers and professionals. From the perspective of socio-epidemiology, the difference between blue-collar work and white-collar work is disappearing.[117] Computerization of work correlates with higher stress levels.[118] During the 1980s, more than 1.5 million middle-management jobs were eliminated, and during the 1990s, upper-middle-management executives were also affected.[119] Many of them had to settle for lesser jobs or no job at all. These trends have been confirmed in the automotive industry and the insurance industry.[120]

Professionals are not escaping this trend either. The restructuring of health care is putting more pressure on health practitioners as a result of stricter external monitoring and administrative control.[121] Gradually, the control medical practitioners have over the services they deliver appears to be slipping in favour of the control exercised by the increasingly computerized health insurance organizations.[122] As levels of stress and job dissatisfaction increase, a growing number of hospitals are turning to stress management seminars, but this end-of-pipe service fails to get to the root of the problem.[123] Among professionals, temporary work is on the rise.[124] Engineers in the U.S. automotive industry are now experiencing layoffs, outsourcing, and diminished benefits as a result of re-engineering.[125] Higher levels of management are also turning to stress management seminars.[126] It would appear that the technique-based connectedness residing in information technology is beginning to dominate entire corporations, although the intensity may decline somewhat in the inner core of the technostructures. Behind this development and its statistics we find countless disrupted lives and a great deal of social breakdown. It is also creating systems of a scale and complexity that make them inaccessible to knowing and doing embedded in experience, thereby requiring a further development of technique.

I will conclude this brief overview by pointing to one last feature that the shift from culture-based connectedness to technique-based connectedness is having on human work. It is what Martin Shain has called the fairness connection.[127] High-strain work characterized by high demand and low control not only produces stress that leads to physical and mental illness. It can also increase the incidence of accidents and injuries on the job. Perhaps even more important is the fact that high-strain work will be experienced in terms of any modern culture as being profoundly unfair. After all, raising demand while reducing control is analogous to increasing the workload of employees while restricting their ability to use their physical and mental resources. When, in addition, their wages and benefits are lowered and their occupational status is reduced, feelings of frustration and resentment will further compound the consequences of stress. The more employees still have a work ethic that is somewhat intact, the more they will suffer from their inability to do a good job, thereby depriving them of what little satisfaction and pride they might have had in their work. Employees may consciously or unconsciously decide that they have a better chance of surviving if they shed any remnant of a work ethic, but this has a terrible price since it requires a partial denial of a person's moral integrity. Since experiences of work are internalized in the brain-mind, they accompany the employees everywhere they go, thus creating the well-documented spillover effects. All this is particularly acute when employees are blamed for an accident or are injured at work because they intuitively sense that, in the majority of cases, this could have been prevented had the balance between demand and control given them a fair chance. The workplace thus intrudes into the lives of employees as a constant uninvited guest in their thoughts as they carry on their daily-life activities. If feelings about their workplace and the role they play in it are profoundly negative, their mental health may be weakened and they may be more vulnerable to stress at work. A destructive cycle is begun in which these negative feelings make it more difficult to go home and eat, relax, and sleep to recover the energies expended at work. If this recovery is interfered with, they will be less ready for the next day's work, thus gradually mining their physical and mental health on the one hand, and their pride and self-respect, and thus their ability to respect their neighbours as themselves, on the other.

Martin Shain sums up current research as follows:

In order to understand how fairness at work affects both the chances of getting sick or injured and recovery, we need to look more closely at what is coming to be called the social biological translation ... This has been defined as a mechanism by which human beings receive messages about the social environment and convert these messages into biological signals that trigger the processes of disease development or health promotion. Key to the social biological translation is the biochemistry of emotion. In recent years, much has been learned about emotions and their effect on the body ... For present purposes, our interest lies in what we may call the 'biochemistry of fairness.' Fairness is a term we encounter or use just about every day but it is nonetheless invested with many different meanings. Here I want to focus on fairness as keeping promises and unfairness as breaking promises. In particular, I want to characterize the employment contract as a set of promises ... When employees perceive that one or more of these express or implied promises have been broken, they are likely to experience a range of negative emotions. If it is correct, and it appears to be, that conditions of work characterized by high demand/ high effort and low control/low reward are seen by many employees as breaches of the employment contract (I did not sign on for this, this is unfair), then a cascade of emotions can be predicted to flow from this perception that includes feeling to one degree or another:

- excluded
- tricked
- rejected/abandoned
- disliked
- unworthy/worthless
- diminished/humiliated
- shamed
- anxious/agitated/insecure
- depressed
- angry/enraged
- suspicious
- helpless.[128]

Shain points out that if such feelings are continuously reinforced and even aggravated during occasional more serious violations of the perceived contract, people may develop a sense that 'nothing and no one can be trusted; there is no order, purpose or meaning in life; things don't make sense; all is not right with the world.'[129] In other words, perceived unfairness at work strikes at our feeling of being whole in a world that is more or less whole, or, to put it in another way, at our ability to live our individual and collective lives. Whatever the ways people find to cope with the situation, the common denominator is destructive of life, be it through finding consolation in the bottle, an escape with drugs, the putting on of a mask of indifference, a hiding behind aggressive violent behaviour, or a retreat from oneself and the world into passivity. In all such cases, the workplace is consciously or unconsciously perceived as the reason for our incompleteness.

This raises the question of the legitimate authority of employers. Does ownership bestow the right of disempowering people by causing 'reduced adaptability, reduced ability to cope with change, impaired learning, impaired memory, increased helplessness, increased passivity, or increased aggression/conflict' and increasing the rates of 'heart/circulatory diseases, immune function disorders, some cancers, mental disorders, and substance abuse.'[130] There is no possibility of obtaining the consent of their employees, with the result that techniques of human relations must be relied on to mitigate strong feelings of broken promises in an end-of-pipe fashion.[131] At one point, the West had moral and cultural resources to resist what in Judaism and Christianity has long been recognized as the threat of wage-labour, leading to employers taking hold of the lives of their employees and even spiritually possessing those lives. However, as technique progressively changed people in its own image, thus causing the weakening of traditions and the secularization of societies, we were left with few moral and cultural resources to resist the way technique-based connectedness shuts us into the 'iron cage' against which Max Weber warned us nearly a century ago. Technique alters our consciousness so as to make us almost unaware of our condition, and weakens any cultural values that would permit our escape from it. (A general understanding of how this is accomplished is the subject of the next chapter.) As far as work is concerned, what this means is that we need to examine the possibility of society interpreting the evolution of work incorrectly. We may have to face the possibility that the most important legacy left to us by technique-based connectedness will include unhealthy and disempowered human beings, societies that must be held together by integration propaganda, and ecosystems that are in decline to a point we cannot ascertain. If this turns out to be the case, then offices and factories may well be the most decisive means by which this technique-based connectedness threatens all life. It is in this context that we need to remind ourselves once again of the warning of the Surgeon-General of the United States about the alarming increase in the incidence of mental illness, reflected in the fact that associated health costs are rising at 7 per cent per annum.[132]

9.5 Living with Information

We began this chapter by showing that most of the new and important phenomena of the second half of the twentieth century were rooted in a rapidly expanding technique-based connectedness within human life

422 Our Third Megaproject?

and society. Nevertheless, each new generation continues to be social-ized into a culture-based connectedness of individual and collective human life, with the result that, as adults, we depend on this cultural orientation. No one directly encounters technique, any more than we encounter entities such as our society or the economy. These concepts help us make sense of daily-life experiences. Since technique is not part of this stock of concepts, we get along without it, possibly in the follow-ing way. We have learned that behind the daily-life world are many experts working on the technical frontier to develop better means for meeting our needs and aspirations. By way of reports from the mass media we hear about their successes of today and their expectations for tomorrow. These reports are part of an extraordinarily fragmented montage of images of our contemporary world. Public opinion and various beliefs weave these images together by means of themes such as the importance of the work of experts to keep our nation competitive in an increasingly global economy, our stake in this via financial mar-kets and their impact on our investments and pension funds, political or environmental threats to peace and security, and the expansion of 'disease care' by new miracle drugs. Were the concept of technique to become a part of this, it would surely refer to something without which our lives and the world would be unthinkable. To most people, it would represent the 'system,' whose workings are largely opaque and whose effects on people's lives might as well be taken for granted since you cannot fight the 'system.' In the higher socio-economic strata, people might identify more closely with this 'system,' because of expectations that as knowers they can make a difference. Since nobody encounters the 'system' directly, many of these opinions and beliefs can readily coexist with a reality that functions quite differently. There is little incentive to deal with this possibility because it would not 'sell' in any case. At present, the concept of capital plays the role that the concept of technique should play, even though the reality to which it refers has long passed us by. The accompanying 'secular theology' permeates the media as an expression of a false consciousness.[133]

On the technical frontier, experts experience their involvement in the 'system' very differently.[134] They have little awareness of the meaning and value of what they do, and even less of the undesired consequences that accompany the desired ones. Since the process of technical desymbolization is largely separated from the process of symboliza-tion, the technical frontier is a primary locus for emerging beliefs and ideologies springing from the extrapolation of what is happening in the laboratory to human life and society. Little or no attention is given to the adjustments that may have to be made to accommodate the new technical advances, or, as we have put it thus far, of 'technique changing people.'

The information scientific and technical specialists work with represents a 'technical shadow': what is left of something after a triple process of abstraction that reveals its place and importance for the local technique-based connectedness. Although such information may be adequate for their work, specialists have to give it some context when communicating with others, and still more context in daily-life settings. A connection with their own lives and the lives of others has to be made, and for this they can only rely on the other part of their structures of experience, that is, their culture. Desymbolized as that culture may be, it registers something of how technique-based connectedness has enveloped culture-based connectedness, thereby making symbolization next to impossible. All that can typically be achieved is some vague reference to strengthening the economy, adding to human well-being, supporting the nation, or creating a future based on a new kind of development that somehow will become sustainable. Public relations firms take the best of these efforts and make them work on the mass media to create the most effective justification of what is happening. This all adds up to what Jacques Ellul has referred to as a technical bluff.[135] It comprises those connections that seek to anchor everything in the greatest good known by a community as symbolized by its cultural unity, although in a greatly weakened form.

As far as technology is concerned, we live from one bluff to another. Simply by calling the latest technologies 'high' (and, by implication, older technologies 'low'), the bluff was that the former would finally be compatible with human life, society, and the biosphere. We would get along with fifth-generation computers. Equipped with artificial intelligence, these computers would help us surmount all our difficulties, as if life were a set of technical problems. When we could not help but be worried about the many mysterious ailments encountered by our friends and neighbours, the mass media assured us that our physicians would have the right drug for each of us. Even today, this pharmaceutical and medical bluff ignores the fact that we produce diseases at a much faster rate than we are able to manufacture cures and there is little chance of catching up, even if we could spend more money on it. Lean production is replacing 'fat' production, creating a kind of 'anorexic' humanity in its workers. The bluff of many material things is that they can satisfy non-material wants. The Internet can deliver friends, information high-ways can decongest regular highways, and SUVs can make us feel more secure when faced with drivers desperately trying to shave a few minutes off their commuting time. The endless talk about human rights and democracy ignores the conformity imposed by technique. Virtually without exception, when governments fear that their citizens may impede the spread of technique-based connectedness, democracy is suspended, as in the case of free-trade negotiations or the non-labelling of GMOs. The new conformity is making the world safe for technique, even when this makes it unsafe for people, communities, and ecosystems.

The contemporary university greatly contributes to this situation. Since, on one side of the campus, the intellectual 'worlds' are full of technology and little else, and on the other side full of everything else and little or no technology, it is next to impossible to understand how so many things reinforce and expand technique-based connectedness as opposed to culture-based connectedness. Underemployment and un-employment should be examined in terms of the 'work niches' created, modified, and eliminated by technique-based connectedness, including the fact that technique feeds on its own problems. To understand why we are running out of time in our lives despite the ever-growing num-ber of time- and labour-saving technologies, it is essential to under-stand that technique-based connectedness not only has enveloped culture-based connectedness but forces the latter to move along at its growing pace. The rate at which we can produce time-saving technolo-gies will be outstripped by the rate at which technique speeds up the pace of life. The evolving architecture of microprocessors must first be understood in terms of its becoming a local manifestation of the larger technique-based connectedness and how this in turn affects culture-based connectedness. Our future will to a significant extent depend on the rate at which technique can resolve certain kinds of difficulties versus the rate at which it creates new ones. Human freedom must be understood in relation to the growing incidence of computer-based monitoring and surveillance. We can all add many more examples of this kind.

Many of the theories that were supposed to shed light on what was happening to humanity in the second half of the twentieth century added to this technical bluff. We could acknowledge the problems of living with the mass media by finding consolation in the belief that, thanks to the mass media and the Internet, we were on the way to the global village, in which we would all have our place. We could acknowledge the problems of industrial societies, in the assurance that something entirely new and better awaited us in the post-industrial information societies of the near future. We could acknowledge the many problems associated with the application of science and technology with the assurance that, simply by advancing a bit further, a qualitative change would follow. The few theories that were truly iconoclastic stripped off some of this bluff. Fortunately, they were so rare that they could easily be dismissed. After all, they denied humanity its much-needed secular religious assurance and justification, thereby sealing their heretical status. It may be argued that again 'there is nothing new under the sun': human life, as far as we can tell, has always been alienated, and the addition of reification changes little. Perhaps this is the case, but, given the growing power of technique and hence the growing magnitude of the consequences, a quantitative difference could translate into a qualitative one.

The most irreversible impact of the shift from experience and culture to information and technique may well be on symbolization, and consequently on what it is to be human. Humanity has always used symbolization, first to impose a culture-based connectedness on human life immersed in nature, and later on human life immersed primarily in society (and secondarily in nature). Can symbolization deal with technique-based connectedness encompassing culture-based connectedness, since it results in a process of desymbolization? Can symbolization deal with experience and culture being flooded by information that comes from a technical approach which filters out much of what is qualitative? Can symbolization survive the computer, which ultimately requires that everything be expressed in terms of zeros and ones, with the result that nothing can be simultaneously zero and one, let alone something more complex? How will this affect our ability to make sense of and live in a socio-cultural world that is frequently contradictory, dialectical, and filled with enfolded wholes? How will the shift from experience to information affect social relations and the ability of people to derive metaconscious images of their social selves and the social selves of others from these relations? Will integration propaganda be able to compensate for any serious negative effects? By its influence on symbolization, is our commitment to technique a gamble with our lives? These kinds of questions bring us to the subject of the last chapter: 'technique changing people.'

chapter 10

Remaking Ourselves in the Image of Technique: Culture within Technique

10.1 Technique as Phenomenon

In the previous chapters, we have seen that 'underneath' the many new phenomena of the latter half of the twentieth century there was an extraordinary stability, namely, a steadily expanding technique-based connectedness within the so-called industrially advanced nations. It gradually acted as a network of the most efficient transformations exchanging inputs and outputs of materials, energy, labour, capital, and information with one another without any decisive reference to their meanings and values for human life, society, and the biosphere. It was as if technique had reorganized the world to increase the efficiency of certain flows of goods and services by ignoring all other forms of connectedness, with a mesmerized humanity looking on, convinced that the undesired consequences would be a small price to pay for the desired ones. In proposing this ideal type, I am not suggesting that technique is some mysterious force that gradually took control in the second half of the twentieth century. The ideal type does not require such an entity. It requires only a recognition that the relationships between people and their surroundings are reciprocal in character, with physical, social, and spiritual dimensions. Everywhere people and societies began to make, break, and transform the relations with one another and the world by means of the technical approach, and to the extent that this changed their world and themselves, a self-reinforcing cycle was created: the more people changed technique, the more technique changed people. This ideal type refers to a particular set of conditions under which the influence people have on technique appears to be less decisive than the influence technique has on people.

Such a situation can be stable if the latter creates an appropriate cultural unity. This cultural unity must support an integration propaganda that makes people keenly aware of how technique is being used to create a better tomorrow, but which simultaneously veils how it undermines that future. Jacques Ellul created the concept of technique to describe this situation.

The validity of this concept can be tested in the following manner. We will examine the characteristics of technique-based connectedness and its evolution. If these characteristics match the cultural values and aspirations of the societies of the time, then technique is under human control and the ideal type misrepresents the situation. If, on the other hand, these characteristics violate these values and aspirations (except those induced by technique), then the ideal type holds. Technique then becomes an entity like the economy or the state, whose influence on human life and society is much more complex than what is intended by the human activities that give rise to it. It is then entirely possible for human beings to (involuntarily) build and enclose themselves in the 'iron cage' of Max Weber or to be possessed by the technical spirit of their age, as suggested by Jacques Ellul (a literal translation of the French title is not *The New Demons* but *The Newly Possessed*).[1] Since this possibility has frequently been dismissed out of hand, we would do well to reflect on how people could end up serving their own creations before we get into the analysis.

Metaphorically speaking, the situation in which technique has a more decisive influence on human life than people have on technique (their own creation) may be compared to one in which an elderly couple on a fixed pension do not have the funds to replace their old furnace. During a particularly severe winter, the furnace runs continuously; and, rightly or wrongly, they believe that by turning down the thermostat and permitting the furnace to stop once in a while, they have a better chance of its lasting the season. The condition of the furnace has now affected the set point on the thermostat, signifying an adjustment of their value of comfort to the state of their heating technology. In the same vein, a society may deem it necessary to adjust its values and aspirations to have a chance in the new global economy, which is decisively shaped by technical development. It may be argued that the second half of the twentieth century saw people and societies go from one adjustment of their values and aspirations to another, without getting around the issues that stood in the way. They continued to brilliantly succeed in the domain of performance values, which remained indissociably linked to difficulties in the domain of (non-technical) context values.[2] People were concerned about the environmental crisis, but despite several decades of an exponential growth in environmental regulations, there was little effect on the trends these were meant to change.[3] People were concerned about the problems of unemployment and underemployment, but many nations had to reduce unemployment benefits to control costs. Similar concerns over public health care, education, a growing dependence on food banks, and homelessness did not translate into effective initiatives. In democratic societies, this ought to have raised concerns about who or what guides and controls what is happening. The traditional explanations to the effect that the transnational corporations had too much influence and that politicians were either ineffective or corrupt simply will not suffice. If most citizens had been really convinced that things were heading in the wrong direction, they could have taken to the streets; even in totalitarian societies this can bring about fundamental changes. When examining the influence and control of any group or institution, it is necessary to analyse the power it exercises and the extent to which people's values and aspirations permit them to tolerate the situation. Hence, I will explore the extent to which 'technique changing people' transforms their values and aspirations so as to legitimate the growth of technique-based connectedness by interpreting it in terms of the culture-based connectedness. If this influence turns out to be considerable, a social and historical situation will have been created in which human life and society evolve as if technique had its way most of the time. Those who exercise power and those who are subjected to it will then be justified. Moreover, such justification must be absolute, that is, anchored in a cultural unity.

The characteristics of technique-based connectedness, and therefore of technique itself, follow directly from our previous analysis. Rationality is the first characteristic. During prehistory and most of history, making things 'better' could not possibly have been conceptualized in terms of efficiency because this would have required a culture-based connectedness that had either been undermined or that conceptualized the world in mechanistic terms. Making things 'better' involved putting them in the widest possible context of human life via experience and culture. As noted, doing so on the basis of technique restricts any consideration of context to the desired outputs and requisite inputs essential for the surrounding technique-based connectedness. The dialectical and enfolded nature of any activity is thus reduced to a rational schema for converting inputs into outputs with the greatest possible efficiency. The technical approach to life focuses exclusively on what is rational in terms of striving for the greatest efficiency and externalizing everything else. The result is, as we have seen, an undermining of the integrality of the activity, a reduced compatibility between it and the local fabric of culturally established relations, a reduced ability of the activity to contribute to this fabric by self-regulation based on negative feedback, and the barring of genuine human participation. The latter makes it progressively more difficult for people to live on the level of experience and culture.

A second characteristic of technique and the evolution of technique-based connectedness is the necessary linking together of techniques. Were it not for this fact, the rationalized relations involved in any activity would have been like loose threads quickly unravelling themselves from the fabric of culturally established relations. I have shown that industrialization involves the necessary linking together of technical improvements by means of a chain-reaction-like process. It ensures that there are no loose threads by making technically improved activities the backbone of the new fabric within a fabric, which we have referred to as a technique-based connectedness within a culture-based connectedness. The cultural unities of the nineteenth century endowed the former with meanings and values that also permeated the latter, thereby integrating the two in a new kind of society that gradually reversed the primacy of culture over technology. I will show how the present cultural unities continue to legitimate this situation.

A third characteristic of technique and its evolution is the absence of human choices based on meanings and values. The evolution of technique-based connectedness is characterized by an automatism of technical choices. The weaving of rationalized relations into the cultural fabric involves no human decisions in the usual sense. Performance values automate human choices. For example, an output of 1.43 tons per hour is 'better' than 1.41 tons per hour. In fact, 1.45 tons per hour would be even better, and there is no reason to stop there. For technique-based connectedness, more is always better. The only real human choice is an implicit one, namely, that performance values (in actual fact, technical ratios) take precedence over any cultural values. All limits traditionally imposed on human activities by values lose their meaning and legitimacy in the relentless drive for ever-greater performance and power. We know how important such limits are in the personality development of children, as they learn to live their individuality and freedom in the face of these limits. This prevents life from deteriorating into a random Brownian movement in which all possibilities are equally meaningful and valuable, which means that they have no meaning and value at all. On the collective plane, desymbolization leads to a weakened role for experience and culture, with the result that a community's cultural unity must be reinforced by integration propaganda. Technical 'choices' automatically decided by performance values guide technical and economic growth on their own terms. In the past, civilizations collapsed when their ways of life were unable to give purpose and direction to the lives of their members. This guidance still occurs, but in a manner that is framed by and derived from the automatism of technical choice. I am aware that technical growth is commonly conceptualized as occurring within economic constraints, and that these in turn are under political control. Such conceptions have more to do with patterns of evolution that preceded the twentieth century than the reality that subsequently emerged. Much of the technical frontier is now located in the large public sectors of the economies of the industrially advanced nations, while the economic revised sequence and its political equivalent permit us to hang on to illusions of freedom and other cultural values to compensate for our ethical irrelevance. In the world of common sense based on experience and culture, we know very well that more is not better (as in the case of food, for example), and that more in one sphere has implications for all others. Hence the need for limits.

Another characteristic of technique and its evolution is its self-augmentation. Every rational thread automatically woven into the cultural fabric of relations engenders others. As it distorts this fabric by rationalizing certain threads and ignoring all others, major consequences for human life, society, and the biosphere are inevitable. The most serious ones must be addressed; and, given the technical orientation of contemporary cultures, what better way to do so than by the technical approach to life? As a consequence, each new category of techniques must be accompanied by others to deal with whatever consequences might interfere with their operation. Only very rarely are problems traced back to their roots because this would involve a calling into question of technique itself. The present intellectual and professional division of labour and its accompanying institutional framework rule this out, and the cultural unity ensures that anyone who criticizes technique is quickly marginalized and pushed to the fringes of society.

The result of all this is that, in the overwhelming majority of cases, technical growth feeds on the problems it creates for human life, society, and the biosphere, thereby trapping us in the labyrinth of technique.[4] For example, the techniques of production led to an environmental crisis, which in turn created the necessity of environmental techniques that compensate in an end-of-pipe fashion without getting to the root of the problem. Techniques of dilution, dispersion, containment, and mitigation led to an entirely new industrial sector.[5] Pollution prevention and many other preventive approaches continue to play a minor role and are not integral to contemporary engineering and management practices.[6] Since the problems do not go away, the public perception of the situation must be managed by public relations techniques.[7] Our techniques for organizing human work necessitate end-of-pipe compensating techniques such as job rotation, job enlargement, ergonomics, industrial relations, human factors, and quality-of-working-life approaches. The only exceptions are socio-technical approaches that attempt to go to the root of the problem, which have led to fascinating but (unfortunately) marginalized innovations. In addition, there has been a proliferation of end-of-pipe social and health services, human relations techniques, stress management approaches, and others. Other techniques act as coping mechanisms, for example, those involving drugs such as Prozac. Yet another set of techniques provides passive leisure-time activities to help us cope with the spillover effects of work on the remainder of life. The urbanization of humanity that was enormously accelerated by the advance of technique is interfering with the ability of people to recover themselves from work by separating daily-life activities in time and space; and here as well new compensating techniques are being developed. Why is this happening? There is no lack of preventive alternatives that partly or wholly go to the root of the problem.[8] It is because technique-based connectedness now so dominates culture-based connectedness that we spontaneously turn to the former to deliver us from all our problems.

There is a second dimension to the self-augmenting character of technical evolution. As mentioned earlier, technical evolution has two primary components. The first involves the usual inventions and innovations as responses to unsatisfactory situations, new opportunities, or imaginative breakthroughs. The second component results from the flow of technical information within the system, causing each area of specialization to act as both a receiver and a transmitter. This is extensively exploited in modern societies that are bent on technical progress. Because this second component is largely independent of human meanings and values, it contributes to the self-augmenting character of technical evolution.

The last characteristic I will briefly mention is the universality of technique and its evolution. The weaving of the new technical relations into the cultural fabric of societies around the globe is having much the same results everywhere, regardless of local cultures and ecosystems. The reasons for this follow directly from what we have already noted. The new relations are first and foremost rational because of the way everything non-technical is excluded by the technical approach to life. In other words, to make things technically 'better' is to make them universal and global, since everything unique to a local culture or ecosystem is excluded in the process. The remaining diversity has two sources. First, a limited technical diversity survives if there is no measure of overall technical superiority.[9] Second, a cultural diversity remains insofar as it does not interfere with the evolution of technique-based connectedness. It compensates for the effects of technique on human life by leaving us some elements of experience and culture, such as traditional foods, customs, and manners, to provide an illusion of a unique identity. However, in the public sphere and the evolution of a way of life, only technique is deemed acceptable.

In setting out some of the characteristics of technique in its evolution, my analysis converges with the characterology of technique, described by Jacques Ellul in *The Technological Society* as rationality, artificiality, automatism of technical choice, self-augmentation, monism, the necessary linking together of techniques, universalism, and autonomy.[10] In *The Technological System*, Ellul refined this characterology by distinguishing between the characteristics of the phenomenon (autonomy, unity, universality, and totalization) and the characteristics of technical progress (self-augmentation, automatism, casual progression, and absence of finality, and the problem of acceleration).[11] I will not review his many examples and arguments except to point out that this characterology, first published over fifty years ago, is even more accurate today than it was at the time. Weber's and Ellul's works are among the very few that, with hindsight, correctly grasped the fundamental changes in human life, and their evolution, that occurred during their time.[12]

It may be helpful to summarize the differences between the characteristics of the technique-based connectedness of human life and those of its culture-based connectedness. The growing dominance of the former over the latter represents a shift from a dependency on experience and culture to a dependency on information and technical rationality; a shift in the reliance on human values to an automatism of technical choice; a shift from an order that evolves from both internal and external challenges to human life to one that mostly feeds on the problems it creates; a shift from activities evolving in the immediate context while enfolding something of a culture's design for living to activities evolving in a technical context; a shift from a culture-based evolution to a self-referential evolution; and a shift from cultural values to performance ratios autonomous from local contexts. Finally, what represents order for the culture-based connectedness of human life is a disorder for technique-based connectedness, and vice versa; and what appears reasonable and rational to the former is irrational for the latter, and vice versa. It is as if technique seizes certain relations from the fabric of human life, society, and the biosphere, transforms them on its own terms, and thereby strengthens itself at the expense of everything else. The 'as if' reflects the fact that there appears to be little or no decisive human intervention in this pattern of events. To the extent that these characteristics of technique describe the behaviour of people and their societies, these actions are no longer guided by human values and aspirations apart from those resulting from the influence of technique on culture.

For the case of engineering and technology (a major component of technique), I have quantitatively confirmed that, in North America, future engineers learn next to nothing about how technology influences human life, society, and the biosphere, and even less on how to use this understanding to adjust design and decision-making to ensure that they get the job done while at the same time preventing or greatly minimizing undesired and harmful effects.[13] In other words, the context undergraduate engineering students encounter in their technical courses is essentially limited to technique-based connectedness. I have also quantitatively verified this for current engineering methods and approaches in the areas of application dealing with materials and production, energy, cities, and the organization of work, while at the same time documenting preventive alternatives in four corresponding annotated bibliographies.[14] For technology (as a major branch of technique), this confirms its autonomy, that is, that the influence technology has on people outstrips the influence people have on technology.

10.2 Technique as Life-Milieu

To continue exploring how technique changes people, it is essential to recognize that technique is not simply an ensemble of means rationally arrived at to obtain the greatest possible efficiency. These means also interpose themselves between human beings and between them and the world, thereby mediating these relationships. In contemporary society, many daily-life contacts with others are mediated by means of one technique or another, as we use the telephone, voice mail, fax machines, cellphones, computers, and the Internet. Much of what happens in our communities, nations, and the world is mediated by the mass media. In factories, machines mediate between workers and the materials of production. In offices, software packages increasingly mediate relations with clients, suppliers, vendors, and other employees. However, the role of technique as intermediary is not limited to machines. Public relations, human relations, group dynamics, organization theory, advertising, therapy, integration propaganda, and the like also mediate between individuals, between organizations and individuals, between individuals and events, and between individuals and the world. There are very few daily-life experiences that are not mediated by techniques.

If these mediations by technique are not neutral, they in effect create a filter between us and others and between us and the world, to the point that it might be argued that we experience ourselves, others, and the world through technique; or, in the extreme, technique might be what prevents us from having experiences in the cultural sense of that term.[15] It is essential to carefully examine how technique mediates our relationships and to what extent it takes over, permeates, or reduces the role of culture. The desymbolizing influence of technique must be understood in relation to the symbolic role of culture.

The non-neutrality of technical mediation can be illustrated by a few examples. When young children first use the telephone, they must learn that certain behaviour, such as pointing, nodding their heads, and even observing what goes on around them, does not pass through the wire. When a face-to-face conversation is mediated by the telephone, the eye etiquette, body language, and conversation distance are filtered out, greatly impoverishing the culture-based connectedness of the two parties. This is why many people do not like to use the telephone for discussing deep personal matters. We gradually learn that we really know nothing about most of our political leaders except the carefully crafted images created by their handlers and public relations experts. Anyone who has ever participated in a demonstration or a similar public event and then watched the media coverage of that event knows the spin that can be put on it simply by decontextualizing images and recontextualizing them in a sequence in order to deliberately create desired impressions. Thanks to carefully crafted store layouts, we tend to walk out with more than we came for; our opinions about people and events are a cause and effect of public opinion; and even our morality is extensively shaped by the media. People who spend a great deal of time with computers may have difficulties with intimate relations;[16] and those who spend a great deal of time on the Internet 'communicating' with others have been found to suffer from a higher degree of loneliness and depression.[17] Much of this kind of evidence would suggest that technically mediated relations with others and the world are far from neutral, and that possibly it is not television that is the message but rather technique.[18] When relations of all kinds are mediated by technique, their culture-based connectedness is greatly impoverished.

Many technical systems now interpose themselves between us and the biosphere to provide us with all inputs of matter and energy and to take from us all outputs of waste generated by the transformations and uses of these inputs. These systems provide us with everything we need to sustain our lives, including food, water, energy, and materials; and they take care of all our wastes, such as garbage, sewage, and waste water. These systems support others that in turn provide us with manufactured goods, transportation, entertainment, information, and much more. It follows from our analysis of production systems in Parts One and Two that this mediation is not neutral either. For example, food systems mediate between us and the cultivation of the earth for sustenance. Of course, grains, vegetables, and fruits continue to be grown. We have seen that the reorganization of agriculture by technique has created agribusiness. That the mediation of agribusiness is far from neutral becomes obvious when we compare the taste, appearance, and nutritional value of something grown by the system with its equivalent, organically grown from traditional seed in a backyard. The gap between the two continues to widen as crops are now genetically modified to better suit the system. Not only the taste of processed food, but also that of most vegetables and fruits, is so poor that a great deal of advertising is required to convince us that much of what we find in our supermarkets is desirable and nutritious. Researchers who seek to impose human values and thereby conclude that some aspects of the food system are not in the public interest risk losing their employment if they persist. Public universities, fearful of losing research funding from agribusiness, are no longer a supportive environment. If, by referring to human values, researchers within the system become convinced that something is wrong, the public relations industry is called upon to undermine their credibility and discredit anything the mass media may pick up.[19] Since nearly half of humanity now lives in large metropolitan areas, such food systems are vitally important, but that does not mean their evolution is guided by human values.[20]

Elsewhere, I have shown how different this mediation could be if the design, construction, and operation of these systems were guided by human values as well as performance values.[21] By using human values, many markets for commodities could be transformed into markets for the services these commodities provide. For example, consumers are not interested in electricity as such but in refrigerating or cooking food, showering with warm water, illuminating rooms at night, and watching television. If markets for electricity were converted into markets for energy services, the interests of utilities and consumers would become aligned in creating a system that is not merely efficient in the production and distribution of electricity but also in the efficiency of its end use. Such a realignment could increase the competitiveness of an economy, make utilities more profitable, raise the standard of living of many people by lowering their energy bills, and substantially reduce the burdens imposed on the biosphere. A comprehensive effort to create genuine service economies would permit the delivery of what people are really interested in with a far lower throughput of matter and energy.[22]

The non-neutrality of the mediation by technique should come as no surprise, since it follows directly from the technical approach to life. Everything is studied for the purpose of making it technically 'better,' as opposed to improving the contribution it makes to human life and the world, assessed in terms of the meanings and values of a culture. After abstracting a 'chunk' of the connectedness of reality, its integrality is ignored in favour of those relations that connect the desired outputs to the requisite inputs, thereby filtering out everything that makes it whole. Whether that is an individual human life, group, community, society, ecosystem, or the biosphere makes no difference. Everything must be reified and commoditized before it can be manipulated by technique and its institutions. The effects of technique as mediation are not uniformly distributed throughout society, because people in different socio-economic strata differentially participate as knowers, doers, or controllers in sets of individual techniques depending on occupation, leisure-time activities, and the urban neighbourhoods in which these activities take place. Nevertheless, there is a substantial common denominator. People who work extensively with computers suffer from technostress and encounter a 'second self.'[23] Living in a modern mass society leads to the 'other-directed personality' in the 'lonely crowd,' the 'minimal self,' personalities fundamentally dependent on integration propaganda, a 'one-dimensional man,' a 'Protean man' whose connections with society and nature are broken, and creates cultures that are narcissistic, cynical, shot through with the magic of technology, and permeated by secular political religions.[24] The permeation of people's lives by work leads to the 'organization man,' 'the technical man,' and the like.[25] It seems to me that all these analyses may be interpreted as recognizing implicitly that experiencing much of daily life through one set of techniques or another has profound effects on human life and culture. A concept of technique as mediation does not require that all individual techniques filter out the same aspects, as Wittgenstein pointed out for the characteristics of games, chairs, and leaves.

There are a number of fascinating but somewhat specific studies of how this or that technology or technique mediates a particular relationship. For example, Giedion has carefully examined the non-neutral mediation of mechanization on activities such as bread-making,[26] and Schivelbusch has meticulously described how trains mediate between the traveller and the landscape in a very different manner than did horse-drawn coaches.[27] All such observations, fascinating as they are, would be relatively trivial were it not for the fact that such effects are endlessly repeated in many daily-life activities in societies with a high density of techniques. Technique as mediation filters out many things that are essential to individual and collective human life and the world in which it is lived, leaving a greatly impoverished culture-based connectedness. When the density of technically mediated relations becomes very high, there may be many things we can no longer experience, learn about, or take action on. This may affect our understanding of all life, since it is being reified by technique. From this perspective, we are becoming an unconscious civilization.[28]

The changes described in previous chapters imply a growing density of technique-mediated relations between people, between people and their society, and between people and the biosphere. As most other relations are gradually crowded out, a qualitative change in human life and society occurs. It is as though technique forms a cocoon around human life through which everything else is experienced, and consequently in relation to which human life evolves. All the necessities of life derive from technique, therefore any technical malfunction can be life-threatening. In other words, technique now becomes the primary life-milieu through which the secondary life-milieu of society is encountered, and in turn through which the tertiary life-milieu of nature is experienced. The pattern is somewhat analogous to what happened when societies began to interpose themselves between prehistoric groups and nature, thereby becoming the primary life-milieu in relation to which human life evolved. As discussed earlier, it is well known how the primary life-milieus of nature in prehistory, and society during history, influenced human consciousness and cultures. Technique can be expected to do the same. Technique, by means of various systems, provides us with all the necessities of life, but it also presents us with entirely new dangers: the threat of total war, 'normal' accidents, poisons creeping up the food-web, unsustainable ways of life, and the impairment of the biosphere's life-support services. Possibly the greatest danger, however, comes from the way the life-milieu of technique interposes itself between human beings and their secondary life-milieu, society, with its diminishing culture-based connectedness. As such, it contributes to desymbolization and the further weakening of the role of culture, creating an entirely new kind of consciousness in the process.

10.3 Technique as Consciousness

As we have seen, the culture-based mediation of any relationship with others and the world is not piecemeal but involves a system of signs that is coextensive with the life of a community and its experience of the world. As far back as we can go, it appears that human beings have distanced themselves from the reality of natural phenomena and their constituents by imagining their potential human significance.[29] This involved the creation of signs, the ability to live in reality in terms of signs, and relating them into a coherent explanatory ensemble. In this manner, human groups, in contrast with other animals, placed themselves in a symbolic universe with the potential of mastery over it. Human beings, although among the weakest creatures and most poorly adapted to living in nature, learned to manipulate it symbolically as opposed to doing so directly and materially.[30] The human story involved first a mastering of the group in nature by means of the symbolic, and later a mastering of society in nature. Had human groups attempted to directly and materially engage in reality, they might not have survived since symbolization is a prerequisite for everything else, including the making of tools. For a very long time, tools and related technologies remained an expression of the symbolic.

The symbolic link with the world appears to be central to what it is to be human, at least until recently. It dialectically associates two aspects related to cultural mediation. First, whatever permeates most experiences of the members of a community profoundly affects the organization of the brain-mind and thus human consciousness and cultures. This opens up the possibility of human life being alienated by a life-milieu. The usual explanation of the sacralization of the primary life-milieu and its corresponding institutionalization in the form of a religion suggests that this inevitably leads to a flight from the world, but it entirely ignores the fact that sacralization first and foremost helps to stabilize a design for its symbolic manipulation. This touches on the second aspect, namely, the creation of a potential for human freedom, since a certain distance is created between the real and the symbolic, in the sense that the latter is an imaginative recreation of reality. It emphasizes some aspects, marginalizes others, and ignores still others, much as a painter symbolizes a landscape on a canvas to get at its deeper meaning and significance for a culture as a whole. Arbitrary and accidental aspects of any sign are eliminated when it is inserted into the symbolic universe, which is absolutized by incorporating the unknown as an extension of the known. Any sign is thus converted into a symbol. Alienation sets in when this absolutization prevents a human group from evolving its design for living in the face of what Toynbee has called challenges.[31] The cultural unity of a society, which results from symbolically mediating our relationship with reality, thus simultaneously creates the possibilities of both human freedom and alienation. Humanity constantly distances itself from a primary life-milieu by symbolizing it to create freedom, while the absolutization involved in this symbolization leads to alienation.

Five aspects of the symbolization of human life in the world are particularly affected when technique becomes the primary life-milieu. The first is what elsewhere I have called a metalanguage for the members of a culture.[32] The symbolization of each and every daily-life experience involves an ongoing expansion and modification of the synaptic and neural connections that make up the brain-mind, to constitute this metalanguage. Such a metalanguage is profoundly affected by the primary life-milieu. It transforms many human experiences, beginning with children when they watch television, and later when they go on the Internet or are socialized into knowing and doing separated from experience and culture. In an information society, a great deal of experience is displaced by information.

The development of a metalanguage prepares the way for the beginning of language in babies and children, as it does for the evolution of a language in general and its stock of concepts in particular. We might say that there is a 'cultural style' by which a metalanguage is brought into words. For example, much has been made of the number of concepts some cultures may have in order to make critically important distinctions, as between different kinds of snow for the Inuit and different kinds of camels for traditional desert peoples. The debate has not given sufficient emphasis to a more fundamental underlying issue. If life in a snowbound environment involves trips by dogsled to go hunting, for example, it is essential to make distinctions between different kinds of snow according to the effect they have on dogsled travel. Such distinctions inevitably become an integral part of people's metalanguages, but how these are or are not translated into linguistic concepts is secondary. When game is far away, a member of the Inuit community may simply observe that this is a bad day to go hunting and everyone will understand why since they have recognized the snow conditions. For desert people, distinguishing between camels according to the distances they can travel is equally essential. It must be recalled that for traditional cultures, as for our own, the purpose for creating new concepts is not to make objective descriptions of particular distinctions, most of which can be taken for granted by the members of a community. Concepts express the deeper meanings and values of the distinctions made in a metalanguage.

Since the metalanguage of each individual member of a cultural community changes as they participate in the evolution of a way of life, the meanings and values of the concepts of the language will also change. For example, during the rapid changes of the initial phases of industrialization, many words in the English language underwent changes in their meanings.[33] The creation of new concepts is particularly important when fundamental changes appear to be happening, which must be expressed in words so that they can be discussed and critically assessed. Those concepts symbolizing deep and broad patterns of change or large areas of human experience must sufficiently generalize a network of distinctions in a metalanguage without blurring or obliterating essential differences. This, as we have seen, is next to impossible because of the presence of a metaconscious sacred and myths, which deflect critical attention from what has been declared to be very good. Hence, an adequate stock of concepts can be created only by an iconoclastic awareness of our defining spiritual commitment to a cultural unity. From this follows a second aspect of symbolization. The influence technique has on contemporary education almost guarantees that few, if any, iconoclastic social scientists, philosophers, and theologians will be available to help change our stock of concepts in order to create a more iconoclastic awareness of ourselves and our world.

A third aspect results from the fact that technique-dependent ways of life of contemporary mass societies greatly increase the distance between the metalanguages of their members on the one hand, and the use and evolution of language on the other. In part, this is the result of science and technique having separated themselves from experience and culture. Even more important is the influence of integration propaganda, including advertising and public relations. There is no longer a speaker in the traditional sense of the term. Someone in an agency uses words as linguistic tools to create a desired effect as efficiently as possible on behalf of a client. There is no communication since no one knows who is speaking, and there is no possibility of responding. Nor are there any listeners because they are reified by these techniques to deliver the desired effects. All this results in what have been called 'plastic words.' In all these processes, the metalanguages of the people involved do not play the role they would in normal human communication.[34] What we have interpreted as technique beginning to distance people's metalanguages from language and culture was recognized by Edward Bernays, one of the founders of public relations, as a unique opportunity, which he expressed as follows: 'If we understand the mechanism and motives of the group mind, is it not possible to control and regiment the masses according to our will without their knowing it? ... theory and practice have combined with sufficient success to permit us to know that in certain cases we can effect some change in public opinion with a fair degree of accuracy by operating a certain mechanism, just as the motorist can regulate the speed of his car by manipulating the flow of gasoline.'[35] No government, ministry, corporation, or charitable organization any longer speaks directly to the public, but through the mediation of public relations. Technique has the ability to fundamentally alter the way we symbolize the living of our individual and collective lives in the world.

A fourth aspect is related to the fact that when science and technique separated themselves from experience and culture, the stock of concepts associated with particular disciplines and sub-specialties was gradually transformed into sets of signs coextensive with scientific disciplines and technical specialties. Such concepts originally reflect the 'technical shadows' of what is technically manipulated but when they enter daily life, frequently through the mass media, they acquire their 'meaning' and 'value' from the context in which they are used. To the extent that such concepts displace or change those arrived at through experience and culture, to that same extent the stock of concepts of a language is desymbolized. Not surprisingly, the technical orientation of contemporary cultures considers desymbolized concepts more real, objective, and valuable than the symbolized ones.

A fifth aspect of how technique influences human consciouness is rooted in the shift from experience to information in daily life. All experience comes from our being embodied in the world, while information distances us from this embodiment. The consequences are far-reaching, as Hubert Dreyfus's study of the Internet shows.[36] The following discussion is based on this study. Once again, we encounter the most unrealistic expectations of this technology. Extreme interpretations see it as ushering in a new stage in human evolution in which we can leave behind our bodies, ignoring the fact that our experience, language, and culture result from our embodiment in the world. Pure minds can get on with the acquisition, communication, and development of information in ways that are unencumbered by the body, ushering in an entirely new kind of civilization. By making back-up copies of these minds, people will even be able to protect themselves from disease, injury, and death, and other problems that come from being 'trapped' in a body. Already, we begin to see signs that a new 'culture' is developing in cyberspace accompanied by mutations in language, communication, and interpersonal interaction.[37] In the face of such grandiose expectations of an Internet-based future, Dreyfus shows that 'if our body goes, so does relevance, skill, reality, and meaning. If that is the trade-off, the prospect of living our lives in and through the Web may not be so attractive after all.'[38] Dreyfus then summarizes the findings of his study as follows:

Chapter 1. The limitations of hyperlinks. The hope for intelligent information retrieval, and the failure of AI. How the actual shape and movement of our bodies play a crucial role in our making sense of our world, so that loss of embodiment would lead to *loss of the ability to recognize relevance*.
Chapter 2. The dream of distance learning. The importance of mattering for teaching and learning. Apprenticeship and the need for imitation. Without involvement and presence *we cannot acquire skills*.

Chapter 3. The absence of telepresence. The body as source of our sense of our grip on reality. How the loss of background coping and attunement endemic to telepresence would lead to the *loss of a sense of the reality of people and things.*

Chapter 4. Anonymity and nihilism. Meaning in our lives requires genuine commitment and real commitment requires real risks. The anonymity and safety of virtual commitments on-line would lead to *life without meaning.*[39]

Enterprise systems and the Internet help to 'wire' the life-milieu of technique. Early studies of these developments confirm the general pattern previously encountered: as technique-based connectedness expands, something is excluded of what until now we have understood it is to be human, individually and collectively. The use of concepts such as relations, communities, and rooms for technique-mediated activities glosses over essential differences. This helps to explain why people who spend a lot of time on the Internet, thereby intensifying the technique-mediated nature of many activities, appear to suffer more from loneliness and depression than those who do not. This is but the tip of the iceberg. The growing use of (desymbolized) information obscures the meaning and value of everything and consequently obscures the culture-based connectedness of human life and society. The life-milieu of technique is not symbolized by means of a culture in terms of its human meanings and values, with the result that we do not distance ourselves from it, thereby exposing ourselves to its direct influence. The question is not whether we will succeed in making computers and the Internet in our own image, but whether we are remaking ourselves in the image of technique. This situation is in sharp contrast with what humanity has done until now: distancing itself from its primary life-milieu by means of symbolization.

In this analysis, I have sought to illuminate a situation that is both very old and very new. Many scholars have observed that human consciousness and cultures were profoundly influenced by nature in prehistory, and by societies during history. Hence, the observation that technique as life-milieu appears to have the same effect is not new. In each of the three situations, human consciousness is an awareness of oneself and the world, experienced bodily and symbolized culturally. There can be no consciousness without a life-milieu and no life-milieu without that consciousness. The brain-mind's organization is modified by each and every moment of a human life and thus by the life-milieu in which it is lived. As I have noted, the usual view that this symbolic link with reality opens up the possibility of alienation forgets that, first and foremost, it creates a measure of freedom, made possible by distancing human life from the life-milieu by symbolizing it. This permits human communities to symbolically dominate their world by imposing their meanings and values, and by absolutizing these, to establish their bearings, roots, and reasons for living. What is radically new is that the present life-milieu is one of our own making and thus no longer mysterious or 'other.' As a result of this 'disenchantment of the world,' human communities may feel little need to distance themselves from it. If human beings cannot symbolize this life-milieu except on its own terms, any margin of freedom with respect to that life-milieu may diminish, to pave the way for an unprecedented level of alienation and reification. Such a possibility raises troubling questions. Will human life as we have known it thus far come to an end, to make way for a mutation to something entirely different? Will the life-milieu of technique be able to sustain whatever new forms of human life may emerge? Will the desymbolizing influence of science and technique irreversibly weaken the role of culture in human life, thereby disabling the symbolic function by which human beings have always lived together in the world? Are we gambling with our humanity and our future?

10.4 Possessed by Technique?

An enquiry into the cultural unity that binds individual and collective human life is essentially asking the question, what does this individual or community live for? To avoid possible misunderstandings, several things must be pointed out. The question is not a prelude to a metaphysical discourse. It is not a matter of ideas, world views, or ideologies but of how people live their lives and how the story of such lives can best be told. Such a narrative must puzzle together what specialists in disciplines such as economics, sociology, psychology, political science, law, religious study, literature, and history might observe about these lives. Since there exists no science of the sciences, we are forced back to a narrative of something that has been, and continues to be, lived. It must take into account the influence of the primary life-milieu on human consciousness and cultures, as well as the extent to which this influence is weakened by the symbolization of this life-milieu. Each member of a community is born into, and socialized to live and help evolve, a fabric of relations. The acquired culture is a design for living that values certain kinds of relations over others, with the result that some parts of the fabric are highly developed and others underdeveloped. The situation is analogous to the human intervention in plant life by means of biotechnology. Since we have not been able to increase the efficiency of photosynthesis, what is essentially happening is that we seek to redistribute the biomass to favour those portions of plants that are useful to human beings, and thereby weaken other portions. In other words, plants become more strongly integrated into the technique-based connectedness of contemporary societies by being less well integrated into ecosystems and the biosphere. In the same vein, technique strengthens certain relations in the fabric of human life by weakening others.[40]

All this has no common denominator with postmodernism, which in essence reflects the influence technique has on individual and collective human life. Under the desymbolizing influence of technique, the role of culture has weakened to the point that philosophical foundationalism has become untenable. Growing up in contemporary societies means being socialized into cultures that depend on a knowing and doing separated from these cultures. People are socialized not to be personally involved in such 'objective' activities, but to act as reified structures or networks. However, in their private lives they can exercise the 'subjective' dimension of their culture, as long as it does not interfere with the collective elaboration of technique-based connectedness guided by performance ratios that express nothing else but the power of the techniques being employed. In literature and the arts, the corresponding metaconscious changes are intuited as the 'death' of the author and the subject, the disappearance of narrative, the impossibility of any human commitment, and the complete relativization of all meanings and values in the face of the exercise of the power of technique. In the same vein, structuralism and its successors reflect the fact that in public life technique reifies and incorporates everything into structures, but they overlook the fact that 'everything else' does not disappear, as is clearly shown from the negative consequences that ensue from the technical approach to life. To claim that this 'everything else' is merely ideology is a projection of the present reality onto the past. The decentring effect of technique on human life and culture is but half the story, since technique-based connectedness can exist only in a parasitic relationship to the cultural and natural orders.

Postmodernism fails to recognize that the desymbolizing influence of technique on culture has two dimensions, corresponding to the double role of the sacred and myths in human life. The first is the elimination of the threat of the unknown by living it as interpolations and extrapolations of the known, thereby putting a community in full contact with reality. Any alternative way of life becomes unliveable. Indeed, postmodernism is non-iconoclastic with regard to its intuitions about technique. However, the sacred and myths did far more than simply eliminate the threat of the unknown. In addition to absolutizing the reality as it was known by a community, whatever was most important in the life of that community was also absolutized as the ultimate good, beyond which nothing more valuable could be imagined. In this way, a culture anchored its values in an absolute, thereby providing its members with the deepest possible meanings and values to guide their lives in the world. Everything in human life was centred on an absolute, deeply hidden in people's metaconscious. It was only this second role that could be affected by the desymbolizing influence of technique. Here we encounter the situation that critics of postmodernism have expressed as 'anything goes.'[41] All this is another way of saying that the universality of technique requires that all cultures be relativized and decentred in order to enthrone technique as the new 'foundationalism' of power. Cultural relativism is possible only if one's own culture has been desymbolized, causing all other cultures to appear as liveable as one's own. In this sense, postmodernism ignores the fact that we also live in a secular sacral world, which is itself a human creation resulting from the influence technique has had on people and their cultures. There is nothing permanent or absolute about this situation, in the same way as there is nothing permanent or absolute about the sacred and myths of other peoples. It was and is a question of people living as if this were the case.

Postmodernism intuits the influence of technique on human consciousness and cultures, interpreted in an entirely non-iconoclastic manner. As such, it is the perfect expression of the spirit of our age, which is in complete submission to technique. All its deconstruction is nothing but a new reductionism, leaving only the relations of power created through technique. Postmodernism reveals a conformity to structures and networks of power relations; and by relegating everything else to ideology, it leaves no bridgehead for any iconoclastic attempt to desacralize technique and regain a measure of human freedom. It is as if technique were the master of this branch of philosophy, instructing enslaved minds on what to think and say. The postmodernist attack on foundationalism is an attack on a straw man that philosophers have created. The presence of a sacred and myths in all human cultures is as close as we have ever come to a 'foundation' for human life; and as we have seen in chapter 7, all the attempts of Western philosophers to create a non-cultural foundation have failed. Once again, it is a question of people and societies during a particular period living as if their sacred and myths are absolute, thereby denying their status as finite and relational beings in need of fabricating a transcendent.

I acknowledge that, well before postmodernism, philosophers who did not have an iconoclastic awareness of their culture's sacred and myths usually ended up articulating the spirit of their age by pushing to the limit what was thinkable in that context. Hegel produced a secular synthesis that in essence translated the Christian incarnation of the Word into the growing determinism of the state. It accurately reflected what was happening in the cultural unities of Western societies. Marx, relying on Hegel to critique Feuerbach and on Feuerbach to critique Hegel, created the concept of a dialectical process of history as a kind of mechanism that would inevitably dethrone capitalism and usher in a secular paradise free from the exploitation of human beings and nature.[42] The acknowledged fact that capitalism alienated the capitalist and proletarian classes alike made the revolutionary role of the latter more than a little mysterious. Both Hegel and Marx saw humanity in the grip of forces beyond its control, but these were no longer the traditional gods. Instead, the new forces were unleashed by what in this narrative I have called 'technology changing people.' Freud created a kind of secular version of neo-Calvinism by explaining that human life was the product of mechanisms deeply buried within us. Nietzsche taught that all that was good in human life was related to the heightening of a feeling of power and the will to power. All that was bad was a result of weakness. Happiness derived from a sense of an increase of power. Worse than any vice was having sympathy for fellow human beings who were weak. I am not aware of a more accurate tracing of what was beginning to happen deep down in the individual and collective metaconscious of people and in the culture of Western societies during Nietzsche's time. Performance ratios were beginning to be turned into values, and human life and society were being reshaped on their terms. A radical break was made with the religious traditions of the West, and this too reflected the new cultural unity based on a secular sacred. Science would not replace religion, according to Nietszche, because people were outgrowing it by showing that there were no objective truths, but only human interpretations required to cope with human weaknesses. For Nietzsche there was only meaninglessness and chaos, and humanity was learning to live with this by its will to power. All this proclaimed the ultimate victory of technology-based connectedness over culture-based connectedness, of 'technology changing people' over 'people changing technology,' and, eventually, of technique over culture.

Each in his own way, Marx, Freud, and Nietzsche created pseudo-scientific theories of human life in the image of technology during the nineteenth and early part of the twentieth century. It was as if everything in human life and society had come down to what I have called the technology-based connectedness. Living a good life in the world by symbolizing everything in terms of meanings and values and making a history by means of them had become little more than expressions of a false consciousness, deep and dark forces, or human finitude and weakness. All other human acts, particularly those that sustained life, were discounted, including acts of self-sacrifice, friendship, and love. It was all but forgotten that these created a measure of freedom within the cultural constraints of a particular epoch and kept the door open to a future other than the one implied in the cultural unity of their societies. Philosophizing as if all of life can be reduced to a play of forces, mechanisms, and powers was a slavish expression of the evolution of the cultural unities of Western societies under the influence of mechanization, rationalization, and, later, technicization. Affirming human life through experience and culture became suspect, and the supersession of which was taken to be a sign of humanity's growing up and coming of age. What was entirely forgotten was that science can take hold only of repeatable phenomena and thus has no access to life beyond the extent to which it is alienated and reified. In the same vein, technique can imitate life only to the extent that it can reconstitute that life through reification. In other words, science and technique can only deal with life as non-life. As people developed science and technique, they simultaneously recreated the rest of human life, but only in terms of those aspects within reach of these creations. Over time, the claims of science and technique became a self-fulfilling prophecy, to the extent that they were not subjected to iconoclasm and freedom.

Until the events described in this narrative, people could live with a relatively quiet conscience, enjoy friendships, and share life with the persons they loved. All this could have meaning and value, and there was the hope of a history that would ensure that all this would not be in vain. Under the influence of desymbolization, the teachings of the above three 'masters of suspicion,' who cast doubts on everything symbolic and cultural, became plausible and hence could penetrate the public psyche. In the past, intellectuals may have had these kinds of thoughts, but they never penetrated the public psyche, sustained as it was by a culture. In the West, people now had to get used to living without any kind of unconditional love, pardon, and salvation. The influence of Christianity on Western civilization was rapidly coming to an end. With Kierkegaard as possibly the only exception, conservative and liberal theologians interpreted the Christian message in a non-iconoclastic manner by reading it through their 'cultural glasses.' An age of disillusion began, which was characterized by a lack of hope.[43] It physically, socially, and spiritually trapped Western civilization in a labyrinth of technique. Marx, Freud, and Nietzsche implied that there was no irreconcilable conflict between the scientific and technical approaches and life itself. Their influence confirmed the validity of their teachings in a self-reinforcing cycle.

However, postmodern philosophers and their precursors were correct on one important issue: there can be no meanings and values, and hence no culture, without a reference point. When the gods of the past fail us, humanity has no choice but to recreate others, or, failing that, act in their place. This would require a new human nature. For Nietzsche this was the free spirit who would outgrow one thing after another. The *Übermensch* in some respects anticipated the other-directed person in a mass society: every day there are many exciting new social contacts, but people outgrow them very rapidly since they are shallow and lack any serious commitment. Every day people are promised new exciting gadgets, medical miracles, and scientific marvels, only to see them discarded shortly after. They are mesmerized by their new palmtop, the new bells and whistles on their cellphone, or a new 'smart' appliance, only to quickly discover that in the final analysis they make no real difference. These people need not make any value judgments. As long as they outgrow the previous fad along with everybody else they will be normal, and that is all they need to worry about. Even their opinions come ready-made and require no symbolic interpretations or commitments. With hindsight, this all becomes painfully obvious.

I will now further distinguish the present narrative from post-modernism by examining what could constitute the cultural unities of contemporary societies. The behaviour of their members is a manifestation of what is individually unique as an expression of their freedom and aspirations, but also of how this individual diversity is constrained by the cultural unities of their communities. Although these are deeply metaconscious, it is possible to search for and critically examine those elements of human experience that appear to be good in themselves, have no limits as to what they can accomplish, or both. Four clusters of such experiences associated with contemporary ways of life suggest themselves. These are related to how we live with science, technique, and the nation state, and how we journey with them in time. We have seen that science implicitly and explicitly showed that underneath experience and culture was a reality like that of a creation, except that the order did not come from a Creator, but from a chaos out of which order emerged. God was replaced by a set of mathematical equations – cold, immutable, and indifferent to the human plight. Their reliability was limited, since a new scientific revolution in any discipline could lead to their partial or complete replacement with others. Nevertheless, we live as if science has no limits, even though it is a knowledge of things out of their full context, hence an ignorance of things in relation to everything else. We find it extremely difficult to come up with things that science will never know, as if it is omnipotent in the domain of knowledge.

In technical doing within technostructures, we have seen a virtuosity out of context with daily life. We can do nothing but marvel at the performance of high-priced sports cars in the abstract because, under the road conditions of daily life, fully engaging the brakes would almost certainly lead to being rear-ended by someone, while all-out acceleration would likely lead to rear-ending someone else. The same is true for the technical virtuosity of modern computers, which is of little use to most people. The examples can be extended almost indefinitely, but the pattern is always that technical virtuosity has little human meaning or significance. It is a part of daily-life experience, but so is the enduring hope that the next generation of techniques will do better. None of the shocks of the twentieth century have been able to derail our fascination with the marvels and power of technique-based connectedness. Civilization deals with these shocks and then continues with its business as usual. Apparently, the omnipotence of technique will fully reveal itself someday. It would appear, then, that technique is a big part of what we live for. After all, our lives and our world are unthinkable without technique, which has made us who and what we are. Whatever our difficulties and whatever our economic, social, political, and religious aspirations, we spontaneously turn to technique in the confidence that it can and will deliver. All we need to do is to imagine how we would deal with anything if we could not use technique, or to probe our knowledge of its limits and our understanding of where not to use it.

We live as if there are no limits to technique. No matter how great the risks of certain techniques, we are confident that we will be able to deal with them thanks to technique. Of course, the techniques of today have limits, but these will be overcome by the techniques of tomorrow. Elsewhere, I have shown that professional education in general, and engineering education in particular, socializes future practitioners to use techniques with little awareness of their limitations.[44]

Another historically unique feature of our time is that everything has become political. Everything can be, and just about has been, politicized. Any suggestion that something is non-political is likely to be branded as naive. We have already encountered some of the reasons for this. Technique eliminates the self-regulating character of all human activities and all spheres of contemporary ways of life, with the result that many of them must be organized, evolved, and controlled by the state. In using the latest techniques, the state reduces everything to the dimension of power and hence to politics. In so doing, it builds the nation state as the primary locus of technique under political control. To live as though everything is political is to live as though the nation state were omnipotent in every domain, thanks to the techniques it harnesses.

Let us consider how we live with science, technique, and the nation state. Is it not obvious that today we know more than yesterday, and that tomorrow we will know even more? Also, can anyone seriously doubt that today we can do more than we could yesterday, and that tomorrow we will be able to do even more? What could possibly stand in the way of our ongoing success? We can readily admit that today's government leaves a lot to be desired, since it is not 'smartly' managing the technical means at its disposal. Such political opinions are lived as temporary setbacks that will come to an end after the next election. After all, we have learned our lessons. Neither communities nor ecosystems should be given the highest priority on the grounds that everything else depends on them. The only priority is to use the latest techniques to stimulate the economy so that it will produce as much wealth as possible, from which we will pay the bills and distribute the rest to education, health care, social security, and environmental cleanup. The best strategy is to put our aspirations and political traditions on the back burner until the time comes to distribute the wealth produced by technique. All this is collapsing the entire political spectrum into a narrow band on the right. How else can we explain the rush into free-trade agreements that surrender everything every political tradition has ever stood for to the invisible hand of the Market, in which the winners are those who excel in the domain of performance values?

Expressions such as 'time will tell' and 'history will judge' point to the way contemporary societies arrange human life in time. No cultural values, beliefs, convictions, or interpretations are required. Having swept the sky clean of the last remaining god and his order, everything now happens within history without any possibility of referring anything to a transcendent. Within that history, technique is proving to be the most decisive phenomenon, which progressively reveals its power over everything as people experience little need to guide or delimit its evolution on the basis of cultural meanings and values. Regardless of their cultures, all societies are engaged in the same journey, thereby reducing their differences to their unequal success on the way. This journey, now euphemistically referred to as sustainable development, provides humanity with a common future. It is a journey created, sustained, and justified by technique. Hence, technique proves itself in time, and history will declare it the victor capable of bringing us everything we desire. In sum, history no longer has any limits since nothing can exist outside of it, with the result that nothing more important than technique can be imagined or expected. At the same time, technique is assured of all the knowledge it could possibly need, since we live as though science were omnipotent in the domain of knowledge. No longer are there any serious endeavours to maintain or evolve culture-based alternatives. Any attempts to delimit scientific research in the name of cultural values or moral principles have been in vain.

The sacred of capital has been knocked from its throne, although it is not yet entirely without power. We have seen that transnational corporations, by exploiting the latest techniques, can raise their own operating capital most of the time; and despite occasional deficits, governments always appear to have adequate funds for advancing the technical frontier since they dare not risk falling behind in the international technical race. Of course, when we get up in the morning, our radios or televisions still inform us whether or not the lesser gods of the markets are smiling on our pension funds that day, and governments frequently defer to their powers. Wealth now flows less from the surplus value of labour than from the flow of information within technique.

To put all this in the language of cultural anthropology, technique and the nation state are the two poles of the sacred today, while history and science are the principal myths.[45] Humanity is no longer required in any essential way. If 'time will tell' and 'history will judge,' then no moral values or judgments are required. Cultures and traditional gods become irrelevant. If history is made by technique and the nation state, both supported by the limitless potential of science, neither experience nor culture is required any longer. The technical system organized by the nation state creates the social roles that human beings must fill in a technical capacity – roles so specialized that they no longer depend on any symbolic taking hold of human life in the world as a context. We simply abstract what we need for any scientific or technical activity, with no regard for the rest. In a traditional society, any intellectual division of labour functioned in the context of the whole as symbolized by means of a culture, and was thus endowed with human meanings and values. Today, performance values point the way, and any other values all but disappear from our lives within technique, thus reducing our social roles to producers and consumers. No morality is required, since what most people do is normal, and what is normal is normative. The meaning of any technical object is symbolized in advertising and integration propaganda. We are no longer called on to symbolize anything in terms of its meaning and value for our lives. The new cultural unity is all in all; that is, technique, the nation state, science, and history take care of everything without any human intervention centred on symbolizing our life in the world. Hence, the nineteenth-century myths of progress and work have mutated into those of history and science, as a result of knowing and doing separating themselves from experience and culture to constitute the phenomenon of technique.

How does the transnational as one locus of technique fit into this structure? Since among the one hundred largest economies in the world we find a majority of transnationals,[46] the question arises of their relation to the nation state, and also of their place within technique-based connectedness. The answers flow from the observation that the development of the most decisive and influential techniques is so uneconomic in the beginning that they can come into being only in the public sectors of the economies of nation states, where the resources of the state and the transnationals can be combined, mostly in military-industrial complexes whose personnel moves in and out through revolving doors between the three entities.[47] This joint effort in the public sector has the additional advantage that the initially uneconomic yet vast outlays of capital can be justified and legitimated in ultimate ways: national security as the struggle between political life and death, economic competitiveness as a struggle between economic life and death, health as the struggle between physical life and death, and success in space as a critical scientific and military frontier.

Politicians and vested interest groups have become very good at converting choices into a single pathway on which everything depends, as if there is no life outside of technique. Similarly, CEOs and government bureaucrats can eloquently explain the ultimate need to provide the 'system' with what it needs. CEOs are now beginning to convince the general public that they have the very best qualifications for public office. Better than anyone else, they can run the nation state as a corporation, having acquired the appropriate technical mindset. All this simply shows the extent to which they are alienated and reified by technique, being unable to impose genuine human meanings and values in the adjudication of alternative choices. In the development of free trade, the state is showing its colours: technical development must proceed regardless of the cost to people and ecosystems. It is loath to put any restrictions on technique in the form of social and environmental standards or restrictions on investment. Have we not learned that decisions about children's work cannot be left to the labour market? Have we forgotten that the conditions required to ensure that free trade is to the advantage of all parties no longer exist? When will we learn to act with the knowledge that the invisible hand of technique is attached to an invisible elbow whose economic effects are causing net wealth production to decline? Does the failure of Keynesian economics mean that the only alternative is to revive the monetarism that brought us the Great Depression and the out-of-control unemployment and inflation that helped to produce so much of the turmoil of the twentieth century? Free trade is free only for technique. It is forced trade for communities and nations unless these can use their sovereignty to impose social and environmental standards. To make the irrational appear rational, integration propaganda is relied upon to command the required loyalty to the 'fatherland,' 'motherland,' or 'our democratic way of life.' Only the state can induce and command this loyalty in a manner that the transnational cannot; and it is for this reason that it holds a more central position in technique-based connectedness than does the transnational.

Gradually, the emergence of mass societies and their reliance on integration propaganda compensated more effectively for the diminished role of culture. Most people became so busy working, consuming, and relaxing that they sensed little need to go much deeper than the symbols that accompanied consumer goods in the confusion of images presented by the mass media.[48] Various crises have from time to time disturbed this fragile superficiality without ever changing anything fundamental. This situation points to the one function of the cultural unities of contemporary societies that appears not to have diminished as a consequence of desymbolization, namely, living as if the reality known through a culture is reality itself, thereby all but eliminating alternative ways of life. It becomes almost impossible to live outside of technique. Many people who may be sceptical about the kind of future that technique is helping to create no longer see any real alternative. There is a growing cynicism which holds that, for better or for worse, we have no choice but to persevere on our present course.

There is little chance that the threat desymbolization poses to human life will attract much attention. Is ours not the first primary life-milieu of our own making, and does this not imply that it must be less hostile than the two previous ones? Hence, there is less chance that people will feel a need to collectively distance themselves from their own life-milieu by symbolizing it. Only a secular conversion (i.e., a change in the intelligence through which we apprehend and live in the world) resulting from an iconoclastic attitude towards technique and the nation state can keep the door open to a genuine human future. This will require a civilization that includes technique but is not dominated by it – a civilization in which culture once again creates a distance between people and their life-milieu by means of symbolization. At present, it is not at all clear how this may happen, given the destructive effect technique continues to have on language and culture. Language plays a central role in the transmission and evolution of a culture, beginning by building on the meta-language acquired during the early stages of primary socialization. Language is now being transformed into 'plastic words'[49] and 'humiliated'[50] by integration propaganda and technical images.[51] The threat does not come from the attacks on language by postmodernist thinkers, but from what is happening to language in daily life.

The above description of the cultural unities of contemporary societies is only half the story. The more the elements of these cultural unities are essential for evolving their ways of life, the less their transgression can be tolerated. A boundary is thus created within which collective life becomes possible by expelling any social chaos. Again, in the language of cultural anthropology, a sacred order is established that must not be violated if the community is to survive as such. Such a sacred order does not 'program' or 'mechanize' the members of a community, thereby rendering impossible any disobedience or, worse, transgression. However, the sacred must provide for a profane in which people can conduct their lives more or less as they see fit, provided that it does not touch the sacred. There must also be a 'sacred of transgression' in order to make the sacred order liveable by providing the members with escape hatches that give access to a domain that converts transgression into a reinsertion into the community. Such a division of the life of a community into the sacred and the profane, with the possibility of a sacred transgression as a carefully delimited 'social chaos' for rejuvenation into the sacred, is well known and has been extensively studied in traditional societies.[52] Religions institutionalized the sacred, thereby binding people in a common unity. What has been generally overlooked is that contemporary societies are indeed secular with respect to traditional religions, but they are not with respect to new kinds of 'secular' religions.

To make contemporary ways of life liveable, no transgression of technique in public life can be tolerated except by means of a sacred transgression, which now takes the form of sex. It is as though people metaconsciously know that they are culturally possessed by the sacred of technique. Freedom and self-expression must therefore be channelled through the biological self, and what better way of doing so than by the sexual transgression of the sacred order of technique?[53] In the 1960s and '70s, 'free love' became the celebration of freedom within ways of life increasingly shaped by technique. Student protests, such as the one in France in 1968, used sex acts to desacralize the places of power within the university. In advertisements, technical objects became sexualized, as many studies have shown.[54] In this manner, strict conformity to their proper (i.e., technical) use was turned into acts of freedom, and consumer goods were endowed with magical powers of many kinds. Profanity, which had invoked the sacred of a bygone era, now invoked the new sacred: sex (as illustrated by the kinds of swear words used in daily life). The effects of desymbolization quickly set in. Using a sex act to prove something, such as the liberation from traditional taboos or the 'conquest' of someone else, is very far from sexual intimacy in a relationship of love. The sex act was now turned into a quest for performance, namely, a search for the greatest possible pleasure and erotic ecstasy. As a result, the sex act became both a transgression of the sacred of technique and a reinsertion into it. The necessity of having to sexually perform in bed led to new anxieties.[55] Surely it is not a coincidence that Freud placed such an enormous emphasis on the sex drive precisely at a time when technology began to have a major influence on human life, and in a place where this influence came more slowly so that it could more easily be observed. Such a major reassessment of the meaning of sexuality in human life made an important contribution to the remaking of human consciousness, first in the image of technology, and later, of technique. As the sacred transgression of technique was gradually desymbolized, alternatives had to be found. Pornography and violence filled the bill. Their explosion on the French Minitel system, and later on the Internet, are signs of our times.

The sacred transgression of the nation state has taken on somewhat different forms depending on the circumstances. For example, in the post–Second World War decades in France, where pseudo-Marxism was in the air, activities seen as helping to pave the way for the Revolution constituted the sacred transgression of the nation state.[56] In the United States, where the influence of technique on society and culture was much more advanced, beating the 'system' became the sacred transgression. It took many forms including feats of vandalism, shoplifting, placing artistic graffiti on technical objects, streaking, and drug or alcohol parties celebrated as transgressions of the social order. Later on, hacking into the sacred places of power or the launching of computer viruses constituted acts of transgression aimed at the sacred of the nation state. The widespread passive (or active) identification with the violence on the mass media is difficult to explain without a reference to a sacred transgression of the social order that so deeply excludes people from their own lives. In addition, there are many more or less politicized social, environmental, and religious protest movements seeking to overthrow specific aspects of the present order. Their revolutionary character is typically very limited because many one-issue causes fail to recognize that they are dealing with mere symptoms of deeper underlying developments.[57] As long as this remains the case, these groups will remain on the fringes of mass societies, unable to bring about badly needed changes.

In all of this, culture becomes something entirely private and personal. Aspects of human conduct of critical importance to traditional ways of life are now irrelevant to the sacred order of technique and the nation state. In fact, some societies are able to encourage multiculturalism, since it no longer represents any threat to public life, in which people participate as knowers, doers, or controllers of technique. People's physical appearance and dress are irrelevant and so is their choice of religion, as long as they are technically competent in their work. (The only exceptions seem to be the positions that demand total conformity to the system, as exemplified by the 'power suit.') In secular societies, issues of conformity or non-conformity must be interpreted in relation to the secular sacred of their cultures. Because of desymbolization, many traditional cultural elements survive, especially in multicultural societies, but their role in public life is mostly decorative. Traditional values, moralities, and religions have essentially become a private matter, and good and evil now have entirely different reference points. Nevertheless, there can be no symbolization without a culture. Were it not for integration propaganda 'speaking' for the many technical loci in society, the situation would be highly unstable. The general public may well be highly distrustful of governments, politicians, and transnational corporations, and sceptical about what they are being told, but this matters very little since it is extremely difficult to symbolize their lives in an alternative way, given the pressures of integration propaganda.[58]

In many parts of the United States but particularly in the Midwest, democracy functions as a secular political religion. In this sacred universe, people have little choice but to love the American way of life or to leave it. It becomes next to impossible to affirm your country but express reservations about some of the things that are going on or about some of the effects it may be having on other countries. What if this way of life is socially unsustainable? What if it has such a large ecological footprint that it deprives many parts of the world of the possibility of going their own way? What if this way of life leads to foreign policies that demonize societies that stand in their paths, as a stepping stone to exercising military power over them? What if the consciousness associated with the American way of life makes it impossible to understand why other nations may be upset about the practices of American transnational corporations, which threaten the livelihood of billions of poor people? Any person who seriously struggles with these kinds of questions quickly becomes suspect. Their status as fellow citizens is all but eliminated when they are branded as tree-huggers, socialists, or, worst of all, communists. Citizens must love the American way of life with all their hearts because otherwise they hate it, and if you hate it you are a threat to the community. As in any sacred universe, there can be no middle ground: either you belong to a community or you don't. Anyone who dares to suggest that possibly the other side's situation must be understood and that only this can lead to genuine political solutions will be treated the way heretics have always been. When the American way of life is regarded as 'Christian' and when people can say that, in a time of crisis, they stand behind their president and their

God, we are surely in the presence of democracy acting as a secular political religion. I have taken this well-known example not because it is an isolated case, but because it is symptomatic of what is happening in many other contemporary nations, including my own. Any faith is under tremendous pressure to turn itself into a religion and to play a more significant role in public life, and when it does the situation could be as explosive as those under the other secular political religions that ravaged the twentieth century.

Along with a new sacred and myths there developed a new hierarchy of values to complete the cultural unities that emerged and evolved with technique. They created a non-ethical domain from which the fading ethical domain was banished. (I am, of course, referring to the values by which people lived, and not to moral principles and rules.) This was the outcome of the effects the separation of knowing and doing from experience had on cultures. Any metaconscious knowledge built up from the practice of science and technique was separated from the metaconscious knowledge built up from experience. As a result, this knowing and doing were not subject to the values and cultural unities of the first generation of industrial societies, and certainly not to traditional values. Initially, knowing and doing separated from experience were guided by performance ratios that evaluated the results on their own terms, namely, to have everything deliver as much output as possible from requisite inputs. When technique-based connectedness became coextensive with culture-based connectedness, performance ratios were turned into performance values and moved to the top of the value hierarchy of every society evolving its way of life by means of technique. What most people did became normal, and what was normal became normative. Public opinions masqueraded as private opinions, even though they required no symbolization or commitment on the part of those who held them. Since 'time would tell' and 'history would judge,' it was no longer required to symbolize events in relation to all others in human life. Everywhere, it was increasingly unnecessary to intervene as an ethical subject. All this and more created a domain of non-morality, in which human beings no longer had to act as ethical subjects.

We see it all around us today. MBAs make business decisions as if people were nothing but items on a budget line, and the results trigger no moral concern of any kind. Worse, if someone raises the issue of the company's having a moral responsibility to an employee who has worked there for twenty-two years, there is a quick and annoyed response to the effect that the company is there to serve its shareholders. We can all repeat examples of this kind from our workplaces, whether they are universities, government offices, hospitals, schools, charities, or non-profit organizations. The Enron scandal should not have come as a surprise. Professional ethics are, as I have noted, purely end-of pipe in their orientation and have no influence whatsoever on the moral and ethical issues integral to professional practice. In other words, the top of the hierarchy of values reflects the necessity of expanding the technique-based connectedness. The lower portion of the hierarchy retains some cultural values, but these exist only to the extent that they relate to a culture-based connectedness that does not interfere with the technique-based connectedness of human life and society. In this way, the hierarchy of values reflects the integration of culture-based connectedness into technique-based connectedness on the latter's terms.

Take biotechnology as an example. To imagine that we have the choice to accept or reject it, or to modify it in some fashion on the basis of ethical criteria, is an illusion that will not last long. First of all, neither the technical specialists involved nor the general public have any idea of what its positive or negative effects will be, since instead of symbolizing this technology they only deal with its 'technical shadow.' How are we to make a judgment about whether it is ethical, unethical, or non-ethical? In the second place, from its very beginning biotechnology has been integral to the necessary linking together of techniques: it was a response to problems created by earlier techniques, and is itself a result of the technical approach to life. In the third place, it is rapidly becoming integral to our life-milieu and to the 'system' (to be examined in the next section). This is not to deny that if, as a society, we began to symbolize biotechnology in terms of its meaning and value for human life, society, and the biosphere, things could be very different. We are on the threshold of unleashing genetic pollution, whose consequences are even more imponderable than those of environmental pollution. The same can be said about other technical frontiers, including information technology and nanotechnology.[59]

The non-ethical is the locus of obtaining the greatest possible outputs for the requisite inputs on the part of one's client, employer, or country, with no tolerance for introducing ethical questions about the meaning and value of such efforts for human life, society, and the biosphere. Increasingly, the only acceptable way in which to raise these kinds of concerns is to apply cost-benefit analysis to determine the level of environmental protection that should be adopted and can be justified.

Traditional morality has become an ideology, especially in the mouths of politicians. No matter how much people long for an environmental ethics, a feminist ethics, an ethics of caring, an ethics of non-violence, and others, these are all limited to specific aspects of human life and thus offer little challenge to technique as a life-milieu and system. Their challenge to technique appears to be limited to the degree that they embody elements of an ethics of non-power as an iconoclasm with respect to technique. However, the practice of an ethics of non-power would force people to the fringes of contemporary societies, in a struggle to reconstitute themselves as ethical subjects by limiting their exposure to their reification by technique. We could not all move to the fringes even if we wanted to because the collapse of our life-milieu would cost billions of lives.

10.5 Technique as System of Non-sense

Because of the reciprocal interaction between people and technique, the developments described in the previous four sections profoundly influenced human consciousness and cultures, while at the same time weakening their roles in individual and collective human life as a consequence of desymbolization. The resulting situation and its evolution up to the present may be described by creating yet another concept, namely, that of technique as system.[60] Such a sociological and historical ideal type has validity only if technique has its own internal dynamic and if the resulting technical developments are given a very high (or even the highest) value by a community, with the result that they are spontaneously adopted with little or no critical and moral intervention. This occurred during the second half of the twentieth century. The sacralization of technique further ensured that society became ethically passive with respect to it. Technique as system is for a mass society what tradition was for earlier societies: the basis for evolving individual and collective human life.

The concept of technique functioning as a system within contemporary societies does not imply that its members have become cogs in a technical mechanism. Technique as system describes a situation in which, paradoxically, the influence of people on technique is mostly an expression of the influence of technique on people, as opposed to the exercise of human freedom expressed through symbolization. A desire for liberation would gradually build very different kinds of cultures, which would permit the creation of a greater distance between people and their primary life-milieu. Instead, human life has been remade in the image of technique, first as *homo economicus* and later as *homo informaticus*. As the role of culture in human life was pushed back, new technical means for integrating communities appeared, of which integration propaganda was a key component. No matter how much 'technique changing people' dominates 'people changing technique,' the relationship remains a reciprocal one, hence any attempt at symbolizing the situation in terms of people having become cogs in a technical mechanism would amount to symbolically closing off the possibility of any human future.

There now exists an unprecedented threat to the possibility of a human future. It comes from the fact that the system of technique is one of non-sense, meaning that its technique-based connectedness operates outside of the domain of experience and culture. Evolving human life and society on the basis of non-sense could close off a genuinely human future by breaking down the cultural order as well as the natural order. It is impossible, however, to reject technique; to do so would be as simplistic as the first human communities deciding that nature was such a threat to their freedom that they were better off to burn it.

The system of technique was organized and institutionalized to perform all necessary functions for its evolution, expansion, and regulation. I will limit myself to a few observations. The advancement of the technical frontier tended to be organized by social structures somewhat similar to the 'invisible colleges' encountered in science, in the sense that a relatively small group of technical experts from around the world collaborated in advancing a particular technical specialty. The flows of information into and out of these social structures encountered institutional and national boundaries. Many attempts were made to stimulate the information flows across these boundaries according to their economic and strategic importance: the creation of military-industrial complexes, the requirement of industrial partners for university research funded by government research agencies, international conferences, and professional and trade literature. The application, innovation, and regulation of a diversity of techniques relevant to particular organizations and institutions tended to occur through social structures resembling the technostructures described earlier. There was a great deal of overlap and interpenetration between the two primary social structures essential for the system of technique. It should be noted that the interface between the portions of the lives of the participants related to their knowing and doing separated from experience, and the remainder embedded in experience, was a source of beliefs, pseudo-values, and ideologies seeking to bridge the gap. Such beliefs, pseudo-values, and ideologies have found their way into society through media interviews and public relations efforts.[61]

Four categories of implications related to technique functioning as a system within human life, society, and the biosphere follow from my analysis.[62] First, the socio-technical entities that emerge as a result of technique-based connectedness dominating the evolution of a particular locus of human activities increase the constraints on human participants as these entities evolve in size and complexity. Participation beyond a well-defined technical role as knower, doer, or controller in these entities must be eliminated as a source of disorder. For example, pilots and passengers flying into or out of a small airport that is not equipped to handle jets have much more discretion and freedom than those using large airports, where everything must be governed by strict procedures if 'normal accidents' are to be avoided.[63] In some cases, such socio-technical entities can be safely run only with rigid military-style discipline.[64] The greater the density of these socio-technical entities in a society, the more they must be technically coordinated and integrated. For the system of technique, the influences of the cultural order of sense and the context-based natural order are sources of disturbance and chaos. Such disturbances of the technical order can be greatly reduced by ensuring that all socio-technical entities are connected to one another. For example, large airports must be integrated into rigidly controlled air traffic control systems and into somewhat less rigidly controlled ground transportation systems for passengers, freight, mail, fuel, and waste disposal. They must also be connected to many services by means of information flows such as those dealing with reservations and tickets, lost or delayed luggage, finance and cost recovery, security and policing, customs and immigration, and public relations. Similar patterns can be observed as the technicization of other spheres advances: factories, offices, hospitals, universities, sports leagues, and much more. During the 1960s and '70s, this was such a striking phenomenon that everything was referred to as a system, with a corresponding growth and emphasis on systems thinking.[65] Everything had to become a system as the technique-based connectedness became dependent on delimiting the influence of culture-based connectedness. The ascendancy of systems thinking came to an end when the proliferation of socio-technical entities in human life came to be taken for granted.

The second consequence for human life and society of technique behaving as a system within them is the growing flexibility derived from technique-based connectedness having become coextensive with culture-based connectedness and having begun its penetration of the connectedness of ecosystems and the biosphere by means of biotechnology and nanotechnology. The accompanying proliferation of socio-technical entities permits the system of technique to respond to whatever problems and needs may arise by smoothly expanding and intensifying technique-based connectedness. For example, when the role of the state in a society was limited to a small number of functions, there was little chance of overlap or of one undermining another. The more the role of the state grew, the more government departments or ministries were compelled to coordinate and systematize their actions. The same holds true for road expansion: as the density of the network of roads increased, each road came to be regarded as a link in that network. As the technical infrastructure of a hospital grew, the more essential it became to use each and every constituent element as productively as possible and to coordinate all their uses. The growing flexibility of the system of technique means that whatever problems or opportunities may arise, they are easily dealt with by one or more existing socio-technical entities expanding the range of existing techniques in their employ or adding new techniques. The evolution of a way of life becomes all but independent from experience and culture, creating a relatively seamless technical order. Intrusions from the cultural and natural orders can be handled almost as a matter of routine. No human intervention (i.e., intervention on non-technical terms) is required, with the result that the system of technique takes on a great deal of autonomy with respect to human life, society, and the biosphere by reifying them.

A third consequence is directly related to the conditions required for technique to operate as a system with respect to human life, society, and the biosphere. Putting a very high value on technique, beyond which nothing more important can be imagined or lived, results in problems and difficulties not attributable to technique itself. These are regarded as having other sources, such as the use people make of it, political interference in its management, or the restrictions of markets. As a consequence, any problems produced by the system of technique will not be traced back to their roots, since this would involve questioning the highest value of contemporary societies. The best solutions to these problems are expected from the application of more technique, as the highest good. Desymbolization does not fundamentally alter this situation because any alternative continues to be hidden by myths. Hence, the system of technique feeds on the problems it produces by not going to the root of any issue but, instead, compensating for negative effects by adding devices or services. We have seen several examples of how chains of compensations for negative effects expanded over time when work or agriculture became technicized. Also, the necessary linking together of techniques involved the proliferation of human techniques as compensation.

The fourth consequence is directly linked to the third. The system of technique is not regulated by negative feedback based on its effects on human life, society, and the biosphere. By compensating for its problems instead of going to their roots, it becomes top-heavy, with the result that from the perspective of experience and culture its desired effects are increasingly undermined by the undesired ones. The expansion of the technical order produces chaos in the social and natural orders, even though ultimately it cannot exist without these.

Contemporary civilization faces enormous risks, well illustrated by the research of J. Tainter.[66] His analysis shows four concepts that can help us understand the collapse of complex civilizations in the past, with the first three being the underpinnings of the fourth:

1. Human societies are problem-solving organizations;
2. sociopolitical systems require energy for their maintenance;
3. increased complexity carries with it increased costs per capita; and
4. investment in sociopolitical complexity as a problem-solving response often reaches a point of declining marginal returns

.... To the extent that information allows, rationally acting human populations first make use of sources of nutrition, energy, and raw materials that are easiest to acquire, extract, process, and distribute. When such resources are no longer sufficient, exploitation shifts to ones that are costlier to acquire, extract, process, and distribute, while yielding no higher returns.

Information processing costs tend to increase over time as a more complex society requires ever more specialized, highly trained personnel, who must be educated at greater cost.

... Sociopolitical organizations constantly encounter problems that require increased investment merely to preserve the status quo. This investment comes in such forms as increasing size of bureaucracies, increasing specialization of bureaucracies, cumulative organizational solutions, increasing costs of legitimizing activities, and increasing costs of internal control and external defense. All of these must be borne by levying greater costs on the support population, often to no increased advantage.

... Thus, while initial investment by a society in growing complexity may be a rational solution to perceived needs, that happy state of affairs cannot last. As the least costly extractive, economic, information-processing, and organizational solutions are progressively exhausted, any further need for increased complexity must be met by more costly responses. As the cost of organizational solutions grows, the point is reached at which continued investment in complexity does not give a proportionate yield, and the marginal return begins to decline.

... As stresses necessarily arise, new organizational and economic solutions must be developed, typically at increasing cost and declining marginal return. The marginal return on investment in complexity accordingly deteriorates, at first gradually, then with accelerated force. At this point, a complex society reaches the phase where it becomes increasingly vulnerable to collapse.

... Once a complex society enters the stage of declining marginal returns, collapse becomes a mathematical likelihood, requiring little more than sufficient passage of time to make probable an insurmountable calamity.[67]

The above explanatory framework deals with an impressive range of collapses of complex societies, including those of the Roman Empire, Mayan society, and Chacoan society. The first concept of the explanatory framework, namely, that societies are problem-solving entities, signals a major departure from earlier classic works such as that of Toynbee.[68] Tainter is writing from a historical vantage point where technique dominates culture, with the result that what Toynbee called challenges are now seen as problems. Observing individual and collective human life through cultural lenses permeated by technique populates the world with problems awaiting technical solutions. In contrast, Toynbee wrote from a historical vantage point where rationality was just beginning to make its mark on culture. Hence, Toynbee's challenges include, but are not limited to, Tainter's problems. For example, the shattering of myths as a consequence of contact with other cultures could lead to a weakening of the values that guided individual and collective human life, the legitimacy of authority, and the metaconscious foundation of a religion. When technique began to undermine the role of culture, especially in public life, the diversity of possible challenges was essentially reduced to those represented by the second, third, and fourth concepts in Tainter's explanatory framework. Toynbee's framework implicitly acknowledges that what appears as a challenge to a society depends on how it symbolizes a particular situation in terms of its culture, with the result that a response will be successful if it contributes to the growth and evolution of a culture, and unsuccessful if it weakens its integrality. In other words, the events that have a significant influence on the growth, stagnation, decline, and possible collapse of a civilization must be interpreted from the perspective of the culture of that civilization. Tainter correctly objects to Toynbee's frequent additional evaluations based on his own cultural vantage point. His explanatory framework gives the impression of being objective and scientific because it excludes a consideration of the cultural 'system' as a whole. It also masks the cultural presuppositions of our time.

The limitations of Tainter's explanatory framework become less and less significant as technique imposes itself on the role of culture, particularly in collective life. Consequently, his observations that our present civilization has entered a period of diminishing returns on its investment in scientific and technical complexity must be taken seriously. The present study has created additional concepts for the analysis of the risk of a collapse. The observation that the system of technique feeds on its own problems and is becoming increasingly top-heavy and ineffective adds to his analysis. What is potentially more serious is that desymbolization makes it increasingly unlikely that our present situation will be adequately diagnosed as a 'non-problem,' in the sense that it cannot be resolved by technique. Tainter's analysis further confirms the likelihood that the 'signal' of the desired effects obtained by means of the system of technique will continue to be undermined by the 'noise' of the undesired effects. Among the latter prominently figures a technical order growing at the expense of the cultural and natural orders. The issue of sustainable development is first and foremost a question of re-establishing a vital human culture capable of intervening in technique in a non-technical way. Such a culture would once again have to distance human life from its life-milieu by means of symbolization. This would usher in a transformation of the system of non-sense.

10.6 Technique as Collective Person

Under U.S. law, the corporation is treated as a legal person with the same rights as other Americans. Joel Bakan compares the 'personality traits' of this legal 'person' to those of human beings, and concludes that we have created a group of psychopaths.[69] We have allowed them

468   Our Third Megaproject?

to be brought into the world to pursue their self-interest; but not having been socialized, they now endanger our world and everyone in it, thanks to the power of the means they use. They have no empathy for anyone, but charm the public by means of public relations techniques that shape public opinion to mask their dangerous self-obsessed personalities. Nor do they feel any remorse when confronted with their illegal behaviour. They go to great lengths to avoid getting caught, pay any fine as just another business expense, and proceed to carry on business as usual. They have no feelings for their victims, born or unborn. The people that serve the corporation lead a kind of schizophrenic lifestyle that prevents them from becoming psychopaths themselves. They contribute their expertise to the many decisions the corporation makes, but they feel little or no personal responsibility for any of these. When they go home, they turn into the opposite character: loving spouses, caring parents, and friends you can count on; and when they fail, they feel remorse and regret. Unfortunately, there is little new under the sun: dictatorships, totalitarian movements, and criminal organizations have always depended on people being able to do just that. If there is anything new here, it is that contemporary cultures make this even easier: the important things have to do with strengthening and expanding technique-based connectedness and the less important things with culture-based connectedness, as long as this does not interfere with the former. All this is justified by the new cultural unities by means of which technique possesses the organization of people's brain-minds. Possessed as they are by the spirit of our age, executives and public relations people can find sense in all of this while engaged in the system of non-sense. Joel Bakan provides us with many informative examples.

Isn't the relentless pursuit of efficiency and the strengthening of technique-based connectedness by disordering the cultural and natural orders a form of collective psychopathic behaviour? Nothing is safe anymore, and no appeal to anything carries any weight. All pretences of corporate social responsibility, business ethics and professional ethics, exist only to the extent that they do not interfere with the psychopathic pursuit of self-interest – this enlightened self-interest only balances the short-term view with a slightly longer-term one. Technique as a system of non-sense has the corporation as its most powerful locus. Oceans of corporate planning to ensure their ultimate self-interest are now engulfing what little is left of (cultural) communities, markets, democracies, and ecosystems. Everything proceeds as described by means of the concepts of technique, technique as life-milieu, technique as consciousness, technique as system of non-sense, and technique as collective 'person.' Technique reigns over us through its most powerful loci, namely, the corporation and the state. The former now regards the latter as a partner, and for a variety of reasons (including the corporation's many political activities) the state is increasingly abandoning its obligation to protect the public interest. Despite the fact that the new emerging technical order corresponds to no religious or political tradition of any kind, nor to any human or social values, technique also reigns through us because of its influence on culture. It has penetrated into the cultural unities of contemporary society and hence the brain-minds of the people who are its members. We are now preparing the way for those who challenge important elements of the reign of technique, such as free trade, to be branded as security risks. People and nations are increasingly seen in secular religious terms, either as belonging to the reign of technique or as comprising threats to that reign. However, this new empire is different from its predecessors in one fundamental way. It is an empire of non-sense in desperate search of sense in the face of growing unrest.

10.7 Living with Non-sense

Human beings have found it extraordinarily difficult to live in the growing chaos created by technique in the socio-cultural and natural orders. Nothing expresses this more dramatically than modern art, music, and literature.[70] The source of this chaos was the fact that technique based on a knowing and doing separated from experience and culture could not deal with anything in the context of a human life, society, or the biosphere. It behaved as if there was simply a chaos of elements that it must reify and incorporate into the technical order. Its separation from experience and culture, and hence from sense, could never be complete, and this condition became the main driving force of technique. It was as though technique wanted to recreate human life and the world in the image of the (information) machine, in spite of the fact that this could not succeed because it ultimately depended on the consent of humanity. In the meantime, technique acted as if everything it touched was nothing but inputs, transformation processes, and outputs, and the sole raison d'être of these transformation processes was efficiency. It was as if there were no generations of human subjects whose stories wove a history and created a future. There was no communication other than the transmission of information, no memory other than the storage of isolated bytes, no language that referred to a beyond, no transcendence.

Before Jacques Ellul began his analysis of the arts (on which my discussion is based), he stressed that it is essential to reject all preconceived ideas of the arts as a quest for the meaning of life, which takes on particular forms according to the available techniques, public tastes, cultural world views, and so on.[71] Nor is art the result of an internal freedom, a refusal to face a disturbing world to escape into a dream, a distraction or a game, a part of the superstructure of society, a dematerialization as a turning away from a materialistic consumer society, or an inner source of joy or pleasure. As an ensemble of signs, art refers itself to a particular time, place, and culture, and that is how it must be interpreted. In this vein, Ellul asked two questions of the arts of his day: how had they become what they were, and what did this mean? In a world where everywhere one encountered techniques of all kinds, many factors combined with technique to create a rich and complex diversity in which the system of technique frequently turned out to be the decisive factor for the evolution of a particular ensemble of factors. The tension between what had been technicized and what had been externalized in the technical approach to life played an essential role, since it was the source of new techniques and of the difficulties of living a life in this new life-milieu.

As technique began to make it impossible for human beings to live in societies evolving by adapting a traditional way of life, it simultaneously exploded all aesthetic traditions integral to it, thereby creating an entirely new situation. Prior to this development, artists participated in traditions that translated into aesthetic styles complete with reference points and norms related to form. On the one hand, this implied certain constraints, and on the other, it offered the possibility of artists using their creativity and imagination to evolve particular aspects in relation to these constraints. Like all traditions, this presented artists with points of departure as well as everything necessary to go further without undermining the integrity of their own aesthetic tradition unless they symbolized the situation as requiring a mutation.

Before technique shattered all traditions, including aesthetic ones, there had been a full encounter between industrialization and the arts in the second half of the nineteenth century.[72] As the effects of industrialization on human life and society expanded in the twentieth century prior to the Second World War, the situation in which artists found themselves became even more difficult because the impact of technology was increasingly felt as a break with the past, with the result that cultural and aesthetic values became incapable of guiding artistic expression. The dead world of machines offered no alternative. New aesthetic expressions had to be invented. At the same time, technology was transforming the means artists had traditionally used, and these means were rapidly becoming everything. For example, architecture could not continue the elaboration of the spatial patterns of traditional cities, which had been shattered by rapid industrialization. Le Corbusier sought to rationalize construction in the image of technology.[73] The new ordering of space had to be suitable to an industrial civilization. Such a transformation could not be accomplished by merely using many new materials. Human life became subdivided into separate activities corresponding to functions, to be spatially allocated and housed in suitable modules. All this was to be connected by the flow of traffic, which had priority over everything else. In this way, the 'radiant city' was an expression of technical authority and the concomitant loss of the integrality of human life in space. Cities now segregated the places where people worked, shopped, relaxed, and lived, just as industry fragmented human life by separating the brain from the hand.

Following the Second World War, technique both accelerated and integrated many developments. As it dealt the final blow to all traditions, desymbolized cultures, and made them into private matters, the situation was at first experienced, as we have seen, as an explosion of human freedom. The new materials, machines, and techniques appeared to make anything possible. Artists began to regard the materials they used as elements of art in themselves, even prior to aesthetic use. New techniques permitted artists to create their own materials, the only limitation being the imagination of the artist. Art could now express pure imagination instead of the real world with its limited materials, possibilities, and traditions. In the absence of such traditions, theories could guide artists to systematically explore all possibilities, and this was further facilitated by the computer. Liberated from the constraints of a material and traditional world, pure fantasy and freedom remained. The problem was that this implied a rejection of all meaning and value, and this was not all. Complete freedom had always implied conflicts with others, but no one appears to have questioned why such conflicts did not arise. With hindsight, it is clear that a liberation from all aesthetic traditions and schools, and from the constraints that come with meaning, value, and history, was no liberation at all. These constraints were merely being shattered by technique and replaced by others. The discourse of artists themselves was significant in the sense that they did not give up talking about freedom, a concept that had no meaning in the absence of constraints. Materially, psychologically, and socially, artists could now do anything. However, this situation did not express freedom but the effects of desymbolization on human life and culture, the disintegration of traditional societies as they were being transformed into mass societies (in which integration propaganda compensated for the loss of culture), and a growing technique-based connectedness that was creating chaos in the social and natural orders.

The conflicts between artists and much of society came soon enough. Ellul humorously asked whether we would put up with chefs who mixed all kinds of foods, not according to culinary tradition but on the basis of some technical criteria. However, when it came to doing violence to our eyes, ears, traditions, and cultures in the name of artistic freedom, the public was supposed to accept this. Art critics informed people how they must experience this art if they were not to be regarded as reactionaries. People could no longer participate in art as human beings except as isolated individuals, because no meaning or message could any longer be communicated or shared. Once again, this was the exact parallel of what was happening in society, where technique was weakening all relationships and social bonds based on kinship and tradition, all but replacing these with integration propaganda.

Since it was impossible for people to become cogs in a technical mechanism, there was both a symbiosis and a contradiction between the system of technique and the mass society dominated by it. A great deal of incoherence and conflict was generated, to which the arts bore witness in their torn character. On the one hand, people found an art totally integrated into the system of technique as a complete reflection of it, including its cold perfection and indifference to everything human; and on the other hand, they saw an art crying out from the disorder of human life in the grip of technique by taking the form of a revolt, although no one was clear as to against whom or what. This latter branch of art called everything into question as being a reflection of technique, disaggregating everything human and social to leave a deeply sensed chaos. It was this contradiction that tore the arts apart into an esoteric formalism on the one hand and an 'art-message' on the other. In turn, this central contradiction engendered others. Ellul gave many examples, including the refusal of avant-garde literature to tell a story with a beginning, a middle, and an end, and modern art no longer painting a subject with a history, although claiming to present a message involving all of society. While 'art for art's sake' was rejected as an explanation or justification, art became 'anti-art for anti-art's sake' because no communication appeared to be possible. Pure spontaneity was glorified, while at the same time the most sophisticated technical means (including computers) were being employed.

When people sensed the break with traditional aesthetic activities, the problem was resolved by claiming that what was happening was non-art. In the theatre, speech was replaced with pure movement, colour, and sound, which reflected what was happening in a technique-dominated mass society. There was an elimination of meaning precisely because technique had no meaning, being nothing but the most efficient transformation of inputs into outputs. What art forms did was to transpose the technical process into the human realm and explain all this in human terms. Since there was no meaning, anything was possible, but this was explained by the freedom of the artist. It is not surprising, therefore, that aesthetic traditions were replaced by incoherence. Technical coherence required the incoherence of culture, including that of the arts and nature. Hence, the arts had to become an epiphenomenon of the system of technique. They existed in the life-milieu of technique and constituted themselves in relation to it, with the result that they had no unity or reality.

The arts thus began to reveal the crisis of living in a mass society dominated and enveloped by technique.[74] They expressed, revealed, and bore the contradictions of these mass societies and even reinforced them. They produced explicit models of the absence of meaning as well as an obsession with death and destruction. The arts were all about self-negation and thus our impotence in dominating the situation. They were the opposite of the magical and religious arts by which human beings had once created a force for organizing and dominating the world. Modern arts, in contrast, prepared people for their subordination to the system of technique. On the one side, Ellul argued, there were artists under the illusion of being outside of society and thus free to talk about it, and on the other side there were artists fully immersed in society. The latter did not ask about its problems and certainly not about the relationship between human beings and technique. They simply translated the universal forms of technique into the arts. Hence, the system of technique spoke through them as they eliminated and transgressed traditional limits. There was a confusion between the different branches of the arts; music, painting, architecture, and writing began to blur together as one branch diffused into another, creating multimedia art.

When technique emerged as the new primary life-milieu, society and nature became secondary in any attempt to explain the situation. Ellul cited Kandinsky as saying that we had to cut all our relations with nature and pull up these roots without any possibility of going back, with the result that artists now had to content themselves with combining pure colours with freely invented forms. These and similar responses did not reflect a realization on the part of artists that technique had in effect become their new nature. They reflected a widespread misunderstanding that because art was no longer rooted in nature, it represented nothing but itself, or that it had been plunged into the subconscious, or that it was without roots, merely producing canvases and texts. Such explanations reflected the difficulties most people, including artists, had in symbolizing themselves apart from the life-milieus of society and nature. For example, cubism could then be seen as the transformation of natural forms. The difficulty with such explanations was that they begged the question of what these transformations were all about and why they were occurring. In the case of surrealism, it was no longer a question of forms at all but one of altering meaning by revealing an invisible world. This also begged the question: the meaning of what, and which invisible world? Abstract art was supposed to call into question the reality in which human beings lived. All such interpretations reflected that, in one way or another, we still felt that we lived in the life-milieu of nature and that it was merely a matter of seeing and understanding nature differently. However, there was neither a disfiguration of reality nor a dematerialization of it, but a distortion that came from seeing new developments with reference to society and nature instead of technique.

Despite the explanations provided, artists continued to create works that faithfully registered the new situation. They, like everyone else, were caught between their traditional symbols and images and the new life-milieu. It was exceedingly difficult to structure aesthetic expression or any other communication in relation to this new situation. Desymbolization further complicated the task of understanding it and learning to live with its effects. For all the generations that had gone before, symbolization had been the way to make sense of and live in any life-milieu by means of a culture, of which language, religion, and the arts had been key components. According to Ellul, humanity now began to face an entirely new situation because technique could not be symbolized, for reasons he explained in several works.[75]

Ellul examined how technique, operating without meanings and values, was reflected in the arts. There was a tendency to portray the domain of love and sexual relations apart from social order and morality, to institute it exclusively on the aesthetic level. He reminded us that Kierkegaard had warned us about this possibility. Since artists were not cognizant of the situation that drove them, Marx would almost certainly have attributed this to a false consciousness. Nevertheless, their creations fully reflected the new life-milieu, which they veiled and mystified when they claimed that they called society into question, or that they were inspired by science, or that they addressed political issues. The general public, not understanding the situation either, by and large rejected this new art because it made no sense to them. For example, people were repulsed by the new 'serious music,' which reminded them of the very reality they were trying to forget. Schoenberg's music traced the image of a world that, for good or evil, had no knowledge of history, being as rigorous as any technical product. Pop music came to the rescue by providing an escape from this unbearable world, which explains its huge success as a purveyor of the irrational and the dream.

Artists transmitted a reflection, an echo, a radar signal of the reality that people refused to live. The 'unity' of the arts could thus be explained by the way technique hypnotized and enslaved people while they believed they dominated technique. The arts were united in expressing this slavery; they were entirely arbitrary except in relation to technique. Their content called everything into question except for the sacred of technique. When technique substituted itself for the symbolic, human beings could only produce the incommunicable and the nonsensical, or manifest the absence of the human in all of this. As a result, the arts in their turn sought to be a complete milieu, structure, or counter-milieu. In other words, they attempted to reproduce in their own way what technique had done in reality: the refusal of any explicit references to a milieu (thus imitating the autonomy of technique). Some sculptors insisted that their works were a complete environment: you should not only look at it, you had to be *in* it. In sum, contemporary arts reflected the life-milieu of technique, including its incoherence, the power of its means, the extreme diversity of its many aspects, and the difficulties of living in it.

Ellul insisted on the fact that the life-milieu of technique was not first and foremost a world of objects but a world of the means by which such objects were created. It was the reflection of a consumer throwaway society. In the arts, it was not the object that was important but the

476 Our Third Megaproject?

process of its creation. Nevertheless, the art object tended to obscure the process and might even dominate it. A painting was reduced to a mere object, except in pop art, where an object made by technique was turned into something aesthetic. In music, people encountered sound objects, except in pop music. The object had to be of pure form because it was without any significance. There also occurred an opposite response to the object, namely, a refusal to produce it any longer. This led to the 'happening,' which sought to dissolve the notion that a finished work was still possible. A musician or orchestra might walk on stage, sit there for a brief time, and then walk off. Clearly, this could not be repeated very often. Yet another response claimed that the arts were nothing but the creation of objects because there was no meaning, subject, or history, which left only the object. In this way, the novel could also become an object. The reasons for the text could only be in the text itself, since there no longer existed anything outside of it, such as a subject, story, or history. There was no greater glorification of the consumer society, which existed only to produce objects. This development was replicated in the arts, which now performed the same function as industry: to create new things. The creativity of artists was thus reduced to techniques producing technical objects.

The justification for all of this centred on the painting in itself, the music in itself, the sculpture in itself, because there was no meaning and no message. The painting was no longer a symbol or a point of departure for reflection – it was a mere object. In music, sound objects were collected from the world around us to be manipulated and assembled, or they might be collected from some electronic device to create new sound objects. The novel was separated from the writer and the context to become a structure that did not say anything. All this corresponded to the technical process that assessed itself on its own terms, namely, performance measures without reference to context. The resulting objects had no real meaning, value, or place in human life. People kept them around for a little while and then threw them away, just as they did with everything else in a consumer society. The triumph of the object was obsolescence. All this was the inevitable consequence of the arts leaving the domain of meaning to create their own empire of non-sense. In this way, the arts were an allegory of a technique-dominated society by adopting its character and orientation.

For Ellul the arts played an important role, having become an essential part of the system of technique. The artist represented a revolt and thus felt compelled to get the public involved, but this helped to veil the difficult conditions under which people had to live. Artists plunged themselves into all the ideologies of the day and expressed them in their own way in order to discover which one compensated the best. For example, at a time when the industrial towns became highly polluted, artists began to pay attention to light and to work out its effects in their paintings while pollution distorted and obscured that light. Paintings began to depict a reality that was no longer there. People went to see this art as a distraction, and then got back to real life. Later, technique trapped the arts in another dilemma. The arts sought to give an account of what was real and true, but they could not represent technique in its raw state. As a way out of this situation, the arts created an ideology that was anti-establishment and revolutionary. Everything had to be overthrown, and its anarchic principle was 'anything goes.' People did not see that this was the exact reflection of technique, which disordered everything cultural and natural. In the 1970s, artists lined up behind the issues of that time: Vietnam, the plight of black people, and the Third World. In the 1980s, it was pollution and the environment. All this might have been a good thing if the issues were referred back to the difficulties posed by technique, but instead everything was limited to politics, which failed to deal with the deeper realities. The taboos that were attacked had already been transgressed by technique. What mattered, apparently, was to provoke and transgress for the sake of doing so, much as performance measures 'valued' the technical process on its own terms. Anything and everything was contested, helping to create chaos in a society on which non-sense had already been imposed.

Attempts were made to create communication by technical means such as drugs, loud music, and strobe lights. Woodstock was the great 'happening' of communion in perfect solitude. It was based on the premise that the art being performed was the point of departure for a transformation. These kinds of developments were so financially successful because of their compensatory value that they became quickly assimilated by the system. Show business, the hit parade, rock concerts, and big musicals all became highly profitable thanks to the support of extensive advertising campaigns.

Ellul also traced the importance of theory for artists. Since the middle of the nineteenth century, the motivation of painters was highly philosophical and idealistic before becoming theoretical. Impressionism and post-impressionism wanted to start with principles of sensory perception in order to acquire a knowledge of the universe. Another development, which led to cubism, wanted to reach the reality hidden behind appearances. Mondrian and Kandinsky wanted to transpose reality into a construction controlled by reason. Others attempted to express the absolute novelty of art in conjunction with the absolute novelty of their epoch. What until then had been an artistic process was now analysed theoretically in technical terms. A systematic and theoretical elaboration had to guide artistic expression and replace sensory impressions, feelings, emotions, and metaphysical experiences. The brain began to replace the heart, leading to an avant-garde art that was highly abstract. For example, dematerialization was mathematically worked out in the design, which resolved the contradictions between subject and object, between interior and exterior, between vital and mechanical, and did so by means of a unified image that was mechanical yet remained an organic reality.

In music we saw the mathematical composition resulting from an organizational process of new sounds that was totally abstract. The theory validated the music. In architecture, the same development occurred. Since the Bauhaus school, there has been an ongoing trend of constructing buildings according to a theory. Architects developed general theories of society and of human beings that could be scientifically expressed. Le Corbusier, and later Iona Friedman, exemplified these developments. Mies van der Rohe derived the principles of architecture from a philosophy of space. These architects worked with preconceived theories of human beings and space, including the relationships between them. The situation was radically different from that of the past, where, for example, cathedrals expressed the medieval conception of the relationship between God and humanity. The intellectual artistic operation had become completely conscious as the work was entirely determined by a theoretical premise. Artists could no longer paint what they saw or felt. Had science not told them to mistrust what they saw? Hence, it became necessary to interpose a theory between the artists' intentions and the paintings. All this accurately reflected the separation of knowing and doing from experience and culture. For some artists, sculpture became the articulation of geometric forms, with conjunctions represented by pieces of steel or other materials. There were all kinds of explanations of what happened in the empty spaces between volumes, which defined zones of tension or action with crossings, and so on. Without some understanding of human life and technique, all of this would appear as little more than elaborate construction games that illustrated geometry. Here, too, art reflected the system of technique, from which it derived its raison d'être; but this could not be explained without some mystical language in order to return to the human being some domination over the object.

In the new novel and in poetry, theory also played a dominant role. There were theories about language, communication, the work within the work, and so on, but no overall theory synthesized all these approaches. There were only fragmentary theories that latched onto one thing or another as keys of interpretation. Aesthetics had to furnish writers and artists with rules of analysis, modes of structuring things, and techniques for programming.

The computer itself became a 'creator' of artistic concepts. The value of any theory depended on the scope of possible applications and its experimental results. Everything became reduced to elements organized by rules. With the aid of the computer, a kind of permutational art was created. In all this, artists sought the kind of freedom denied to human beings. In the process, they created a world that excluded experience and culture, that is, a world of non-sense. Since there were no theoretical criteria for the selection of theories, the public's reaction was taken as a criterion for valuing a theory.

This situation led to a fundamental contradiction. Since art was no longer a question of an emotion, an aesthetic feeling, a story, a conception, or an aspiration, it was incomprehensible to the non-expert. Art became hermetically sealed from everything human, and therefore uninteresting. If there were to be any appreciation by the public, long explanations had to be furnished by the artist or the art critic. This 'pure' art, which had broken all bonds with its context, was frequently interpreted as being 'spiritual.' In this way, it also reflected the system of technique, which liberates technical advances from any moral scruples or questioning in terms of its meaning or value.

Within the system of technique, artists played the role of the 'specialists of freedom.' However, as I have said, they experienced little real freedom with respect to the system, and they certainly did not exercise their imagination over it. They played this role because the public could not renounce their humanity and need for freedom. Modern societies did not need to deal with artists the way revolutionaries were usually dealt with. Like everyone, artists were conformists to the system and performed the role of a safety valve within it. According to Ellul, all the talk about subversion and revolution became the new conformist vocabulary of that time, much as order, beauty, and morality had been the conformist vocabulary in France in the 1860s. Artists could even be famous, respected, and wealthy, and avant-garde art was simply a reflection of what technique had already accomplished. Extensive commercialization, using the latest marketing and public relations techniques, removed the last element of doubt as to what was happening here.

When the message disappeared from the arts, an explanatory discourse became essential. What else could be done with a painting of two colour bars than to talk about it? Would anyone pay any attention to something that was so dehumanized and technicized, except to talk about the process that led up to it (including the theory)? The ambiguous role of art criticism was to have the object say something to reveal the sense of non-sense. It was not easy to produce a discourse about work that was stripped of all meaning. Without art criticism, it was unlikely that the public would give many artworks a moment of their attention. In fact, it is highly doubtful that modern art could have thrived without art criticism, because it and it alone attempted to give art some sense. Art criticism was just another technique. It was only in this way that the efficiency of art could be maximized in terms of its effect on the public. Art criticism guaranteed the artist an eminent social role; without it, art would have had no connection to society. In the same vein, literary works had to be considered on their own terms independent of all biographical, social, ideological, and historical aspects. The work should not be studied in relation to the author but in relation to the person who read it. It was the reading that constituted the work and that had to be guided by literary criticism. Criticism substituted for lack of meaning in literature, to become a kind of metaphysics of language.

Ellul summed up his analysis of the arts with the remark that for as far back as we can go, human beings have been the 'talking animal,' and all human creations have been dictated by the centrality of *logos*, including art. Contemporary arts manifested the end of this primacy of the word. They represented a break with the whole cultural ensemble born out of the centrality of the word. Artistic expression no longer had anything to do with the word – it was derived exclusively from the means of action. The *logos* and the word were finished. It was now the act, no longer personal or heroic in character, but mechanical.

Today, we must either consider that everything good, beautiful, and human (gradually constructed over a very long time by means of culture) is worth defending, or that it all belongs to the past and we have to start over. This was the question for Ellul. For example, what we call music kills our musical emotions and the reality of our being, and we do not even know if it serves anyone. In the past, music was tied to a particular historical epoch that gave it its expression and content as well as its value and meaning. The best of contemporary arts is an expression of the almost inexpressible suffering of people in our society. The crucifixion of humanity by the inhumanity of technique is a suffering that art can no longer transfigure and symbolize; it can only express it in its nudity and crudity. It throws the suffering in the face of the spectator and is even a cause of anxiety and exasperation. All this because the arts have become ahistorical and inauthentic, and have created a rigid hierarchy by reinforcing and confirming the reality of a technique-dominated society, as opposed to being a protest against it. Instead of becoming aware of the situation, the arts close the last door and give reality over to the illusory. This prevents people from understanding the technique-dominated world while fixing their attention on the insignificant and on appearances. The elimination of the meaning of the arts is the elimination of the 'why' humanity has lived until now. It is as though humanity itself has been eliminated. Breaking with all traditions to get back to an origin is in effect the detachment of human beings from everything that we have lived by thus far, only to immerse us into the unknown of the system of technique. There can no longer be a reference to a past that might permit us to judge what is happening thanks to a point of reference of a moral, aesthetic, or ideological character, which would permit human beings to remain masters of such an explosion of techniques. There were pure 'raw' sounds in the beginning, but human beings found it necessary to go beyond them to create something else. Why were people not content with these basic sounds, feeling a need to symbolize them? Why is modern music giving up on interpreting and elaborating these sounds?

The attack on language in particular is the culmination of our integration into the system of technique. The calling into question of art and language arises because humanity itself is being called into question by technique. When human beings are on the verge of non-existence, they create a vacuum in aesthetics and an absence of any message. The very being of humanity is being extinguished, leaving little more than mechanisms. The arts attempt to explain this non-existence, with the only meaning being the calling of everything into question, which goes hand in hand with our perfect conformity to technique. The arts are an accomplice to the annihilation of what it is to be human and the reification of the human being.

Is there any hope left? If we continue to endlessly repeat that humanity is an accident born of hazards and necessities, that everything is political, that everything boils down to class struggle, that everything is relative, and so on, then there is nothing to do and nothing to hope for. The arts must break with the system of technique. Different decisions and choices will have to be made to help us transcend our situation, take back meaning and value in the face of non-sense, and challenge the system of technique. It is only then that artists can stop being seismographs registering the earthquakes that come from successive waves of technical advances. Our task is to find meaning in this chaos left by technical rationality and to transcend it. If we succeed in symbolizing reality once again, we will create a force that will permit artists to play the double role of procreators first and midwives next, to a mutation that will constitute an escape from the mathematics of destiny. It is with these words (paraphrased and slightly shortened) that Jacques Ellul concluded his analysis of the arts. In it, he held out a model of learning and scholarship – not mere information-gathering in the context of a discipline, but a quest to understand who we are and where we are heading so as to create knowledge and wisdom capable of challenging the empire of non-sense, which has so fragmented the contemporary university.[76]

Humanity's third megaproject has undermined the accomplishments of the first two. We once again face a choice: to veil our present situation in the usual ways, or to symbolize it and thereby regain a measure of freedom with respect to technique. It is a choice for life and a human future.

Epilogue

According to the spirit of our age, this narrative should end with an effective and efficient prescription for our woes. Failing that, there should at least be some affirmation to the effect that the human spirit will triumph in the long run; that there is a political solution; that biotechnology, nanotechnology, information technology, and other developments are producing something genuinely new and better; or that there is some new 'ism' to the effect that if only we believe or do this or that, humanity and our world will be saved. I cannot subscribe to any of this.

There is no 'solution' since life is not a 'problem.' I shudder when I think of the 'solution' offered by National Socialism or by communism for ushering in a new creation. I also fear the 'solutions' of the one remaining superpower, whose presidents keep proclaiming freedom and democracy while their policies continue to achieve the opposite. For example, the capital flows required to maintain this 'empire of freedom' and its 'free' trade initiatives are distorting economies and harming the poorest and most vulnerable part of the human family. Among the industrially advanced nations, it leads in almost every negative statistic, including the level of inequality among its people, the proportion of people incarcerated, and the size of the private police force required to supplement the public police. Current tax-reduction proposals have reached a new level of public immorality. I trust that my readers understand by now that this is not an anti-American agenda but a concern for a people who have possibly suffered more than any other from an extreme accommodation to the necessities of technique. We in Canada are going in the same direction, and, barring a decisive intervention, much of this lies in store for us.

None of this offers any hope. What we need to do instead is to look at the lives of our children. What kind of future are we helping to create for them? They require more and more education for a life wasted by more and more underemployment and unemployment. They are particularly vulnerable to the effects of technique-based connectedness dominating culture-based connectedness in their lives, making it so much more difficult to grow up and make some sense of their lives and the world. Many require technical means to help them cope. A significant proportion of young people cannot look forward to gradually being able to surmount these difficulties and live a reasonably happy and healthy life. This is not a moral judgment that political and religious leaders are so fond of making, but a sociological observation. Technique is penetrating deeply into the fabric of relationships woven by human life in the world. Each new generation born into and growing up in this fabric must find its way, sustained by an ever more desymbolized culture. Torn between a knowing and doing embedded in experience and a knowing and doing separated from experience, we can frequently extrapolate what goes on in the latter to maintain some semblance of hope and courage of what we can expect for tomorrow in the former. If my narrative has any merit, this hope is entirely misplaced. We know all too well what happens to the bodies of kids growing up on junk food; therefore, there should be no surprise when their brain-minds are grown from technique-mediated 'experiences.' This mental junk food will lead to similar results in the psychic, social, and spiritual dimensions of their lives.

Where does that put us as adults? Imagine that contemporary ways of life really turned out to be socially and environmentally unsustainable, resulting in the collapse of our civilization. Suppose that the survivors held a kind of 'Nuremberg trial' of who was responsible. We can hear the arguments for the defence already. We could all say that because of our intellectual and technical division of labour we were simply busy improving this or that aspect, and that we had no overall responsibilities. No one would be responsible, including our politicians, who would simply claim to have followed the wishes of the people.

What then is there left to say? There are other avenues, and our future is by no means a foregone conclusion. We need to do something that initially will appear to be entirely ineffective and certainly inefficient, namely, to awaken from our technique-induced slumber. We need to take back something of our lives and our world. For myself, I believe the most decisive issue is that of human freedom, because without freedom there can be no committed human relationships of love and friendship, and no 'system' that serves our values and aspirations instead of our serving it. This is not the freedom of philosophers or politicians, but a lived freedom that is created within the constraints generated by the fabric of relationships woven by human life in the world. For more and more people this means the constraints resulting from living in our life-milieu and system, of which I have attempted to give an account in this narrative. Their influence on our awareness of ourselves and the world has realigned our values and aspirations with what this life-milieu and system can deliver. Otherwise there would be no such system. In other words, the first step on the road to a lived freedom is a recognition of the extent to which our individual and collective life is *possessed* by our life-milieu and system. This will bring us face to face with the unsettling realization that, contrary to what we are told by integration propaganda, our values and aspirations do not really correspond to what we yearn for most: a more meaningful, satisfying, and purposeful life, achieved by pushing back the constraints imposed by technique. This in turn requires healthy and viable communities engaged in a collective journey that expresses freely chosen values and aspirations.

What I am advocating is a human freedom lived in the tension between 'people changing technique' and 'technique changing people.' This requires a great deal of scepticism, if not an outright iconoclasm towards our secular sacred universe. There is not nearly as much new under the sun as we were led to believe. All the talk about secularization, coming of age, and our ability to be objective participants in science and technique have hidden what I have attempted to uncover by this narrative. It is one way of explaining the obvious, namely, that individually and collectively those who are employed work increasingly hard to better our lives; but in many respects these efforts are woven together by the present fabric of relationships, to worsen them instead. Our individual and collective efforts are being turned against us by the system we have created, which means that whatever theories, ideologies, beliefs, habits, and experiences we live by are not being confirmed by our daily lives. On the professional level we see the same thing: engineers, managers, economists, policy-makers, doctors, lawyers, social workers and many others are serving a system that undoes our best intentions. Out of this gap between what we believe is happening and what is actually happening an iconoclasm should emerge, and out of this should grow small changes in the conduct of our lives. Out of these in turn should come a confirmation of this gap, to create a self-reinforcing development. To the extent that this begins to create some free play within the system, we will be able to intervene out of a modest measure of freedom.

All this is not nearly as abstract as it may sound. For example, for those who labour in the 'intellectual backyards' of our universities a measure of freedom must be exercised with respect to the constraints of their discipline-based structure. Elsewhere, I have shown how this can be done for the engineering, management, and regulation of technology, with considerable benefits for all sectors of society. I have also shown how the relevance of many of these disciplines can be greatly improved by reducing the knowledge externalities embedded in them. This can be done through intellectual round tables of two kinds, which could be institutionalized in a new kind of faculty. In this way, the relevance of contemporary universities would undoubtedly be greatly improved. I am sure that my readers could invent similar initiatives in their own 'backyards.' My point is that there is not likely to be a shortage of genuinely creative innovations that could deliver much more of what human beings are genuinely interested in than the kinds of things achieved by our contemporary ways of life. The real difficulty lies in the fact that such initiatives are iconoclastic and hence heretical with respect to our secular sacred world. We have seen all too clearly what has happened to those heretics who dared to question artificial intelligence, Star Wars, or genetic engineering, or who stood up for their patients while under contract to a drug company.

Instead of suggesting 'solutions' to our 'problems' or, heaven forbid, a 'Solution' to our 'Problem,' I am suggesting conversations that would focus on what is happening to our lives and what we can and should do about it. I am thinking about what the Polish people did when their lives were not turning out the way they wanted and how grass-roots conversations led to the Solidarity movement. Perhaps if they had been able to keep these conversations going, Poland might now be one of the creative centres of contemporary civilization. It is a sign of our time that the traditional religious and political communities, as well as the many one-issue groups, are not engaged in these conversations. I have no illusions about the possibility of such conversations beginning in our schools and our universities, since the almost exclusive purpose of our educational system is to socialize people into a knowing and doing separated from experience and culture. Under the enormous pressure of integration propaganda, it is not likely that we will start getting together with our neighbours in our kitchens, church basements, union halls, community centres, or clubs to begin these conversations. Nevertheless, I see this as a first step towards exercising a modest measure of freedom.

Individually, we could begin by questioning the global significance of our local actions, such as shopping at a big-box store. Are we taking the bread out of our neighbours' mouths to put crumbs into the mouths of others? When a 'depot' gets the most competitive bids for their annual suppliers, will building fixtures become junk for lack of spare parts the next year? Are our clothing purchases increasing 'the gap' between rich and poor even further? To what extent is the problem also within us when we sacrifice our humanity for (supposed) convenience and price? To what extent are our employers undermining our humanity, and when and how should we voice our concerns? Can we do anything about our pension funds adding to the speculative bubbles? How can we avoid creating more one-issue organizations that fail to see any issue as only a symptom of deeper underlying patterns of change? How can we begin to reverse banishing our values and aspirations to our private lives? To fully live a life is to persistently ask these kinds of questions of ourselves and our neighbours; if answers are difficult to come by, we can endeavour to help one another. Maybe, just maybe, a public conversation will emerge about where we are going with science and technology and where they are taking us. People will then be able to participate in a growing grass-roots democracy that insists on a civil society asking fundamental questions about our individual and collective lives. The answers to these questions are in our lives themselves, and not in some moral, philosophical, political, or religious statements.

Constructing a narrative of where we think we are going, and where we could and should be going, is another step – one that could lead us to live the tension between who we are and who we would like to be. Otherwise, our narrative as I have attempted to describe it in this work will go on with more chapters of the same. It is a question of freedom and iconoclasm, and that is what I profess.

Notes

Introduction

1 Joseph Stiglitz, 'The Roaring Nineties,' *Atlantic Monthly* (October 2002): 76–89.
2 By culture I mean the totality of human creations made by a group or society and passed on from generation to generation. This usage must be distinguished from culture being the artistic and esthetic decoration on the 'social cake.'
3 This point is most forcefully made by Eric Voegelin. See his *Plato* (Baton Rouge: Louisiana State University Press, 1966).
4 Bertell Ollman, *Alienation: Marx's Conception of Man in Capitalist Society* (Cambridge: Cambridge University Press, 1971).
5 Emile Durkheim, *Selected Writings*, ed. and trans. with an introduction by Anthony Giddens (London: Cambridge University Press, 1972).
6 Max Weber, 'Science as a Vocation,' in *From Max Weber: Essays in Sociology*, ed. H.H. Gerth and C. Wright Mills (New York: Oxford University Press, 1963), 129–56. See also Rogers Brubaker, *The Limits of Rationality: An Essay on the Social and Moral Thought of Max Weber* (London: Allen & Unwin, 1984).
7 Arnold Toynbee, *A Study of History*, abridgement of vols. 1–10 by D.C. Somervell (London: Oxford University Press, 1946).
8 Jacques Ellul, *The Technological Society*, trans. John Wilkinson (New York: Alfred Knopf, 1964); Jacques Ellul, *The Technological System*, trans. Joachim Neugroschel (New York: Continuum, 1980); Jacques Ellul, *The Technological Bluff*, trans. Geoffrey W. Bromiley (Grand Rapids, MI: W.B. Eerdmans, 1990).
9 Hubert Dreyfus, *What Computers Still Can't Do: The Limits of Artificial Intelligence* (Cambridge, MA: MIT Press, 1992).

10 David Healy, *Let Them Eat Prozac* (Toronto: James Lorimer, 2003).
11 Cited as an opening quotation by Michael Jacobs, *The Green Economy: Environment, Sustainable Development and the Politics of the Future* (Vancouver: UBC Press, 1993), viii.

Chapter 1

1 W.H. Vanderburg, *The Growth of Minds and Cultures: A Unified Theory of the Structure of Human Experience* (Toronto: University of Toronto Press, 1985).
2 The concept of a metaconscious will be explained in detail in chapter 2. It is the structure of experience, which represents the transcendance of each individual experience.
3 Cornelius Castoriades, *The Imaginary Institution of Society*, trans. K. Blamey (Cambridge, MA: MIT Press, 1987).
4 Ibid.
5 Vanderburg, *The Growth of Minds and Cultures.*
6 Mathis Wackernagel and William Rees, *Our Ecological Footprint: Reducing Human Impact on the Earth* (Gabriola Island, BC: New Society Publishers, 1996).
7 'Class' appears to be an 'essentially contested concept' in the social sciences. Other than grouping people according to income distribution or occupational category, which yields considerable statistical information, social scientists seem to disagree about class analysis on almost all fronts. Even among Marxist practitioners the disagreements are widespread. See, for instance, Ralf Dahrendorf, *Class and Class Conflict in Industrial Society* (Stanford, CA: Stanford University Press, 1981); Ralph Miliband, *The State in Capitalist Society* (London: Quartet Books, 1976); Ralph Miliband, *Divided Societies: Class Struggle in Contemporary Capitalism* (New York: Oxford University Press, 1989); T.B. Bottomore, *Political Sociology* (London: Pluto Press, 1993); T.B. Bottomore, *Citizenship and Social Class* (London: Pluto Press, 1992); T.B. Bottomore, *Elites and Society* (London: Routledge, 1993); and Goran Therborn, *Science, Class and Society: On the Formation of Sociology and Historical Materialism* (London: New Left Books, 1977). Galbraith argued that with the birth of the modern corporation and the ensuing separation of ownership and control of capitalist organizations, the notion of class is empty baggage. John Kenneth Galbraith, *The New Industrial State* (Boston: Houghton-Mifflin, 1978). For other attempts at class analysis, see also Amitava Kumar, ed., *Class Issues: Pedagogy, Cultural Studies and the Public Sphere* (New York: New York University Press, 1997); Marshal Gordon, *Repositioning Class: Social Inequality in Industrial Societies* (London:

Sage, 1997); and Michael Savage, *Class Analysis and Social Transformation* (Philadelphia: Open University Press, 2000).

8 Karl Polanyi, *The Great Transformation: The Political and Economic Origins of Our Time* (Boston: Beacon Press, 1957). See also David S. Landes, *The Unbound Prometheus: Technological Change and Industrial Development in Western Europe from 1750 to the Present* (Cambridge: Cambridge University Press, 1969).

9 Environmental bookkeeping is, in essence, no different from keeping track of your bank account in terms of the financial flows that go into and come out of your account. Environmental bookkeeping tracks the flows of materials and energy. The product, process, or system to be analysed is enclosed in an imaginary spatial boundary across which input and output flows of matter and energy are tracked in order to monitor the 'account.' The location of this control surface, which is fixed in space, is usually indicated by means of a dashed line on a diagram showing the product, process, or system for which environmental bookkeeping is required. It separates the 'account' from the remainder of the world. Its location must be carefully chosen to ensure useful results. As a general rule of thumb, the control surface is placed where most flows are known in order to permit the calculation of unknown flows by 'balancing the books.' In this way, engineers can obtain useful results of an overall nature before the detailed design of a product, process, or system or before the full implications of a complex decision are worked out.

The 'balancing of the books' is accomplished by applying the first and second laws of thermodynamics. The first law holds that matter and energy can neither be created nor destroyed by society or the biosphere. They can only be transformed. This conservation principle implies that for a control volume drawn around a product, process, or system, the inputs of matter or energy equal the outputs minus any storage. If we draw the control volume around the economy of a society, it means that all matter and energy in the economy are temporarily borrowed from the biosphere, to which they ultimately return. For the biosphere, which receives no external material inputs, it means that all transformations of matter are organized in terms of closed cycles so that the outputs of one transformation become the inputs into the next and so on, with the result that there are no pollutants. The economy borrows energy from the biosphere and transforms it into useful forms whose usage converts them into low-temperature heat that finds its way back into the biosphere. I will include the solar energy input into the biosphere because the ozone layer 'prepares' it for its many functions. The second law of thermodynamics holds

that energy cannot flow in cycles because any transformation is irreversible. Hence, linear chains of energy transformations carried out by society
originate in the biosphere and end there.

10 For further details, see W.H. Vanderburg, *The Labyrinth of Technology*
(Toronto: University of Toronto Press, 2000), ch. 6.

11 Polanyi, *The Great Transformation*.

12 Adam Smith, *An Inquiry into the Nature and Causes of the Wealth of Nations*,
2 vols. (Chicago: University of Chicago Press, 1976).

13 Ibid. I have examined this in some detail in Vanderburg, *The Labyrinth of
Technology*, ch. 10.

14 Gilbert Simondon, *Du mode d'existence des objets techniques* (Paris: Aubier,
1989).

15 Leo Huberman, *Man's Worldly Goods: The Story of the Wealth of Nations*
(New York: Monthly Review Press, 1968). See especially chapter 16.

16 Lewis Mumford, *The Myth of the Machine: Technics and Human Development*
(New York: Harcourt, Brace, Jovanovich, 1968).

17 Smith, *The Wealth of Nations*.

18 Robert Karasek and Töres Theorell, *Healthy Work: Stress Productivity and
the Reconstruction of Working Life* (New York: Basic Books, 1990).

19 There is no intention here of promulgating an alternative class theory to
that of Marx.

20 This development is a significant factor in explaining why engineering
schools have great difficulty teaching design, and why DFX (Design for X)
is added on to the design process to consider issues such as quality and
sustainability.

21 Emile Durkheim, *Selected Writings*, ed. and trans. with an introduction by
Anthony Giddens (London: Cambridge University Press, 1972).

22 Max Weber, 'Science as a Vocation,' in *From Max Weber: Essays in Sociology*,
ed. H.H. Gerth and C. Wright Mills (New York: Oxford University Press,
1963), 129–56.

23 Jacques Ellul, *The Technological Society*, trans. John Wilkinson (New York:
Alfred Knopf, 1964), 111–12.

24 Landes, *The Unbound Prometheus*, 2–3.

25 See chapter 4.

26 Peter Chapman, *Fuel's Paradise: Energy Options for Britain*, (Harmonds-
worth: Penguin, 1975); Mohan Munasinghe and Peter Meier, *Energy Policy,
Analysis and Modelling* (London: Cambridge University Press, 1993); David
Foley, *The Energy Question* (London: Penguin, 1992); Robert Hill, Phil
O'Keefe, and Colin Snape, *The Future of Energy Use* (London: Earthscan,
1995).

27 A preliminary analysis has been undertaken in Vanderburg, *The Labyrinth of Technology*, ch. 9.
28 Polanyi, *The Great Transformation*.
29 Smith, *The Wealth of Nations*. I will capitalize the word 'Market' when it is used in the context of its acting as an organizing principle of society.
30 Mumford, *The Myth of the Machine*.
31 Vanderburg, *The Labyrinth of Technology*.
32 Jack P. Manno, *Privileged Goods: Commoditization and Its Impact on Environment and Society* (Boca Raton, FL: Lewis, 2000).
33 As a person of his time, place, and culture, Marx did not question the cultural myths of his own era, including progress and work. As a consequence, transient values and characteristics that were unique to the nineteenth century became the immutable basis for interpreting the past and the future. All of human history could thus be divided into five successive stages, each characterized by a unique economic base (comprising the forces and relations of production) and a superstructure. As a result, Engels could claim that Marx had discovered the laws of history. Later, this 'truth' became the foundation of the secular religion of communism. In the following chapters, I will seek to show that the distancing of the economy from society occurred during the eighteenth and nineteenth centuries, as technology-based connectedness gradually began to dominate culture-based connectedness.
34 The most influential religious tradition of the time, namely, Christianity, warned against the possibility of money becoming a spiritual power capable of possessing human life. See, for example, Jacques Ellul, *Money and Power*, trans. LaVonne Neff (Downers Grove, IL: Inter-Varsity Press, 1984).
35 Vanderburg, *The Labyrinth of Technology*, chs. 6 and 8.
36 Garrett Hardin, 'The Tragedy of the Commons,' *Science* 162 (13 December, 1968): 243–8.
37 Michael Jacobs, *The Green Economy: Environment, Sustainable Development and the Politics of the Future* (Vancouver: UBC Press, 1993).
38 Ibid.
39 Smith, *The Wealth of Nations*.
40 Jacobs, *The Green Economy*.
41 Herman Daly and John Cobb, *For the Common Good: Redirecting the Economy toward Community, the Environment, and a Sustainable Future* (Boston: Beacon, 1989).
42 Manno, *Privileged Goods*.
43 Daly and Cobb, *For the Common Good*.

44 Landes, *The Unbound Prometheus*.

45 Susan George, *Food for Beginners* (New York: Writers and Readers Publishing, 1982).

46 Huberman, *Man's Worldly Goods*.

47 For a good discussion of this, see Heinz Lubasz, *The Development of the Modern State* (New York: Macmillan, 1964), especially the introduction.

48 I have extensively examined these issues in Vanderburg, *The Labyrinth of Technology*.

49 Sarah Anderson and John Cavanagh, *Top 200: The Rise of Corporate Global Power* (Washington, DC: Institute for Policy Studies, 2000).

50 Vanderburg, *The Growth of Minds and Cultures*.

51 David Rothenberg, *Hand's End: Technology and the Limits of Nature* (Berkeley: University of California Press, 1993).

52 Landes, *The Unbound Prometheus*.

53 See chapter 2.

54 Roberto Vacca, *The Coming Dark Age* (New York: Doubleday, 1973).

55 Charles Perrow, *Normal Accidents* (New York: Basic, 1984). Scott Douglas Sagan, *The Limits of Safety: Organizations, Accidents, and Nuclear Weapons* (Princeton, NJ: Princeton University Press, 1993), chapter 1.

56 Perrow, *Normal Accidents*, 27.

57 Amory Lovins, *World Energy Strategies: Facts, Issues and Options* (New York: Harper Colophon, 1980).

58 P. Sainath, *Everybody Loves a Good Drought* (London: Headline Books, 1996).

Chapter 2

1 W.H. Vanderburg, *The Growth of Minds and Cultures: A Unified Theory of the Structure of Human Experience* (Toronto: University of Toronto Press, 1985).

2 For a good discussion, see Rogers Brubaker, *The Limits of Rationality: An Essay on the Social and Moral Thought of Max Weber* (London: George Allen and Unwin, 1984).

3 Vanderburg, *The Growth of Minds and Cultures*.

4 Ibid.

5 The current fashion of representing the organization of the brain-mind as a neural net is a substantial improvement over the previous fashion of likening it to a computer. Nevertheless, it is flawed. First, the name does not permit a distinction between the organization of the brain-mind and what is simulated in a computer as a neural net. Second, the limited success of neural nets in mimicking the behaviour of some organisms cannot be extrapolated to their becoming *homo logos* with a unique con-

sciousness and culture. The outputs of similarly functioning structures in the brain-mind must be presented to an 'inner' consciousness, thus implying higher symbolic structures that are very different in kind.

6 I am aware that alienation is an 'essentially contested concept,' but this only points to its importance to our culture. William Connolly has given us a classic account of this in his essay 'Essentially Contested Concepts,' in *The Terms of Political Discourse* (Princeton, NJ: Princeton University Press, 1984), 10–44. For Marxist interpretations of alienation, see Bertell Ollman, *Alienation: Marx's Conception of Man in Capitalist Society* (Cambridge: Cambridge University Press, 1971) and István Mészáros, *Marx's Theory of Alienation* (London: Merlin Press, 1970). Richard Schacht, *Alienation* (Garden City, NY: Doubleday, 1970) is a good introductory text that explores other existentialist and sociological meanings of the term. For the Judaeo-Christian notion of alienation, see Jacques Ellul, *Ethics of Freedom*, trans. Geoffrey W. Bromiley (Grand Rapids, MI: Eerdmans, 1976), 1–70.

7 My use of the terms 'prehistory' and 'history,' to distinguish the times when human beings lived in groups and in societies respectively, implies no value judgment. It is now widely accepted that what the nineteenth century regarded as 'primitive cultures' are in fact as complex as our own. In fact, we may argue that these cultures succeeded in what we have been unable to accomplish, namely, to create appropriate technologies and sustainable ways of life.

8 W.H. Vanderburg, prod. and ed., *Perspectives on Our Age: Jacques Ellul Speaks on His Life and Work* (Concord, ON: House of Anansi Press, 1997).

9 I acknowledge that for some readers the term 'prehistoric' suggests that human life was much more 'primitive' than ours. Ever since cultural anthropology, depth psychology, and the sociology of religion have rehabilitated the role of myths in human life, it has become apparent that there is nothing 'primitive' about prehistoric cultures. These exhibited a highly *enfolded* complexity, while contemporary cultures exhibit a highly *unfolded* complexity, which each have positive and negative consequences for human life. In this work, the term 'prehistory' will refer to that period in our past when food-gathering and hunting peoples lived in groups as opposed to societies. Many of these were the first, and probably the only, 'leisure societies' humanity has enjoyed.

10 The following sections are based on Vanderburg, *The Growth of Minds and Cultures*. For full details and references, please consult this work.

11 Ibid.

12 This and the following paragraphs are adapted from Vanderburg, *The Growth of Minds and Cultures*, pp. 56–8.

13 Vanderburg, *The Growth of Minds and Cultures*.

14 Repressed experiences can also form metaconscious knowledge, which would be subconscious as well. However, an examination of this form of metaconscious knowledge falls beyond the scope of this work.

15 See Thomas S. Kuhn, *The Structure of Scientific Revolutions*, 2nd ed. (Chicago: University of Chicago Press, 1970).

16 See J.R. Milton, 'The Origin and Development of the Concept of the "laws of nature,"' *Archives Européennes de sociologie* 22 (1981): 173–95.

17 See, for example, Roger Caillois, *Man and the Sacred*, trans. M. Barash (New York: Free Press, 1959); Mircea Eliade, *The Sacred and the Profane*, trans. Willard R. Trask (New York: Harper and Row, 1961); Mircea Eliade, *Patterns in Comparative Religion*, trans. Rosemary Sheed (Cleveland: World, 1970); Jacques Ellul, *The New Demons*, trans. C. Edward Hopkin (New York: Seabury Press, 1975); Claude Levi-Strauss, *The Raw and the Cooked*, trans. John and Doreen Weightman (New York: Harper and Row, 1969); Richard Stivers, *Evil in Modern Myth and Ritual* (Athens: University of Georgia Press, 1982); Paul Ricoeur, *The Symbolism of Evil*, trans. Emerson Buchanan (New York: Harper and Row, 1967); Vanderburg, *Growth of Minds and Cultures*.

18 Perhaps it is now possible to understand why both the Judaic and Christian traditions did not read the first few chapters of Genesis as an explanation of the origin of the world (which is taken care of in the first verse), but as a setting out of the fundamental relations between God, people, and the land (biosphere). From this perspective, the debate between science and religion regarding the origin of the world is the result of reading these ancient texts with different 'cultural glasses' shaped by, among other things, Darwin's replacement of 'the great clockmaker' with evolution.

19 The role of the sacred in a society is often not clearly distinguished from that of its myths. As a result, some of the works cited in note 17 include much of what we have called the sacred of a society. The distinction between myths and a sacred in this chapter follows the work of Ellul, *The New Demons*, chs. 3 and 4.

20 Góran Therborn, *Science, Class and Society* (London: Verso, 1980).

21 See Vanderburg, *Perspectives on Our Age*. Chapter 4 develops the distinction between religion (of the dominant culture) and faith (related to communications from a transcendent God who is holy, i.e., separated from that culture).

22 This situation does not preclude the possibility of individuals striving for an alternative defining commitment, as encountered in the works of

Kierkegaard, for example. I am grateful to Professor Hubert Dreyfus for his lectures on this topic. The term 'defining commitment' is his.

23 See Arnold Toynbee, *A Study of History*, abr. and ed. D.C. Somervell (New York: Dell, 1978); Pitirim Sorokin, *Social and Cultural Dynamics*, rev. and abr. in 1 vol. by the author (Boston: Porter Sargent, 1957).

24 Therborn, *Science, Class and Society*.

25 Jacques Ellul, *Reason for Being: A Meditation on Ecclesiastes*, trans. Joyce Main Hanks (Grand Rapids, MI: Wm. B. Eerdmans, 1990).

26 Milan Kundera, *The Unbearable Lightness of Being*, trans. Michael Henry Heim, 1st ed. (New York: HarperPerennial, 1991).

27 Paul Ricoeur, *Hermeneutics and the Human Sciences: Essays on Language, Action and Interpretation*, ed. and trans. John B. Thompson (New York: Cambridge University Press, 1981), 6.

28 Translated by Rita Vanderburg from 'Rapport de Pariset' à la Chambre de Commerce de Lyon, 1867.

29 Translated by Rita Vanderburg from Jerome Blanqui, *Des classes ouvrières en France pendant l'année 1848* (Paris: Pagnerre, 1849).

30 Ibid.

31 D. Furia and P.-Ch. Serré, *Techniques et sociétés* (Paris: Armand Colin, 1970).

32 Giedion has provided some interesting illustrations of what constituted medieval comfort. See S. Giedion, *Mechanization Takes Command: A Contribution to Anonymous History* (New York: Norton, 1969), 258–304.

33 Lewis Mumford, *The Myth of the Machine: Technics and Human Development* (New York: Harcourt, Brace, Jovanovich, 1968).

34 W.H. Vanderburg, *The Labyrinth of Technology* (Toronto: University of Toronto Press, 2000).

35 Ellul, *The New Demons*.

36 The situation in the Middle East would surely be less hopeless if Jews, Christians, and Muslims all engaged in some soul-searching about the extent to which their beliefs and political engagements reflect the religious creations of their cultures rather than divine revelation. Even a recognition that the former may be a critical component would undermine extreme positions and possibly create greater tolerance. I recognize that, as in the case of nature and nurture, religion and revelation become enfolded into individual and collective human life.

37 For a good discussion see Robert H. Nelson, *Reaching for Heaven on Earth: The Theological Meaning of Economics* (Lanham, MD: Rowman and Littlefield, 1993).

38 Jacques Ellul, *Métamorphose du bourgeois* (Paris: Calmann-Lévy, 1967).

39 Giedion, *Mechanization Takes Command.*
40 Alexander Solzhenitsyn, *The Gulag Archipelago 1918–1956: An Experiment in Literary Investigation*, trans. Thomas P. Whitney (New York: Harper and Row, 1975).
41 Jacques Ellul, *Changer de révolution: L'inéluctable prolétariat* (Paris: Éditions du Seuil, 1982).
42 Some of these characteristics are based on Raymond Aron, *Eighteen Lectures on Industrial Society*, trans. M.K. Bottomore (London: Weidenfield and Nicolson, 1967).

Chapter 3

1 The terminology is ironic. I do not mean that the people involved in technology become mere 'software.' The use of this term anticipates the findings in Part Three, which will show that the computer allows contemporary technology to take on the properties of a system that involves human beings on its own terms.
2 Thomas P. Hughes, *Networks of Power: Electrification in Western Society, 1880–1930* (Baltimore: Johns Hopkins University Press, 1993); George Besalla, *The Evolution of Technology* (Cambridge: Cambridge University Press, 1989).
3 For an alternative view, see Wiebe E. Byker, Thomas P. Hughes, and Trever J. Pinch, *The Social Construction of Technological Systems: New Directions in the Sociology of Technology* (Cambridge, MA: MIT Press, 1987). I am much more comfortable with the approach taken by William Ray Arney, *Experts in the Age of Systems* (Albuquerque: University of New Mexico Press, 1991).
4 Silvano Arieti, *Creativity: The Magic Synthesis* (New York: Basic Books, 1976).
5 For an interesting exception to this case, see Clive Ponting, *A Green History of the World* (London: Penguin, 1991).
6 One of the least effective components of this internal mechanism of technological development results from the huge research investments of the military-industrial complexes supposedly aimed at stimulating the remainder of the economies in which they are embedded. No doubt such stimulation occurs, but it has been greatly exaggerated to justify the enormous dependence of many industrially advanced nations on the design, manufacture, and sales of weapons of mass destruction. See, for example, Victor Perlo, *Militarism and Industry: Arms Profiteering in the Missile Age* (New York: International Publishers, 1963); Herbert I. Schiller

and Joseph D. Phillips, eds., *Super-state: Readings in the Military-industrial Complex* (Urbana: University of Illinois Press, 1970); Sidney Lens, *The Military-industrial Complex* (Philadelphia: Pilgrim, 1970). I will return to this subject in chapter 6.

7 If there is any doubt about this situation, one should only examine how governments have all but compelled universities to build much closer linkages with industry for the purpose of transferring scientific and technological knowledge and innovations into the economy. As a consequence, the public interest that these institutions were commited to serve is continually redefined in economic terms. For a critique of the underlying economic view of society, see W.H. Vanderburg, *The Labyrinth of Technology* (Toronto: University of Toronto Press, 2000), ch. 1.

8 Steven Vogel, *Cat's Paws and Catapults: Mechanical Worlds of Nature and People* (New York: Norton, 1998).

9 See, for example, Denis Goulet, *The Uncertain Promise* (New York: IDOC/North America, 1977); E.F. Schumacher, *Small Is Beautiful* (New York: Harper & Row, 1973).

10 Vanderburg, *The Labyrinth of Technology*, ch. 10.

11 Ibid., ch. 11.

12 David S. Landes, *The Unbound Prometheus* (London: Cambridge University Press, 1969).

13 Jacques Ellul, *The Technological Society*, trans. J. Wilkinson (New York: Knopf, 1964). See especially ch. 2.

14 Ellul, *The Technological Society*.

15 Ibid., ch. 2; see also Jacques Ellul, *Autopsy of Revolution*, trans. Patricia Wolf (New York: Knopf, 1971).

16 D.J. Dijksterhuis, *The Mechanization of the World Picture*, trans. C. Dikshorn (London: Oxford University Press, 1961).

17 Mumford, *The Myth of the Machine*; Lynn White Jr, *Medieval Technology and Social Change* (New York: Oxford University Press, 1962).

18 Mumford, *The Myth of the Machine*.

19 Landes, *The Unbound Prometheus*.

20 David Rothenberg, *Hand's End: Technology and the Limits of Nature* (Berkeley: University of California Press, 1993).

21 David A. Hounshell, *From the American System to Mass Production, 1800–1932* (Baltimore: Johns Hopkins University Press, 1984).

22 Ibid.

23 John Kenneth Galbraith, *The Nature of Mass Poverty* (Cambridge, MA: Harvard University Press, 1979).

24 Jacques Ellul, *Changer de révolution: L'inéluctable prolétariat* (Paris: Éd. du

Seuil, 1982); Barrington Moore Jr, *Social Origins of Dictatorship and Democracy* (New York: Beacon Press, 1967). The situation is also well dramatized by Aleksandre Solzhenitsyn, *The Gulag Archipelago, 1918–1956* (London: Collins, 1974).

25 Mumford, *The Myth of the Machine.*
26 Claude Lévi-Strauss, *The Savage Mind* (Chicago: University of Chicago Press, 1966).
27 Jacques Ellul, private communication.
28 Goulet, *The Uncertain Promise.*
29 W.H. Vanderburg, 'Political Imagination in a Technical Age,' in *Democratic Theory and Technological Society*, ed. Richard B. Day, Ronald Beiner, and Joseph Masciulli (New York: M.E. Sharpe, 1988), 3–35.
30 Schumacher, *Small Is Beautiful.*

Chapter 4

1 I have examined the process of primary socialization as culture acquisition, by which babies and children learn to make sense of and live in the world, in W.H. Vanderburg, *The Growth of Minds and Cultures: A Unified Theory of the Structure of Human Experience* (Toronto: University of Toronto Press, 1985). I am now combining this model with the five-stage skill acquisition model described in Hubert Dreyfus and Stuart Dreyfus with Tom Athanasiou, *Mind over Machine: The Power of Human Intuition and Expertise in the Era of the Computer* (New York: Free Press, 1986). For an update, see the special issue of the *Bulletin of Science, Technology and Society*, June 2004, which is entirely devoted to this model and its applications.
2 Dreyfus and Dreyfus, *Mind over Machine.*
3 Hubert Dreyfus and Stuart Dreyfus, 'What Is Morality? A Phenomenological Account of the Development of Ethical Expertise,' in *Universalism vs. Communitarianism: Contemporary Debates in Ethics*, ed. David M. Rasmussen (Cambridge, MA: MIT Press, 1990), 237–64.
4 Hubert Dreyfus, *On the Internet* (London: Routledge, 2001).
5 Dreyfus and Dreyfus, *Mind over Machine*, 32.
6 Vanderburg, *The Labyrinth of Technology.*
7 Shlomo Avineri, *The Social and Political Thought of Karl Marx* (Cambridge: Cambridge University Press, 1968).
8 For further details, the reader is referred to chapters 1 and 4 of Vanderburg, *The Labyrinth of Technology.*
9 Vanderburg, *The Growth of Minds and Cultures.*
10 Ibid.

11 M. McCloskey, 'Intuitive Physics,' *Scientific American* 248 (April 1983): 122–30. For evidence of an 'intuitive arithmetic,' see J. Lave, 'The Values of Quantification,' in *Power, Action and Belief*, ed. J. Law (London: Routledge & Kegan Paul, 1986), 88–111.

12 Pravin Varaiya, 'Productivity in Manufacturing and the Division of Mental Labor,' in *Knowledge and Industrial Organization*, no. 2, ed. Ake E. Andersson, David F. Batten, and Charlie Karlsson (New York: Springer-Verlag, 1989), 21.

13 Shoshana Zuboff, *In the Age of the Smart Machine: The Future of Work and Power* (New York: Basic Books, 1988).

14 Ken C. Kusterer, *Know-How on the Job: The Important Working Knowledge of 'Unskilled' Workers* (Boulder, CO: Westview Press, 1978); George Sturt, *The Wheelwright's Shop* (Cambridge: Cambridge University Press, 1923).

15 Vanderburg, *The Labyrinth of Technology*, ch. 10.

16 A continuum is a mathematical abstraction representing a material with uniformly distributed properties.

17 Vanderburg, *The Labyrinth of Technology*.

18 A.E. Musson and Eric Robertson, *Science and Technology in the Industrial Revolution* (Toronto: University of Toronto Press, 1969).

19 David S. Landes, *The Unbound Prometheus* (Cambridge: Cambridge University Press, 1969).

20 Thomas S. Kuhn, *The Structure of Scientific Revolutions*, 2nd ed. (Chicago: University of Chicago Press, 1970).

21 David S. Landes, *Revolution in Time: Clocks and the Making of the Modern World* (Cambridge, MA: Harvard University Press, 1983).

22 E.P. Thompson, 'Time, Work, Discipline and Industrial Capitalism,' *Past and Present* 38 (December 1967): 56–97; Carl J. Couch, 'The Mechanization of Time,' in *Constructing Civilizations* (Greenwich, CT: JAI Press, 1984), 345–69.

23 Neil Evernden, *The Natural Alien: Humankind and Environment* (Toronto: University of Toronto Press, 1985); Neil Evernden, *The Social Creation of Nature* (Baltimore: Johns Hopkins University Press, 1992).

24 J.H. Van den Berg, *The Changing Nature of Man*, trans. H.F. Croes (New York: Norton, 1961).

25 Wolfgang Schivelbusch, *The Railway Journey: The Industrialization and Perception of Time and Space in the 19th Century* (New York: Berg, 1986).

26 Georg Simmel, 'The Metropolis and Mental Life,' in *The Sociology of Georg Simmel*, ed. Kurt Wolff (Glencoe, IL: Free Press, 1950).

27 Stephen Kern, *The Culture of Time and Space, 1880–1918* (Cambridge, MA: Harvard University Press, 1983).

28 E.A. Wrigley, 'The Process of Modernization and the Industrial Revolution in England,' *Journal of Interdisciplinary History* 3 (1972): 225–60.
29 John R. Milton, 'The Origin and Development of the Concept of the "Laws of Nature,"' *Archives Européennes de sociologie XXII* (1981): 173–95.
30 Ibid.
31 M. Kline, 'The Mathematization of Science,' in *Mathematics: The Loss of Certainty* (New York: Oxford University Press, 1980).
32 M. Kline, 'The Authority of Nature,' in *Mathematics: The Loss of Certainty* (New York: Oxford University Press, 1980).
33 Ibid.
34 Harry Redner, *The Ends of Science: An Essay in Scientific Authority* (Boulder, CO: Westview Press, 1987).
35 G. Simondon, *Du mode d'existence des objets techniques* (Paris: Aubier-Montaigne, 1969).
36 G. Gusdorf, 'Past, Present and Future in Interdisciplinary Research,' *International Social Science Journal* 29, no. 4 (1977): 580–600.
37 Ibid.
38 See, for instance, Charles Jencks, *What Is Postmodernism?* (London: Academy Edition, 1986); Edward Lucie-Smith, *Movements in Art Since 1945* (London: Thames and Hudson, 1992).
39 A. Keller, 'Mathematics, Mechanics and the Origins of the Culture of Mechanical Invention,' *Minerva* 23, no. 3 (1985): 348–61.
40 Auguste Comte, *Auguste Comte and Positivism: The Essential Writings*, ed. G. Lenzer (Chicago: University of Chicago Press, 1983).
41 A. Mark Smith, 'Knowing Things Inside Out: The Scientific Revolution from a Medieval Perspective,' *American Historical Review* 95 (1990): 726–44.
42 Ibid., 742.
43 Ibid.
44 Frederick Binkerd Artz, *The Development of Technical Education in France, 1500–1800* (Cambridge, MA: MIT Press, 1966); R.A. Buchanan, 'The Rise of Scientific Engineering in Britain,' *British Journal for the History of Science* 18, Part 2, no. 59 (1985): 218–33; Monte A. Calvert, 'School versus Shop: Conflict and Compromise,' in *The Mechanical Engineer in America: Professional Cultures in Conflict* (Baltimore: Johns Hopkins Press, 1967), 63–85; A.R. Hall, 'On Knowing, and Knowing How To ...,' *History of Technology* 3 (1978): 91–103.
45 E.T. Layton, 'Mirror Twin Images: The Communities of Science and Technology in Nineteenth Century America,' *Technology and Culture* 12 (1971), 562–80.
46 Eda Kranakis, *Constructing a Bridge: An Exploration of Engineering Culture,*

Design and Research in Nineteenth Century France and America (Cambridge, MA: MIT Press, 1997).

47 For some alternative perspectives, see G. Bohme, 'The "Scientification" of Technology,' in *The Dynamics of Science and Technology*, ed. W. Krohn, Edwin T. Layton Jr, and Peter Weingart (Dordrecht: D. Reidel, 1978), 219–50. This volume contains many excellent case studies. See also M. Fores, 'Constructed Science and the Seventeenth Century "Revolution,"' *History of Science* 22 (1984): 217–44; M. Fores, 'The History of Technology: An Alternative View,' *Technology and Culture* 20 (1979): 853–61; Thomas S. Kuhn, 'Mathematical versus Experimental Traditions in the Development of Physical Science,' *Journal of Interdisciplinary History* 7 (1976): 1–31; David Landes, 'The Creation of Knowledge and Technique: Today's Task and Yesterday's Experience,' *Daedalus* 109 (Winter 1980): 111–20; Layton, 'Mirror Twin Images,' 562–80; E.T. Layton, 'Technology as Knowledge,' *Technology and Culture* 15 (January 1974): 31–41; J. Sebestik, 'The Rise of the Technological Science,' *History and Technology* 1, no. 1 (1983): 25–44; D.E. Whisnant, 'The Craftsman: Some Reflections on Work in America,' in *Technology as Institutionally Related to Human Values*, ed. P. Ritterbush (Washington: Acropolis Books, 1974), 109–26.

Chapter 5

1 Michael Polanyi, *Personal Knowledge* (Chicago: University of Chicago Press, 1962).

2 The controversy over T.S. Kuhn's concept of a scientific paradigm can in part be accounted for by the fact that it simultaneously appeared to designate a scientific discipline as a way of making sense of and dealing with the world and as a practice that was internalized by individual practitioners. Thomas S. Kuhn, *The Structure of Scientific Revolutions*, 2nd ed. (Chicago: University of Chicago Press, 1970). See also Margaret Masterman, 'The Nature of a Paradigm,' in *Criticism and the Growth of Knowledge*, ed. Imre Lakatos and Alan Musgrave (Cambridge, MA: Cambridge University Press, 1970).

3 This will be examined in detail in Part Three.

4 Jacques Ellul, 'Remarks on Technology and Art,' *Bulletin of Science, Technology and Society* 21, no. 1 (February 2001): 26–37.

5 Given the highly enfolded substructure of experience scientists acquire from secondary socialization, it is not difficult to understand why Kuhn, describing their behaviour, found himself using the concept of a paradigm in a diversity of ways. The situation is analogous to what we find in daily

life. The mind manifests itself in human behaviour in a multitude of ways, including someone's personality, mood, know-how, intuition, beliefs, values, and prejudices. This has not led to the rejection of the concept of mind but to a set of analytical distinctions between particular manifestations of that mind and human behaviour.

6 The myth of science will be examined in Part Three.

7 Jacques Ellul, *The New Demons*, trans. C. Edward Hopkin (New York: Seabury Press, 1975).

8 Diana Crane, *Invisible Colleges: Diffusion of Knowledge in Scientific Communities* (Chicago: University of Chicago Press, 1972).

9 Kuhn, *The Structure of Scientific Revolutions*.

10 Ibid.

11 Emile Durkheim, *Selected Writings*, ed. Anthony Giddens (London: Cambridge University Press, 1972).

12 W.H. Vanderburg, *The Growth of Minds and Cultures: A Unified Theory of the Structure of Human Experience* (Toronto: University of Toronto Press, 1985), ch. 8; Stephen P. Turner, *The Social Theory of Practices: Tradition, Tacit Knowledge and Presuppositions* (Cambridge, MA: Polity Press, 1994).

13 W.H. Vanderburg, *The Labyrinth of Technology* (Toronto: University of Toronto Press, 2000).

14 Ibid.

15 W.H. Vanderburg, 'STS as a Vital Intellectual Ecumenism,' *Bulletin of Science, Technology and Society* 20 (February 2000): 3–9.

16 Ken C. Kusterer, *Know-how on the Job: The Important Working Knowledge of 'Unskilled' Workers* (Boulder, CO: Westview Press, 1978).

17 Vanderburg, *The Labyrinth of Technology*.

18 Kusterer, *Know-how on the Job*.

19 Robert Karasek and Töres Theorell, *Healthy Work: Stress, Productivity, and the Reconstruction of Working Life* (New York: Basic Books, 1990).

20 Michael Floyd, Margery Povall, and Graham Watson, *Mental Health at Work* (London: Jessica Kingsley, 1994); Robert T. Golembiewski, Robert A. Boudreau, Robert F. Munzenrider, and Huaping Luo, *Global Burnout: A Worldwide Pandemic Explored by the Phase Model* (Greenwich, CT: JAI Press, 1996); R. Bourbounais, 'Job Strain and Psychological Distress in White Collar Workers,' *Scandinavian Journal of Work, Environment and Health* 22, no. 2 (1996): 139–45; James House, *Occupational Stress and the Mental and Physical Health of Factory Workers* (Ann Arbor: Institute for Social Research, University of Michigan, 1980).

21 Robert Blauner, *Alienation and Freedom: The Factory Worker and His Industry* (Chicago: University of Chicago Press, 1964); Michael Argyle, *The Social*

Psychology of Work (London: Penguin, 1989); Simon Marcson, ed., *Automation, Alienation and Anomie* (New York: Harper & Row, 1970); Harry Braverman, *Labor and Monopoly Capital: The Degradation of Work in the Twentieth Century* (New York: Monthly Review Press, 1974).

22 Kusterer, *Know-how on the Job.*

23 Richard G. Frank and Willard G. Manning Jr, eds., *Economics and Mental Health* (Baltimore: Johns Hopkins University Press, 1992); Office of Economic Affairs, American Psychiatric Association, *Economic Factbook for Psychiatry* (Washington, DC: American Psychiatric Press, 1987).

24 Kusterer, *Know-how on the Job.*

25 E.T. Layton, 'Mirror Twin Images: The Communities of Science and Technology in Nineteenth Century America,' *Technology and Culture* 12 (1971): 562–80; J. Sebestik, 'The Rise of the Technological Science,' *History and Technology* 1, no. 1 (1983): 25–44.

26 Ibid.

27 Benson Snyder, 'Literacy and Numeracy: Two Ways of Knowing,' *Daedalus* 119 (1990): 233–56; Pepper White, *The Idea Factory: Learning to Think at MIT* (New York: Penguin, 1991).

Chapter 6

1 John Kenneth Galbraith, *The New Industrial State*, 3rd ed. (New York: New American Library, 1979).

2 Rogers Brubaker, *The Limits of Rationality: An Essay on the Social and Moral Thought of Max Weber* (London: George Allen and Unwin, 1984).

3 Jacques Ellul, *The Technological Society*, trans. John Wilkinson (New York: Alfred Knopf, 1964).

4 Radovan Richta, *Civilization at the Crossroads: Social and Human Implications of the Scientific and Technological Revolution* (Prague: International Arts and Sciences Press, 1969). The analysis by Michael Hardt and Antonio Negri in *Empire* (Cambridge, MA: Harvard University Press, 2000) fails to come to grips with the developments described by Richta and Galbraith.

5 The facts in the following paragraphs are drawn from Galbraith, *The New Industrial State*, ch. 2.

6 The statistics cited here are based on ibid., 11–18.

7 Ibid.

8 Ibid., 11–12.

9 Ibid., excerpted from 12n4.

10 Ibid., 59.

11 Ibid., 58.

12 William Ray Arney, *Experts in the Age of Systems* (Albuquerque: University of New Mexico Press, 1991).
13 Galbraith, *The New Industrial State.*
14 Michael J. Piore and Charles F. Sabel, *The Second Industrial Divide: Possibilities for Prosperity* (New York: Basic Books, 1984); W. Sengenberger, G. Loveman, and M. J. Piore, eds., *The Re-emergence of Small Enterprises: Industrial Restructuring in Industrialized Countries* (Geneva: International Institute for Labour Studies, 1991).
15 This has been the subject of many popular works. See, for example, Vance Packard, *The Status Seekers* (Richmond Hill, ON: Simon and Schuster, 1961); Vance Packard, *The Pyramid Climbers* (Greenwich, CT: Fawcett, 1964); William Whyte, *The Organization Man* (New York: Simon and Schuster, 1956).
16 Galbraith, *The New Industrial State.* See especially chapters 7, 10, and 15.
17 AnnaLee Saxenian, *Regional Advantage: Culture and Competition in Silicon Valley and Route 128* (Cambridge, MA: Harvard University Press, 1994).
18 See, for instance, 'Vivendi Shareholders Sue for "False" Data,' *Globe and Mail* (22 August 2002), B8.
19 'The Great CEO Pay Heist,' *Fortune Magazine* (25 June 2001).
20 Ibid. See also 'The Greedy Bunch,' *Fortune Magazine* (2 September 2002).
21 Ibid. See also 'System Failure,' *Fortune Magazine* (24 June 2002).
22 Sarah Anderson and John Cavanagh, 'Top 200: The Rise of Corporate Global Power' (Baltimore: Institute for Policy Studies, 2000).
23 B. Commoner, 'The Environmental Cost of Economic Growth,' in *Energy, Economic Growth and the Environment: Papers Presented at a Forum Conducted by Resources for the Future Inc.,* ed. Sam H. Schurr (Baltimore: Johns Hopkins University Press, 1972), 30–65; W.H. Vanderburg, *The Labyrinth of Technology* (Toronto: University of Toronto Press, 2000), 218–20.
24 Galbraith, *The New Industrial State.* See particularly chapter 3.
25 Ernst von Weizsäcker, Amory B. Lovins, and L. Hunter Lovins, *Factor Four: Doubling Wealth By Halving Resource Use, The New Report to the Club of Rome* (London: Earthscan, 1997).
26 For a time, General Motors extended the approach of vertical integration to the procurement of personnel educated in knowledge separated from experience, by means of the General Motors Institute.
27 Later, Japanese lean production companies would perfect this to an art, creating families of companies, many of which may have been spun off from the parent company. See James P. Womack, Daniel T. Jones, and Daniel Roos, *The Machine That Changed the World* (New York: Macmillan, 1990).

28 W.H. Vanderburg, 'STS as a Vital Intellectual Ecumenism,' *Bulletin of Science, Technology and Society* 20 (February 2000): 3–9.

29 W.H. Vanderburg, 'Johannesburg, Kyoto, and the Need for Knowledge Infrastructure Renewal,' *Bulletin of Science, Technology and Society*, 22, no. 6 (December 2002): 419–25.

30 Herman E. Daly and John B. Cobb Jr, *For the Common Good: Redirecting the Economy toward Community, the Environment, and a Sustainable Future* (Boston: Beacon Press, 1989).

31 David Reisman with Nathan Glazer and Reuel Denney, *The Lonely Crowd: A Study of the Changing American Character* (Garden City, NY: Doubleday Anchor, 1950).

32 Galbraith, *The New Industrial State*, chs. 18, 19.

33 Stephen Bayley, *Sex, Drink and Fast Cars* (New York: Pantheon, 1986).

34 Jacques Ellul, *The New Demons*, trans. C. Edward Hopkin (New York: Seabury, 1975), ch. 4.

35 Dan Silverman, 'Fantasy and Reality in Nazi Work-Creation Programs, 1933–1936,' *Journal of Modern History* 65 (1993): 113–51; Richard Overy, *War and Economy in the Third Reich* (Oxford: Clarendon Press, 1994), chs. 6, 7.

36 In Canada it is estimated that the taxes one pays each year amount to one's earnings over eight months.

37 Victor Perlo, *Militarism and Industry: Arms Profiteering in the Missile Age* (New York: International Publishers, 1963); Herbert I. Schiller and Joseph D. Phillips, eds., *Super-state: Readings in the Military-industrial Complex* (Urbana: University of Illinois Press, 1970); Sidney Lens, *The Military-industrial Complex* (Philadelphia: Pilgrim, 1970).

38 Scott D. Sagan, *The Limits of Safety: Organizations, Accidents and Nuclear Weapons* (Princeton, NJ: Princeton University Press, 1993).

39 Perlo, *Militarism and Industry*; Schiller and Phillips, *Super-state*; Lens, *The Military-industrial Complex*.

40 Ibid.

41 W.H. Vanderburg, *The Labyrinth of Technology* (Toronto: University of Toronto Press, 2000).

42 Galbraith, *The New Industrial State*, ch. 29.

43 Shlomo Avineri, *The Social and Political Thought of Karl Marx* (Cambridge: Cambridge University Press, 1968).

44 Brubaker, *The Limits of Rationality*.

45 Ellul, *The Technological Society*.

46 Robert H. Nelson, *Reaching for Heaven on Earth: The Theological Meaning of Economics* (Savage, MD: Rowman and Littlefield, 1991).

47 Noam Chomsky, *Profit over People: Neoliberalism and Global Order* (New York: Seven Stories Press, 1999), 23–4.

48 There is no intention here of defining two new social classes.

49 Alexis de Tocqueville, *Democracy in America*, trans. George Lawrence (Garden City, NY: Doubleday, 1969).

50 Avineri, *The Social and Political Thought of Karl Marx.*

51 Georg Simmel, 'The Metropolis and Mental Life,' in *The Sociology of Georg Simmel*, ed. and trans. Kurt H. Wolff (Glencoe, IL: Free Press, 1950), 409–24.

52 Avineri, *The Social and Political Thought of Karl Marx.*

53 A.A. Brill, ed., *Sigmund Freud: Basic Writings* (New York: Random House, 1973).

54 Raymond Aron, *Progress and Disillusion: The Dialectics of Modern Society* (New York: F.A. Praeger, 1968).

55 Reisman, *The Lonely Crowd.*

56 For some classic works on a mass society, the reader is referred to de Tocqueville, *Democracy in America*; Simmel, 'The Metropolis and Mental Life,' 409–24; Karl Mannheim, *Man and Society in an Age of Reconstruction: Studies in Modern Social Structure*, trans. Edward Shils (London: K. Paul, Trench, Trubner, 1940); Robert Ezra Park and Ernest W. Burgess, *Introduction to the Science of Sociology* (Chicago: University of Chicago Press, 1922); Hannah Arendt, *The Origins of Totalitarianism* (New York: Harcourt, Brace, 1951); Erich Fromm, *Escape from Freedom* (New York: Farrar & Rinehart, 1941); José Ortega y Gasset, *The Revolt of the Masses* (London: G. Allen & Unwin, 1932); Raymond Williams, *Culture and Society, 1780–1950* (New York: Columbia University Press, 1958); Robert A. Nisbet, *The Quest for Community* (London: Oxford University Press, 1969); Daniel Bell, *The End of Ideology: On the Exhaustion of Political Ideas in the Fifties* (Glencoe, IL: Free Press, 1960); Reisman, *The Lonely Crowd*; T.S. Eliot, *Notes Towards the Definition of Culture* (London: Faber and Faber, 1948); Whyte, *The Organization Man*; C. Wright Mills, *White Collar: The American Middle Classes* (New York: Oxford University Press, 1951); C. Wright Mills, *The Power Elite* (New York: Oxford University Press, 1956); Arthur J. Vidich and Joseph Bensman, *Small Town in Mass Society: Class, Power, and Religion in a Rural Community* (Princeton, NJ: Princeton University Press, 1968).

57 In this and the following paragraphs, I am relying on Jacques Ellul, *Propaganda: The Formation of Men's Attitudes*, trans. Konrad Kellen and Jean Lerner (New York: Vintage, 1965). See especially chapter 2.

58 Daniel J. Boorstin, 'Statistical Morality,' in *The Americans: The Democratic Experience* (New York: Vintage, 1974), 238–44.

59 Daniel J. Boorstin, 'How Opinion Went Public,' in *Democracy and Its Discontents: Reflections on Everyday America* (New York: Random, 1974), 12–21.
60 Ibid. See also Walter Lippmann, *Public Opinion* (New Brunswick, NJ: Transaction, 1997).
61 Ellul, *Propaganda*.
62 Jacques Ellul, *The Technological Bluff* (Grand Rapids, MI: Wm. B. Eerdmans, 1990).
63 Jacques Ellul, *Betrayal of the West* (New York: Continuum, 1978).
64 Galbraith, *The New Industrial State*, chs. 26, 27; Edwin R. Black, 'Politics on a Microchip,' *Canadian Journal of Political Science* 16, no. 4 (1983): 675–90; W.H. Vanderburg, 'Political Imagination in a Technical Age,' in *Democratic Theory and Technological Society*, ed. Richard B. Day, Ronald Beiner, and Joseph Masciulli (New York: M.E. Sharpe, 1988), 3–35.
65 Black, 'Politics on a Microchip'; Jacques Ellul, *The Political Illusion*, trans. Konrad Kellen (New York: Knopf, 1967); Ellul, *The New Demons*.
66 Vanderburg, *The Labyrinth of Technology*, ch. 9.
67 In this and the following paragraphs, I am indebted to Ellul, *Propaganda*. See especially chapter 3.
68 Vanderburg, *The Labyrinth of Technology*.
69 Adam Smith, *An Inquiry into the Nature and Causes of the Wealth of Nations*, 2 vols. (Chicago: University of Chicago Press, 1976).
70 Avineri, *The Social and Political Thought of Karl Marx*.
71 Jacques Ellul, *Métamorphose du bourgeois* (Paris: Calmann-Lévy, 1967); Max Weber, *The Protestant Ethic and the Spirit of Capitalism* (New York: Charles Scribner's, 1958).
72 Vanderburg, *The Labyrinth of Technology*. See especially ch. 10.
73 Ellul, *Propaganda*; Ellul, *The Political Illusion*. See also Lippmann, *Public Opinion*; Noam Chomsky, *Letters from Lexington: Reflections on Propaganda* (Monroe, ME: Common Courage, 1993); Noam Chomsky, *Necessary Illusions: Thought Control in Democratic Societies* (Toronto: Anansi, 1991); Edward S. Herman and Noam Chomsky, *Manufacturing Consent: The Political Economy of the Mass Media* (New York: Pantheon, 1988).
74 Vanderburg, *The Labyrinth of Technology*; Jacques Ellul, 'De la signification des relations publiques dans une société technicienne,' *L'année sociologique* 13 (1963): 69–152; Noam Chomsky, *Media Control: The Spectacular Achievements of Propaganda* (New York: Seven Stories, 1997); Chomsky, *Letters from Lexington*; Chomsky, *Necessary Illusions*; Herman and Chomsky, *Manufacturing Consent*. See also Stuart Ewen, *PR! A Social History of Spin* (New York: Basic Books, 1996); Stuart Ewen and Elizabeth Ewen, *Channels of Desire: Mass Images and the Shaping of American Consciousness* (Minneapolis:

University of Minnesota Press, 1992); Stuart Ewen, *Captains of Consciousness: Advertising and the Social Roots of the Consumer Culture* (New York: McGraw-Hill, 1976).

75 Sheldon Rampton and John Stauber, *Trust Us, We're Experts! How Industry Manipulates Science and Gambles with Your Future* (New York: Jeremy P. Tarcher / Putnam, 2001); Chomsky, *Media Control*; Chomsky, *Letters from Lexington*; Chomsky, *Necessary Illusions*; Herman and Chomsky, *Manufacturing Consent*. See also Ewen, *PR!*; Ewen and Ewen, *Channels of Desire*; Ewen, *Captains of Consciousness*.

76 Ellul, *Propaganda*, 145–6.

77 Ibid.; Ellul, *The Political Illusion*; Herman and Chomsky, *Manufacturing Consent*.

78 Ellul, *Propaganda*, 160.

79 Ibid., 159.

80 Ellul, *The Political Illusion*.

81 Barry Commoner, 'The Environmental Costs of Economic Growth,' in *Energy, Economic Growth and the Environment*, ed. Sam Shurr (Baltimore: Johns Hopkins University Press, 1971), 30–65; Theo Colborn, Dianne Dumanoski, and John Peterson Myers, *Our Stolen Future: Are We Threatening our Fertility, Intelligence and Survival? A Scientific Detective Story* (New York: Dutton, 1996); Deborah Cadbury, *The Feminization of Nature: Our Future at Risk* (London: Hamish Hamilton, 1997); David Weir and Mark Schapiro, *Pesticides and People in a Hungry World* (Oakland, CA: Food First Books, 1981).

82 Juliet B. Schor, *The Overworked American: The Unexpected Decline of Leisure* (New York: Basic Books, 1991); J. Gershuny, 'Are We Running Out of Time?' *Futures* (January/February 1992): 3–22; Benjamin Hunnicutt, *Work without End: Abandoning Shorter Hours for the Right to Work* (Philadelphia: Temple University Press, 1988).

83 Hubert L. Dreyfus, *On the Internet* (London: Routledge, 2001); Scott Lash, *Critique of Information* (London: Sage, 2002); Laura Pappano, *The Connection Gap: Why Americans Feel So Alone* (New Brunswick, NJ: Rutgers University Press, 2001); Pippa Noris, *Digital Divide: Civic Engagement, Information Poverty and the Internet Worldwide* (New York: Cambridge University Press, 2001); Sherry Turkle, *Life on the Screen: Identity in the Age of the Internet* (New York: Simon & Schuster, 1995); Sherry Turkle, *The Second Self: Computers and the Human Spirit* (New York: Simon and Schuster, 1984); Craig Brod, *Technostress: The Human Cost of the Computer Revolution* (Reading, MA: Addison-Wesley, 1984); C.A. Bowers, *Let Them Eat Data: How Computers Affect Education, Cultural Diversity and the Prospects for Ecological*

Sustainability (Athens: University of Georgia Press, 2000); Orrin E. Klapp, *Overload and Bordom: Essays on the Quality of Life in the Information Society* (Westport, CT: Greenwood Press, 1986).

84 Turkle, *The Second Self*.

85 James Howard Kunstler, *The Geography of Nowhere: The Rise and Decline of America's Man-made Landscape* (New York: Simon & Schuster, 1993); James Howard Kunstler, *Home from Nowhere: Remaking Our Everyday World for the 21st Century* (New York: Simon & Schuster, 1998); James Howard Kunstler, *The City in Mind: Meditations on the Urban Condition* (New York: Free Press, 2001).

86 Faye Duchin, *The Future of the Environment: Ecological Economics and Technological Change* (New York: Oxford University Press, 1994); Michael Redclift, *Wasted: Counting the Cost of Global Consumption* (London: Earthscan, 1996). See also Commoner, 'The Environmental Costs of Economic Growth,' 30–65, and Colborn, Dumanoski, and Myers, *Our Stolen Future*.

87 Altieri, *Genetic Engineering in Agriculture*; de la Perrière, Ali, and Seuret, *Brave New Seeds*. See also Vandana Shiva, *Tomorrow's Biodiversity* (London: Thames and Hudson, 2000); Vandana Shiva, *Monocultures of the Mind: Perspectives on Biodiversity and Biotechnology* (London: Zed Books, 1993).

88 Robert Jay Lifton and Richard Falk, *Indefensible Weapons: The Political and Psychological Case Against Nuclearism* (Toronto: CBC Enterprises, 1982); Peter R. Beckman, *The Nuclear Predicament: Nuclear Weapons in the Cold War and Beyond* (Englewood Cliffs, NJ: Prentice-Hall, 1992).

89 Neil Freeman, *The Politics of Power: Ontario Hydro and Its Government, 1906–1995* (Toronto: University of Toronto Press, 1996); Jesse H. Ausubel and Cesare Marchetti, 'Electron: Electrical Systems in Retrospect and Prospect,' in *Technological Trajectories and the Human Environment*, ed. Jesse H. Ausubel and H. Dale Langford (Washington, DC: National Academy Press, 1997).

90 Freeman, *The Politics of Power*.

91 Schor, *The Overworked American*.

92 Herman Daly and John B. Cobb Jr, *For the Common Good: Redirecting the Economy toward Community, the Environment, and a Sustainable Future* (Boston: Beacon, 1989); Clifford Cobb, Tel Halstead, and Jonathan Rowe, 'If the GDP Is Up, Why Is America Down?' *Atlantic Monthly* (October 1995): 59–78.

93 Braden R. Allenby and Deanna J. Richards, eds., *The Greening of Industrial Ecosystems* (Washington, DC: National Academy Press, 1994), Introduction.

94 Robert Karasek and Töres Theorell, *Healthy Work: Stress, Productivity, and the Reconstruction of Working Life* (New York: Basic Books, 1990), 11.
95 Vanderburg, *The Labyrinth of Technology*.
96 Karasek and Theorell, *Healthy Work*.
97 U.S. Department of Health and Human Services, *Mental Health: A Report of the Surgeon General* (Rockville, MD, 1999).
98 Vanderburg, *The Labyrinth of Technology*, particularly ch. 5.
99 Michael Jacobs, *The Green Economy: Environment, Sustainable Development and the Politics of the Future* (Vancouver: UBC Press, 1993).

Chapter 7

1 Max Weber, 'Science as a Vocation,' in *From Max Weber: Essays in Sociology*, ed. H.H. Gerth and C. Wright Mills (New York: Oxford University Press, 1963), 129–56.
2 Hubert Dreyfus and Stuart Dreyfus, *Mind over Machine: The Power of Human Intuition and Experience in the Era of the Computer* (New York: Free Press, 1986). The July 2004 issue of the *Bulletin of Science, Technology and Social Development* is dedicated to the skill acquisition model and its applications.
3 Eric Voegelin, *Order and History*, vol. 2, *The World of the Polis* (Baton Rouge: Louisiana State University Press, 1987); Eric Voegelin, *Order and History*, vol. 3, *Plato and Aristotle* (Baton Rouge: Louisiana State University Press, 1987).
4 Robert Graves, *The Greek Myths* (London: Penguin, 1965).
5 Dreyfus and Dreyfus, *Mind over Machine*, 2.
6 Plato, *The Apology*, trans. Benjamin Jowett (Danbury, CT: Grolier, 1980); I.F. Stone, *The Trial of Socrates* (Boston: Little, Brown, 1988).
7 Jacques Ellul, *Histoire des institutions* (Paris: Presses Universitaires de France, 1956).
8 I am indebted to the lectures given by Professor Hubert Dreyfus in his course Existentialism in Literature and Film, taught at Berkeley in 2002. Much of this section reflects what I learned from him and from my French mentor, Jacques Ellul.
9 John Horgan, *The End of Science: Facing the Limits of Knowledge in the Twilight of the Scientific Age* (Reading, MA: Helix Books, 1996); Bernard d'Espagnat, *In Search of Reality* (New York: Springer-Verlag, 1983).
10 For as far as we can go back, Jewish culture regarded itself as historical. Also, it surely is no accident that the *Song of Songs*, which extols the human body, was incorporated into the Jewish Bible. The Jewish orienta-

tion to life had an influence on Christian thought, but this rapidly became undermined by Greek philosophy. Western civilization never quite recovered from this confusion.

11 Alvin Toffler, *Future Shock* (Toronto: Bantam, 1971).

12 Max Weber, *The Theory of Social and Economic Organization*, ed. T. Parsons (New York: Oxford University Press, 1947), 160–1.

13 Ibid., 161.

14 Jacques Ellul, *The Technological Society*, trans. John Wilkinson (New York: Vintage Books, 1964), xxv. For an overview of Ellul's theory of contemporary society, see W.H. Vanderburg, ed., *Perspectives on Our Age: Jacques Ellul Speaks of His Life and Work* (Toronto: CBC Enterprises, 1981). For further details, see Jacques Ellul, *The Technological System*, trans. Joachim Neugroschel (New York: Continuum, 1980). The reader should be aware that some translators have rendered the French word *technique* as technology. This is erroneous since there is no term equivalent to technology in the French language. The French *technologie* refers to the thinking about, or philosophy of, technique.

15 In the following paragraphs, I am indebted to Jeremy Campbell, *The Improbable Machine: What the Upheavals in Artificial Intelligence Research Reveal about How the Mind Really Works* (New York: Simon and Schuster, 1989). See also Arthur Koestler, *The Ghost in the Machine* (London: Hutchinson, 1967).

16 W.H. Vanderburg, *The Labyrinth of Technology* (Toronto: University of Toronto Press, 2000), ch. 4.

17 Hubert Dreyfus, *What Computers Still Can't Do: The Limits of Artificial Intelligence* (Cambridge, MA: MIT Press, 1992).

18 Dreyfus and Dreyfus, *Mind over Machine*.

19 Campbell, *The Improbable Machine*, 140.

20 Ibid., 101.

21 Ibid., 103.

22 Ibid., 140.

23 Didier Nordon and Hubert Nyssen, eds., *Les mathématiques pures n'existent pas* (Paris: Actes Sud, 1981).

24 Bill Joy, 'Why the Future Does Not Need Us,' *Wired Magazine* 8 (April 2000); Ray Kurzweil, *The Age of Spiritual Machines: When Computers Exceed Human Intelligence* (New York: Viking, 1999).

25 I will show that technique-based connectedness emerges out of the technology-based connectedness of a society.

26 Dreyfus, *What Computers Still Can't Do*, 126–8.

27 William E. Connolly, 'Essentially Contested Concepts,' in *The Terms of*

Political Discourse (Princeton, NJ: Princeton University Press, 1984), 10–44.

28 Michael Polanyi, *Personal Knowledge* (Chicago: University of Chicago Press, 1962).

29 Thomas S. Kuhn, *The Structure of Scientific Revolutions*, 2nd ed. (Chicago: University of Chicago Press, 1970).

30 Hubert L. Dreyfus and Paul Rabinow, *Michel Foucault: Beyond Structuralism and Hermeneutics* (Chicago: University of Chicago Press, 1983).

31 Jean-Luc Porquet, *Jacques Ellul, l'homme qui avait (presque) tout prévu* (Paris: Le Cherche Midi, 2003).

Chapter 8

1 William E. Dunstan, *Ancient Greece* (Fort Worth, TX: Harcourt College Publishers, 2000).

2 Jacques Ellul, *The Technological Society*, trans. John Wilkinson (New York: Knopf, 1964), 382.

3 J.H. Van den Berg, *The Changing Nature of Man* (New York: Delta, 1961).

4 Jacques Ellul, *La technique, ou, L'enjeu du siècle* (Paris: Economica, 1990). For the English translation see note 2 above.

5 Jacques Ellul, *Propaganda: The Formation of Men's Attitudes*, trans. Konrad Kellen and Jean Lerner (New York: Vintage Books, 1973); Jacques Ellul, *The Political Illusion*, trans. Konrad Kellen (New York: Knopf, 1967).

6 Ellul, *The Technological Society*, 344.

7 Ida Hoos, *Systems Analysis in Public Policy* (Berkeley: University of California Press, 1983). See also James William Gibson, *The Perfect War: The War We Couldn't Lose and How We Did* (Boston: Atlantic Monthly Press, 1986).

8 This is well illustrated by Shulamit Reinharz in *On Becoming a Social Scientist: From Survey Research and Participant Observation to Experiential Analysis* (San Francisco: Jossey-Bass, 1979).

9 George Devereux, *From Anxiety to Method in the Behavioral Sciences* (Paris: Mouton, 1967).

10 Paul Goodman, *Growing up Absurd: Problems of Youth in the Organized System* (New York: Random House, 1960).

11 Neil Postman, *Amusing Ourselves to Death: Public Discourse in the Age of Show Business* (New York: Viking Penguin, 1986); Neil Postman, 'Learning by Story,' *Atlantic Monthly* 264 (December 1989): 119–24.

12 Ibid. See also Ellen Rose, *Hyper Texts: The Language and Culture of Educational Computing* (London, ON: Althouse Press, 2000).

13 Postman, *Amusing Ourselves to Death*; Postman, 'Learning by Story.'

14 Craig Brod, *Technostress: The Human Cost of the Computer Revolution* (New York: Addison-Wesley, 1986). See also Sherry Turkle, *The Second Self: Computers and the Human Spirit* (New York: Simon and Schuster, 1984).

15 Fictionalized but nevertheless well-researched accounts may be found in Steven Pressfield, *Gates of Fire* (New York: Bantam, 1999); Mary Renault, *The Last of the Wine* (New York: Vintage, 1975); Robert Graves, *I, Claudius* (Manchester: Carcanet, 1998).

16 Heinz Lubasz, *The Development of the Modern State* (New York: Macmillan, 1964).

17 Victor Perlo, *Militarism and Industry: Arms Profiteering in the Missile Age* (New York: International Publishers, 1963); Herbert I. Schiller and Joseph D. Phillips, eds., *Super-state: Readings in the Military-Industrial Complex* (Urbana: University of Illinois Press, 1970); Sidney Lens, *The Military-Industrial Complex* (Philadelphia: Pilgrim, 1970).

18 Deborah Shapley and Rustum Roy, *Lost at the Frontier: U.S. Science and Technology Policy Adrift* (Philadelphia: ISI Press, 1985).

19 In turn, the more popular these video games become, the more likely they are to be made into big-budget Hollywood productions such as *Mortal Kombat* and *Tomb Raider*.

20 Perlo, *Militarism and Industry*; Schiller and Phillips, *Super-state*; Lens, *The Military-Industrial Complex*.

21 Ibid.

22 Vance Packard, *The Hidden Persuaders* (London: Longmans, Green, 1957).

23 Ibid.

24 Sharon Beder, *Global Spin: The Corporate Assault on Environmentalism* (White River Junction, VT: Chelsea Green Publishing, 1998); William Greider, *Who Will Tell the People: The Betrayal of American Democracy* (New York: Simon and Schuster, 1992); Stuart Ewen, *PR! A Social History of Spin* (New York: Basic Books, 1996); Stuart Ewen and Elizabeth Ewen, *Channels of Desire: Mass Images and the Shaping of American Consciousness* (Minneapolis: University of Minneapolis Press, 1992); Stuart Ewen, *Captains of Consciousness: Advertising and the Social Roots of the Consumer Culture* (New York: McGraw-Hill, 1976).

25 Jacques Ellul, *The Humiliation of the Word*, trans. Joyce Main Hanks (Grand Rapids, MI: Eerdmans, 1985). See also Guy Debord, *The Society of the Spectacle* (New York: Zone Books, 1994); Richard Stivers, *Technology as Magic: The Triumph of the Irrational* (New York: Continuum, 1999), ch. 4.

26 O. Duane Weeks and Catherine Johnson, eds., *When All the Friends Have Gone: A Guide for Aftercare Providers* (Amityville, NY: Baywood, 2001); Philippe Ariès, *The Hour of Our Death*, trans. Helen Weaver (New York:

516 Notes to pages 356–66

Knopf, 1981). See also J.J. Farrell, *Inventing the American Way of Death*
(Philadelphia: Temple University Press, 1980); H. Feitel, ed., *New Meanings
of Death* (New York: McGraw-Hill, 1977); Edwin S. Shneidman, ed., *Death:
Current Perspectives* (Palo Alto, CA: Mayfield Publishing, 1976).

27 Mary Bradbury, *Representations of Death: A Social Psychological Perspective*
(New York: Routledge, 1999); Glennys Howarth, *Last Rites: The Work of the
Modern Funeral Director* (Amityville, NY: Baywood, 1996).

28 Robert Rhodes, ed., *Imperialism and Underdevelopment* (New York: Monthly
Review Press, 1970).

29 Susan George, *Food for Beginners* (New York: Writers and Readers Publishing, 1982).

30 Dan Morgan, *Merchants of Grain* (New York: Viking, 1979).

31 Susan George, *How the Other Half Dies: The Real Reasons for World Hunger*
(London: Penguin, 1989).

32 Miguel A. Altieri, *Genetic Engineering in Agriculture: The Myths, Environmental Risks and Alternatives* (Oakland, CA: Food First Books, 2001).

33 Siegfried Giedion discusses the mechanization of bread-making in *Mechanization Takes Command* (New York: Oxford University Press, 1948),
179–208.

34 Vandana Shiva, *Poverty and Globalization* (London: BBC Reith Lectures,
2000).

35 Ibid. See also Altieri, *Genetic Engineering in Agriculture*.

36 Sheldon Rampton and John Stauber, *Trust Us, We're Experts: How Industry
Manipulates Science and Gambles with Your Future* (New York: Jeremy P.
Tarcher / Putnam, 2001). See also Daniel Charles, *Lords of the Harvest:
Biotech, Big Money, and the Future of Food* (Cambridge, MA: Perseus Books,
2001).

37 To date, this is the case in the United States and Canada. The governments
in both countries are encouraging producers of GM foods to provide
voluntary labelling.

38 Shiva, *Poverty and Globalization*.

39 William Whyte, *The Organization Man* (New York: Simon and Schuster,
1956); Vance Packard, *The Pyramid Climbers* (Greenwich, CT: Fawcett,
1964); Christopher Lasch, *The Minimal Self: Psychic Survival in Troubled
Times* (New York: W.W. Norton, 1984); David Riesman with Nathan
Glazer and Reuel Denney, *The Lonely Crowd: A Study of the Changing
American Character* (New Haven, CT: Yale University Press, 1963); Good-
man, *Growing up Absurd*; Robert Jay Lifton, *The Broken Connection* (New
York: Simon and Schuster, 1979); Ellul, *Propaganda*; Christopher Lasch, *The
Culture of Narcissism: American Life in an Age of Diminishing Expectations*

(New York: W.W. Norton, 1978); Richard Stivers, *The Culture of Cynicism: American Morality in Decline* (Cambridge, MA: Blackwell, 1994); Turkle, *The Second Self*; Craig Brod, *Technostress: The Human Cost of the Computer Revolution* (Reading, MA: Addison-Wesley, 1984); William Ray Arney, *Experts in the Age of Systems* (Albuquerque: University of New Mexico Press, 1991); Joel Bakan, *The Corporation* (Toronto: Viking Canada, 2004).

40 Michel Foucault, 'On Power,' in *Michel Foucault: Politics, Philosophy, Culture; Interviews and Other Writings 1977–1984*, ed. Lawrence D. Kritzman (London: Routledge, 1990).

41 Herman E. Daly and John B. Cobb Jr, *For the Common Good: Redirecting the Economy Toward Community, the Environment, and a Sustainable Future* (Boston: Beacon Press, 1989).

42 Michael Jacobs, *The Green Economy: The Environment, Sustainable Development and the Politics of the Future* (Vancouver: UBC Press, 1993).

43 Lester R. Brown, *Eco-Economy: Building an Economy for the Earth* (New York: W.W. Norton, 2001).

44 Max Weber, 'Science as a Vocation,' in *From Max Weber: Essays in Sociology*, ed. H.H. Gerth and C. Wright Mills (New York: Oxford University Press, 1963), 129–56; Ellul, *The Technological Society*.

45 John Kenneth Galbraith, *The New Industrial State* (New York: New American Library, 1979).

46 Jack P. Manno, *Privileged Goods: Commoditization and Its Impact on Environment and Society* (Boca Raton, FL: Lewis, 2000).

47 W.H. Vanderburg, 'STS in Seattle, New York and Baltimore,' editorial, *Bulletin of Science, Technology and Society* 21 (June 2001): 1–4. In Vanderburg, *The Labyrinth of Technology*, I have examined in detail the case for technology itself.

48 Devereux, *From Anxiety to Method in the Behavioral Sciences*.

Chapter 9

1 Daniel Bell, *The Coming of Post-Industrial Society: A Venture in Social Forecasting* (New York: Basic Books, 1976); Joel Jay Kassiola, *The Death of Industrial Civilization: The Limits to Economic Growth and the Repoliticization of Advanced Industrial Society* (Albany: State University of New York Press, 1990); Krishan Kumar, *From Post-industrial to Post-modern Society: New Theories of the Contemporary World* (Cambridge, MA: Blackwell Publishers, 1995); Yoneji Masuda, *The Information Society as Post-industrial Society* (Tokyo: Institute for the Information Society, 1980); Seymour Martin Lipset, ed., *The Third Century: America as a Post-industrial Society* (Stanford,

CA: Hoover Institution Press, 1979); Bo Gustafsson, ed., *Post-industrial Society* (New York: St Martin's Press, 1979); Alain Touraine, *The Post-Industrial Society; Tomorrow's Social History: Classes, Conflicts and Culture in the Programmed Society*, trans. Leonard F.X. Mayhew (New York: Random House, 1971); Peter F. Drucker, *Post-capitalist Society* (New York: HarperBusiness, 1994); David Ramsay Steele, *From Marx to Mises: Post-capitalist Society and the Challenge of Economic Calculation* (La Salle, IL: Open Court, 1992); Martyn J. Lee, ed., *The Consumer Society Reader* (Malden, PA: Blackwell, 2000); Jean Baudrillard, *The Consumer Society: Myths and Structures* (London: Sage, 1998); Neva R. Goodwin, Frank Ackerman, and David Kiron, eds., *The Consumer Society* (Washington, DC: Island Press, 1997); Erich Fromm, *Escape from Freedom* (New York: Farrar & Rinehart, 1941); José Ortega y Gasset, *The Revolt of the Masses* (London: G. Allen & Unwin, 1932); Ray B. Browne and Marshall W. Fishwick, eds., *The Global Village: Dead or Alive?* (Bowling Green, OH: Bowling Green State University Popular Press, 1999); Marshall McLuhan, *The Global Village: Transformations in World Life and Media in the 21st Century* (New York: Oxford University Press, 1989); Guy Debord, *The Society of the Spectacle* (New York: Zone, 1994); John Kenneth Galbraith, *The New Industrial State*, 3rd ed. (New York: New American Library, 1979); Zbigniew Brzezinski, *Between Two Ages: America's Role in the Technetronic Era* (New York: Viking Press, 1970); Zbigniew Brzezinski, *America in the Technetronic Age* (New York: School of International Affairs, 1967); John P. Rasmussen, *The New American Revolution: The Dawning of the Technetronic Era* (New York: Wiley, 1972).

Some authors of these theories obviously did not bother checking what was happening. For example, computers were a signpost of the new smokestackless industries and of dematerialization. One statistic will suffice: to make a single 32 MB DRAM computer chip requires 32 kg of water, 1.6 kg of fossil fuels, 700 grams of elemental gases (mainly nitrogen), and 72 grams of chemicals (hundreds are used, including lethal arsine gas and corrosive hydrogen fluoride). *Japan Times* (23 January 2003) 15.

2 Bell, *The Coming of Post-Industrial Society*, 7.

3 Claude Shannon cited in Tom Forester, ed., *The Microelectronics Revolution: The Complete Guide to the New Technology and Its Impact on Society* (Oxford: Basil Blackwell, 1980), 508.

4 Edward Tenner, *Why Things Bite Back: Technology and the Revenge of Unintended Consequences* (New York: Knopf, 1996).

5 Chellis Glendinning, *When Technology Wounds: The Human Consequences of Progress* (New York: William Morrow, 1990).

6 Winner, *Autonomous Technology.*
7 Jacques Ellul, *The Technological Bluff*, trans. Geoffrey W. Bromiley (Grand Rapids, MI: W.B. Eerdmans, 1990).
8 This is illustrated by the growing role of mechatronics in mechanical engineering.
9 Jean-Luc Porquet, *Jacques Ellul, l'homme qui avait (presque) tout prévu* (Paris: Le Cherche Midi, 2003).
10 Bell, *The Coming of Post-Industrial Society.*
11 Ibid.
12 Noam Chomsky, *Profit over People: Neoliberalism and Global Order* (New York: Seven Stories Press, 1999), 23–4.
13 Ibid.
14 Tim Jackson, *Material Concerns: Pollution, Profit and Quality of Life* (New York: Routledge, 1996).
15 Touraine, *The Post-Industrial Society.*
16 Peter Ferdinand Drucker, *Post-Capitalist Society* (New York: Harper Business, 1993).
17 Notable exceptions are John Kenneth Galbraith, *The New Industrial State* (New York: New American Library, 1979) and Radovan Richta, *Civilization at the Crossroads: Social and Human Implications of the Scientific and Technological Revolution* (Prague: International Arts and Sciences Press, 1969).
18 W.W. Rostow, *The Economics of Take-off into Sustained Growth: Proceedings of a Conference Held by the International Economic Association* (London: Macmillan, 1963).
19 Deborah Shapley and Rustum Roy, *Lost at the Frontier: U.S. Science and Technology Policy Adrift* (Philadelphia: ISI Press, 1985).
20 Herman Daly and John B. Cobb Jr, *For the Common Good: Redirecting the Economy toward Community, the Environment, and a Sustainable Future* (Boston: Beacon, 1989); Clifford Cobb, Ted Halstead, and Jonathan Rowe, 'If the GDP Is Up, Why Is America Down?' *Atlantic Monthly* (October 1995): 59–78.
21 Harvey Cox, 'The Market as God: Living in the New Dispensation,' *Atlantic Monthly* (March 1999): 18–23.
22 Maude Barlow and Tony Clarke, *Global Showdown: How the New Activists Are Fighting Global Corporate Rule* (Toronto: Stoddart, 2002).
23 Robert Bocock, *Consumption* (London: Routledge, 1993).
24 McLuhan, *The Global Village.*
25 Jerry Mander, *Four Arguments for the Elimination of Television* (New York: Quill, 1978).
26 James Burnham, *The Managerial Revolution* (Harmondsworth, UK: Penguin Books, 1962).

27 James R. Beniger, *The Control Revolution: Technological and Economic Origins of the Information Society* (Cambridge, MA: Harvard University Press, 1986).

28 David Riesman with Nathan Glazer and Reuel Denney, *The Lonely Crowd: A Study of the Changing American Character* (New Haven, CT: Yale University Press, 1963).

29 William Whyte, *The Organization Man* (New York: Simon and Schuster, 1956).

30 Debord, *The Society of the Spectacle*.

31 Herbert Marcuse, *One Dimensional Man: Studies in the Ideology of Advanced Industrial Society* (Boston: Beacon, 1964).

32 Ulrich Beck, *Risk Society: Towards a New Modernity*, trans. Mark Ritter (London: Sage Publications, 1992).

33 Arnold Toynbee, *A Study of History*, abridgement of vol. 1–10 by D.C. Somervell (London: Oxford University Press, 1946).

34 Jacques Ellul, *La technique, ou, L'enjeu du siècle* (Paris: Economica, 1990).

35 Theodore Roszak, *The Making of a Counter Culture: Reflections on the Technocratic Society and Its Youthful Opposition* (Garden City, NY: Doubleday, 1968).

36 Lewis Mumford, *The Myth of the Machine: Technics and Human Development* (New York: Harcourt, Brace & World, 1967); Lewis Mumford, *The Myth of the Machine: The Pentagon of Power* (New York: Harcourt, Brace, Jovanovich, 1970).

37 Michel Foucault, 'On Power,' in *Michel Foucault: Politics, Philosophy, Culture; Interviews and Other Writings 1977–1984*, ed. Lawrence D. Kritzman (London: Routledge, 1990).

38 Daniel Bell, 'The Social Framework of the Information Society,' in *The Microelectronics Revolution*, ed. Forester, 500–49.

39 Joseph Weizenbaum, 'Once More, the Computer Revolution,' in *The Microelectronics Revolution*, ed. Forester, 550–70.

40 Daniel Bell, 'A Reply to Weizenbaum,' in *The Microelectronics Revolution*, ed. Forester, 571–4.

41 Hubert Dreyfus, *What Computers Still Can't Do* (Cambridge, MA: MIT Press, 1992).

42 Hubert Dreyfus and Stuart Dreyfus with Tom Athanasiou, *Mind over Machine: The Power of Human Intuition and Expertise in the Era of the Computer* (New York: Free Press, 1986).

43 Herbert A. Simon, 'What Computers Mean for Man and Society,' in *The Microelectronics Revolution*, ed. Forester, 419–33.

44 Joseph Weizenbaum, 'Where Are We Going? Questions for Simon,' in *The Microelectronics Revolution*, ed. Forester, 434–52.

45 Richta, *Civilization at the Crossroads.*
46 Jacques Ellul, *L'idéologie Marxiste Chrétienne* (Paris: Le Centurion, 1979);
    Robert V. Andelson and James M. Dawsey, *From Wasteland to Promised
    Land: Liberation Theology for a Post-Marxist World* (Maryknoll, NY: Orbis
    Books, 1992); Michael R. Candelaria, *Popular Religion and Liberation: The
    Dilemma of Liberation Theology* (Albany: State University of New York
    Press, 1990).
47 Günter Friedrichs and Adam Schaff, eds., *Microelectronics and Society:
    For Better or for Worse: A Report to the Club of Rome* (New York: Pergamon
    Press, 1982).
48 For example, 2 B or not 2 B (To be or not to be); N E 1 there? (Anyone
    there?); 4 U R ... (For you are ...); Lol (Laugh out loud); Cu (See you);
    :) (Happy); Omg (Oh my God); Aso (And so on).
49 Craig Brod, *Technostress: The Human Cost of the Computer Revolution* (Read-
    ing, MA: Addison-Wesley, 1984); Sherry Turkle, *The Second Self* (New
    York: Simon and Schuster, 1984); Stephen L. Talbot, *The Future Does Not
    Compute: Transcending the Machines in Our Midst* (Sebastopol, CA: O'Reilly
    and Associates, 1995).
50 A. Harmon, 'Researchers Find Sad, Lonely World in Cyberspace,' *New
    York Times* (30 August 1998): 22.
51 For an excellent entry point into the literature on the emergence of the
    lean production system, see James P. Womack, Daniel T. Jones, and Daniel
    Roos, *The Machine That Changed the World* (New York: Macmillan, 1990). Its
    ideological slant stems from the fact that it studied lean production from
    the perspective of 'people changing technology' while largely ignoring
    how technology simultaneously changes people. I have relied on this
    work for the basic facts about lean production.
52 Vanderburg, *The Labyrinth of Technology*, ch. 10.
53 Robert Karasek and Töres Theorell, *Healthy Work: Stress, Productivity and
    the Reconstruction of Working Life* (New York: Basic Books, 1990), 11. See
    also the two Special Issues of the *Bulletin of Science, Technology & Society*,
    24, nos. 4–5
54 Womack, Jones, and Roos, *The Machine That Changed the World.*
55 Ibid., 13–14.
56 Robert Karasek, 'Labor Participation and Work Quality Policy: Require-
    ments for an Alternative Economic Future,' *Scandinavian Journal of Work,
    Environment and Health* 23, supplement 4 (1997): 60.
57 Karasek and Theorell, *Healthy Work*
58 Boye Lafayette de Mente, 'Karoshi: Death from Overwork,' *NTC's Dictio-
    nary of Japan's Cultural Code Words* (Lincolnwood, IL: National Textbook

Company, 1994); Takashi Haratani, 'Karoshi: Death from Overwork,' *Encyclopedia of Occupational Health and Safety*, 4th ed. (Geneva: International Labour Organization, 1998).

59 Womack, Jones, and Roos, *The Machine That Changed the World*, 13.

60 K. Matsushita, quoted in Peter J. Denning, 'Beyond Formalism,' *American Scientist* 79 (January–February 1991): 19.

61 Vanderburg, *The Labyrinth of Technology*, ch. 10.

62 For an overview of these and other details, the reader is again referred to Womack, Jones, and Roos, *The Machine That Changed the World*.

63 Henry Mintzberg, 'The Fall and Rise of Strategic Planning,' *Harvard Business Review* 72 (January-February 1994): 107–14.

64 W.H. Vanderburg, *The Growth of Minds and Cultures: A Unified Theory of the Structure of Human Experience* (Toronto: University of Toronto Press, 1985).

65 Hubert L. Dreyfus, *What Computers Still Can't Do* (Cambridge, MA: MIT Press, 1992); Hubert Dreyfus and Stuart Dreyfus with Tom Athanasiou, *Mind over Machine: The Power of Human Intuition and Expertise in the Era of the Computer* (New York: Free Press, 1986); Russell L. Ackoff and Fred E. Emery, *On Purposeful Systems* (Chicago: Aldine-Atherton, 1972).

66 W. Edwards Deming, *Elementary Principles of the Statistical Control of Quality: A Series of Lectures* (Tokyo: Nippon Kagaku Gijutsu Remmei, 1952).

67 Thomas H. Davenport, *Process Innovation: Reengineering Work through Information Technology* (Boston: Harvard Business School Press, 1993).

68 Michael Hammer and James Champy, *Reengineering the Corporation: A Manifesto for Business Revolution* (New York: Harper Collins, 1993).

69 For a critical overview, see E. Mumford and R. Hendricks, *Reengineering Rhetoric and Reality: The Rise and Fall of a Management Fashion*. Manuscript in preparation.

70 Thomas K. Landauer, *The Trouble with Computers: Usefulness, Usability and Productivity* (Cambridge, MA: MIT Press, 1995).

71 Ibid.

72 Peter Brödner, *The Shape of Future Technology: The Anthropocentric Alternative* (New York: Springer-Verlag, 1990).

73 Hammer and Champy, *Reengineering the Corporation*, 32.

74 Ibid., 35.

75 I am indebted to my student, Oscar Guerra, for the following examples. For full details see Jorge Oscar Guerra Gutiérrez, 'Extending the Precautionary Principle to Work Redesign: The Implications of Business Process Reengineering for Low-Level White-Collar Work' (M.A.Sc. thesis: University of Toronto, 2001).

76 Hammer and Champy, *Reengineering the Corporation*, 36–8 and 204 ff.

77 Ibid.

78 Michael Hammer, 'Re-engineering Work: Don't Automate, Obliterate,' *Harvard Business Review* 68 (July-August 1990): 104–12.

79 Ibid.

80 J.R. Caron, S.L. Jarvenpaa, and D.B. Stoddard, 'Reengineering at CIGNA Corporation: Experiences and Lessons Learned from the First Five Years,' *MIS Quarterly* 18 (September 1994): 233–50.

81 Alternative names used in the literature include Enterprise Systems and Enterprise Resource Planning.

82 Thomas Davenport, 'Putting the Enterprise into the Enterprise System,' *Harvard Business Review* 22 (July-August 1998): 123.

83 E. Brown, 'The Best Software Business Bill Gates Doesn't Own,' *Fortune* (29 December 1997): 243.

84 Shoshana Zuboff, *In the Age of the Smart Machine: The Future of Work and Power* (New York: Basic, 1988); Shoshana Zuboff, 'Problems of Symbolic Toil: How People Fare with Computer-Mediated Work,' *Dissent* (Winter 1982): 51–61; Shoshana Zuboff, 'Automate/Informate: The Two Faces of Intelligent Technology,' *Organizational Dynamics* (1985): 5–18.

85 Zuboff, *In the Age of the Smart Machine*, 62.

86 Ibid., 63.

87 Ibid., 64.

88 Ibid., 131.

89 Ibid., 132.

90 Ibid., 134.

91 Ibid., 135.

92 Ibid., 136.

93 Ibid., 138.

94 Davenport, *Process Innovation*.

95 Zuboff, 'Problems of Symbolic Toil.'

96 Zuboff, 'Automate/Informate.'

97 See, for example, Peter Josef Benda, 'Extending the Preventive Engineering Paradigm to the Analysis and Design of Discrete Product Manufacturing Systems: The Role of Functional Integration in the Production of Social and Psychological Outputs' (M.A.Sc. thesis, University of Toronto, 1997).

98 Gutiérrez, *Extending the Precautionary Principle to Work Redesign*.

99 Karasek and Theorell, *Healthy Work*.

100 Alternative models focus on other work-related effects. See, for example, Martin Shain, 'Stress and Satisfaction,' *Occupational Health and Safety*

Canada 15 (April-May 1999): 38–47; Martin Shain, 'The Fairness Connection,' *Occupational Health and Safety Canada* 16 (June 2000): 22–8.
101 T.D. Wall, P.R. Jackson, and S. Mullarkey, 'Further Evidence on Some New Measures of Job Control, Cognitive Demand and Production Responsibility,' *Journal of Organizational Behaviour* 16, no. 5 (1995): 432.
102 P.R. Jackson, T.D. Wall, R. Martin, and K. Davids, 'New Measures of Job Control, Cognitive Demand and Production Responsibility,' *Journal of Applied Psychology* 78, no. 5 (1993): 753–62; Wall, Jackson, and Mullarkey, 'Further Evidence,' 431–55.
103 Zuboff, 'Automate/Informate.'
104 Brödner, *The Shape of Future Technology*.
105 Mumford and Hendricks, *Reengineering Rhetoric and Reality*.
106 Jeremy Rifkin, *The End of Work: The Decline of the Global Market Force and the Dawn of the Post-Market Era* (New York: Putnam, 1995), 151.
107 Ibid., 148.
108 Ibid., 151.
109 OTA Report cited in Rifkin, *The End of Work*, 188.
110 K. Naughton, 'CyberSlacking,' *Newsweek* (29 November 1999), 62–5; Enid Mumford, *Systems Design: Ethical Tools for Ethical Change* (London: Macmillan, 1996), 62.
111 Charles Perrow, *Normal Accidents* (New York: Basic Books, 1984).
112 Rifkin, *The End of Work*, 190–1.
113 Juliet Schor, *The Overworked American*.
114 Landauer, *The Trouble with Computers*, 47–9.
115 Mike Parker and Jane Slaughter, 'Managing by Stress: The Dark Side of the Team Concept,' *International Labor Relations Report* 24 (1988), 19–23; Mike Parker and Jane Slaughter, *Choosing Sides: Unions and the Team Concept* (Boston: South End Press, 1988).
116 F.E. Emery, E.L. Hilgendorf, and B.L. Irving, *The Psychological Dynamics of Smoking* (London: Tobacco Research Council, 1969).
117 Karasek and Theorell, *Healthy Work*, 42; M. Wallace, M. Levens, and G. Singer, 'Blue Collar Stress,' in *Causes, Coping and Consequences of Stress at Work*, ed. Cary L. Cooper and Roy Payne (New York: Wiley, 1988), 74–5.
118 J.P. MacDuffie, 'The Changing Status and Roles of Salaried Employees in the North American Auto Industry,' in *Broken Ladders: Managerial Careers in the New Economy*, ed. Paul Osterman (New York: Oxford University Press, 1996), 82.
119 Rifkin, *The End of Work*, 171.
120 MacDuffie, 'Changing Status and Roles of Salaried Employees in the North American Auto Industry,' 81; E.D. Scott, K.C. O'Shaughnessy, and

P. Cappelli, 'Management Jobs in the Insurance Industry: Organizational Deskilling and Rising Pay Inequity,' in Osterman, ed., *Broken Ladders*, 128.

121 David Mechanic, *Mental Health and Social Policy: The Emergence of Managed Care*, 4th ed. (Needham Heights, MA: Allyn and Bacon, 1999).

122 Steven L. Sauter and Lawrence R. Murphy, eds. *Organizational Risk Factors for Job Stress* (Washington, DC: American Psychological Association, 1995).

123 Z.E. Neuwirth, 'The Silent Anguish of the Healers,' *Newsweek* (13 September 1999): 79.

124 Rifkin, *The End of Work*, 192.

125 MacDuffie, 'The Changing Status and Roles of Salaried Employees in the North American Auto Industry,' 81.

126 B. Stone, 'Get a Life,' *Newsweek* (7 June 1999): 68.

127 Martin Shain, 'Returning to Work after Illness or Injury: The Role of Fairness,' *Bulletin of Science, Technology and Society* 21 (October 2001): 361–8.

128 Ibid.

129 Ibid.

130 Ibid.

131 Vanderburg, *The Labyrinth of Technology*, ch. 10.

132 U.S. Department of Health and Human Services, *Mental Health: A Report of the Surgeon General* (Rockmille, MD: Department of Health and Human Services, U.S. Public Health Services, 1999).

133 Robert H. Nelson, *Reaching for Heaven on Earth: The Theological Meaning of Economics* (Lanham, MD: Rowman and Littlefield, 1993).

134 William Ray Arney, *Experts in the Age of Systems* (Albuquerque: University of New Mexico Press, 1991).

135 Ellul, *The Technological Bluff*.

Chapter 10

1 Jacques Ellul, *The New Demons*, trans. C. Edward Hopkin (New York: Seabury Press, 1975).

2 W.H. Vanderburg, *The Labyrinth of Technology* (Toronto: University of Toronto Press, 2000).

3 Donella H. Meadows, Dennis L. Meadows, and Jørgen Randers, *Beyond the Limits: Confronting Global Collapse, Envisioning a Sustainable Future* (Post Mills, VT: Chelsea Green Publishing, 1992).

4 I have examined a microcosm of this problem in the engineering, management, and regulation of contemporary technology. See Vanderburg, *The Labyrinth of Technology*.

5 Lester R. Brown, *Eco-Economy: Building an Economy for the Earth* (New York: W.H. Norton, 2001); Storm Cunningham, *The Restorative Community: The Greatest New Growth Frontier* (San Fransisco: Berrett-Koehler, 2002).

6 Vanderburg, *The Labyrinth of Technology*.

7 Sheldon Rampton and John Stauber, *Trust Us, We're Experts! How Industry Manipulates Science and Gambles with Your Future* (New York: Penguin Putnam, 2001).

8 Vanderburg, *The Labyrinth of Technology*.

9 Abraham Charnes, ed., *Data Envelopment Analysis: Theory, Methodology, and Application* (Boston: Kluwer Academic Publishers, 1994).

10 Jacques Ellul, *The Technological Society*, trans. John Wilkinson (New York: Vintage Books, 1964).

11 Jacques Ellul, *The Technological System*, trans. Joachim Neugroschel (New York: Continuum, 1980).

12 Jean-Luc Porquet, *Jacques Ellul: L'homme qui avait (presque) tout prévu* (Paris: Le Cherche-Midi, 2003).

13 Vanderburg, *The Labyrinth of Technology*.

14 A. Jürgensen, N. Khan, and W.H. Vanderburg, *Sustainable Production: An Annotated Bibliography* (Lanham, MD: Scarecrow Press, 2001); N. Khan and W.H. Vanderburg, *Sustainable Energy: An Annotated Bibliography* (Lanham, MD: Scarecrow Press, 2001); N. Khan and W.H. Vanderburg, *Healthy Cities: An Annotated Bibliography* (Lanham, MD: Scarecrow Press, 2001); N. Khan, N. Nakajima, and W.H. Vanderburg, *Healthy Work: An Annotated Bibliography* (Lanham, MD: Scarecrow Press, 2004).

15 As a character proclaims in Max Frisch's *Homo Faber*: 'technology ... the knack of so arranging the world that we don't have to experience it.' Max Frisch, *Homo Faber: A Report* (Orlando: Harvest Books, 1994).

16 Craig Brod, *Technostress: The Human Cost of the Computer Revolution* (Reading, MA: Addison-Wesley, 1984).

17 A. Harmon, 'A Sad, Lonely World is Discovered in Cyberspace, Surprising Researchers,' *New York Times* (30 August 1998), 22.

18 Marshall McLuhan, *Understanding Media: The Extensions of Man* (Cambridge, MA: MIT Press, 1994). For an interpretation of McLuhan's work as seen from a similar vantage point to my own, see Ellul, *The Technological System*, introduction. See also Jacques Ellul, *The Humiliation of the Word*, trans. Joyce Main Hanks (Grand Rapids, MI: Eerdmans, 1985).

19 Rampton and Stauber, *Trust Us, We're Experts!*

20 Eric Schlosser, *Fast Food Nation: The Dark Side of the All-American Meal* (New York: Perennial, 2002).

21 Vanderburg, *The Labyrinth of Technology*.

22 Tim Jackson, *Material Concerns: Pollution, Profit and Quality of Life* (New York: Routledge, 1996).

23 Brod, *Technostress*; Sherry Turkle, *The Second Self: Computers and the Human Spirit* (New York: Simon and Schuster, 1984).

24 Brod, *Technostress*; Sherry Turkle, *The Second Self*; David Reisman with Nathan Glazer and Reuel Denney, *The Lonely Crowd: A Study of the Changing American Character* (New Haven, CT: Yale University Press, 1950); Christopher Lasch, *The Minimal Self: Psychic Survival in Troubled Times* (New York: W.W. Norton, 1984); Jacques Ellul, *Propaganda: The Formation of Men's Attitudes*, trans. Konrad Kellen and Jean Lerner (New York: Vintage Books, 1973); Herbert Marcuse, *One-Dimensional Man: Studies in the Ideology of Advanced Industrial Society* (Boston: Beacon Press, 1991); Robert Jay Lifton, *The Protean Self: Human Resilience in an Age of Fragmentation* (New York: Basic Books, 1993); Christopher Lasch, *The Culture of Narcissism: American Life in an Age of Diminishing Expectations* (New York: W.W. Norton, 1978); Richard Stivers, *The Culture of Cynicism: American Morality in Decline* (Cambridge, MA: Blackwell, 1994); Richard Stivers, *Technology as Magic: The Triumph of the Irrational* (New York: Continuum, 1999); Robert H. Nelson, *Reaching for Heaven on Earth: The Theological Meaning of Economics* (Lanham, MD: Rowman and Littlefield, 1993).

25 William Whyte, *The Organization Man* (Garden City, NY: Doubleday, 1957); Victor C. Ferkiss, *Technological Man: The Myth and the Reality* (New York: G. Braziller, 1969).

26 S. Giedion, *Mechanization Takes Command* (New York: Oxford University Press, 1948).

27 Wolfgang Schivelbusch, *The Railway Journey: The Industrialization and Perception of Time and Space in the 19th Century* (New York: Berg, 1986).

28 John Ralston Saul, *The Unconscious Civilization* (New York: Free Press, 1997).

29 Ernst Cassirer, *An Essay on Man* (New Haven, CT: Yale University Press, 1944).

30 Ibid. See also Paul Ricoeur, *The Symbolism of Evil*, trans. E. Buchanan (New York: Harper and Row, 1967); Jacques Ellul, 'Symbolic Function: Technology and Society,' *Journal of Sociological and Biological Structures* 1 (1978): 207–18.

31 Arnold Toynbee, *A Study of History*, abridgement of vol. 1–10 by D. C. Somervell (London: Oxford University Press, 1946).

32 W.H. Vanderburg, *The Growth of Minds and Cultures: A Unified Theory of the Structure of Human Experience* (Toronto: University of Toronto Press, 1985).

33 Bill Bryson, *The Mother Tongue: English and How It Got That Way* (New York: Avon Books, 1991).

34 Jacques Ellul, *Les relations publiques* (Paris: L'année Sociologique, Presses Universitaires de France, 1965).
35 Edward L. Bernays, *Propaganda* (New York: Liverwright, 1928), 47, 48.
36 Hubert L. Dreyfus, *On the Internet* (New York: Routledge, 2001).
37 For example: 2 B or not 2 B (To be or not to be); N E 1 there? (Anyone there?); 4 U R ... (For you are ...); Lol (Laugh out loud); Cu (See you); :) (Happy); Omg (Oh my God); Aso (And so on).
38 Dreyfus, *On the Internet*, 7.
39 Ibid.
40 I have shown this in considerable detail for the domain of technique, referred to as technology, in *The Labyrinth of Technology* (Toronto: University of Toronto Press, 2000).
41 Margaret A. Rose, *The Post-modern and the Post-industrial: A Critical Analysis* (Cambridge: Cambridge University Press).
42 Jacques Ellul, *La pensée Marxiste*, posthumously edited by Michel Hourcade, Jean-Pierre Jezequel, et Gérard Paul (Paris: La Table Ronde, 2003).
43 Jacques Ellul, *Hope in Time of Abandonment*, trans. C. Edward Hopkins (New York: Seabury, 1973).
44 Vanderburg, *The Labyrinth of Technology*, ch. 3.
45 Jacques Ellul, *The New Demons*; Richard Stivers, *Evil in Modern Myth and Ritual* (Athens: University of Georgia Press, 1982).
46 Sarah Anderson and John Cavanagh, *Top 200: The Rise of Corporate Global Power* (Washington, DC: Institute for Policy Studies, 2000).
47 Herbert I. Schiller and Joseph D. Phillips, eds., *Super-state: Readings in the Military-industrial Complex* (Urbana: University of Illinois Press, 1970).
48 Stuart Ewen, *Captains of Consciousness: Advertising and the Social Roots of Consumer Culture* (New York: McGraw-Hill, 1976); *PR! A Social History of Spin* (New York: Basic Books, 1996); Stuart Ewen with Elizabeth Ewen, *Channels of Desire: Mass Image and the Shaping of American Consiousness* (Minneapolis: University of Minnesota Press, 1992).
49 Uwe Poerksen, *Plastic Words: The Tyranny of a Modular Language*, trans. Jutta Mason and David Cayley (University Park: Pennsylvania State University Press, 1995).
50 Ellul, *The Humiliation of the Word*.
51 Ellul, *Propaganda*.
52 Paul Garwood, ed., *Sacred and Profane: Proceedings of a Conference on Archaeology, Ritual and Religion* (Oxford: Oxford Committee for Archaeology, 1991); Thomas J.J. Altizer, *Mircea Eliade and the Dialectic of the Sacred* (Philadelphia: Westminster Press, 1963).

53 Ellul, *The New Demons*; Stivers, *Evil in Modern Myth and Ritual*.
54 Stephen Bayley, *Sex, Drink and Fast Cars* (New York: Pantheon Books, 1986).
55 William H. Masters, Virginia E. Johnson, and Robert C. Kolodny, *Heterosexuality* (New York: HarperCollins, 1994); William H. Masters and Virginia E. Johnson, *Human Sexual Response* (Boston: Little, Brown, 1966).
56 Ellul, *The New Demons*.
57 Neil Carter, *The Politics of the Environment: Ideas, Activism, Policy* (Cambridge: Cambridge University Press, 2001).
58 This interpretation has been forcefully argued by Ellul in *Propaganda*. See particularly chapters 3, 4, and 5, dealing with life in a mass society.
59 W.H. Vanderburg, 'Some Reflections on Teaching Biotechnology, Nanotechnology and Information Technology,' editorial, *Bulletin of Science, Technology and Society* 24 (February 2004): 5–9.
60 Ellul, *The Technological System*.
61 Jacques Ellul, *The Technological Bluff*, trans. Geoffrey W. Bromiley (Grand Rapids, MI: W.B. Eerdmans. 1990); Rampton and Stauber, *Trust Us, We're Experts*; Stuart Ewen, *Captains of Consciousness*; Stuart Ewen, *PR!*
62 Ellul, *The Technological System*.
63 Charles Perrow, *Normal Accidents* (New York: Basic Books, 1984).
64 Ibid.
65 F.E. Emery, ed., *Systems Thinking: Selected Readings* (Harmondsworth, UK: Penguin, 1981).
66 Joseph A. Tainter, *The Collapse of Complex Societies* (Cambridge: Cambridge University Press, 1988).
67 Ibid., 194–5.
68 Arnold Toynbee, *A Study of History*, abridgement of vol. 1–10 by D.C. Somervell (London: Oxford University Press, 1946).
69 Joel Bakan, *The Corporation* (Toronto: Viking Canada, 2004).
70 Jacques Ellul, *L'empire du non-sens: l'art et la société technicienne* (Paris: Presses universitaires de France, 1980); W.H. Vanderburg, 'Comments on the Empire of Non-Sense: Art in a Technique-Dominated Society,' *Bulletin of Science, Technology and Society* 21 (February 2001): 38–54; Jacques Ellul, 'Remarks on Technology and Art,' *Bulletin of Science, Technology and Society* 21 (February 2001): 26–37; S. Giedion, *Mechanization Takes Command* (New York: Oxford University Press, 1948).
71 Unless otherwise indicated, the remainder of this section is based on Ellul, *L'empire du non-sens*. Since this work has not been translated into English, this section will provide the reader with a paraphrase of its principal arguments. I will omit the detailed discussions of specific artistic creations

or the dialogues with many alternate interpretations included in Ellul's study. The focus is on the interpretation of the arts as a reflection of the system of technique.

72   Ellul, 'Remarks on Technology and Art,' 26–37.

73   Ibid.

74   Ellul, *Propaganda*. See particularly chapters 2–4.

75   Ellul, *The Technological System*, 176–8; Ellul, *The Humiliation of the Word*; Ellul, 'Symbolic Function: Technology and Society,' 207–18.

76   It should be noted that most of the critics of Jacques Ellul give no evidence of having read more than one or two of his works. As such, their criticism is somewhat analogous to placing one or two pieces of a jigsaw puzzle on a table and then criticizing the entire picture. Apart from his studies of technique, Ellul has examined human life in a mass society, democracy transformed by technique into an illusion, Western values betrayed in the service of technique, our possession by a secular sacred, myths and the secular political religions built on these, the commonplaces, an autopsy of the Marxist revolution, and the contours of a genuine mutation. Permeated throughout his analysis is a refusal to give in but instead a determination, as an activist, to create play in the system and to live in Hope.

Index

abstraction: science, 307; technology, 301–2
advanced industrial society, 381
advertising, 264, 267–70, 285–6, 354
agribusiness, 358, 360, 435
agriculture, 357–8
alienation, 81, 106, 366; technique, 454
American Academy of Engineers, 306
analytical exemplar, 232
anamnesis: Plato, 320
apprentice, 176–8, 180
Aquinas, St Thomas, 323
Aristotle, 318–9, 321–3; rationality, 321
army, 346
Aron, Raymond, 277
art: culture, 470–80; technique, 473, 480–2
artificial intelligence, 89, 328–9, 331–3
Augustine of Hippo, St, 323
autonomy; technostructure, 254
Aztec culture, 162

Bakan, Joel, 467

Barth, Karl, 101
Bauhaus school, 478
Bell, Daniel, 378–9, 387–8
Bernays, Edward, 441
biology-based connectedness, 8–9, 59
biotechnology, 358–60
black box: corporation, 260, 263
Blue Cross, 306
Boolean algebra; logic, 329
bottlenecks, 35–7
brain-mind, 18, 78–9, 85–6, 108; culture, 71
Brödner, Peter, 401, 416
business process re-engineering, 400–5, 416

capital, 117; agriculture, 359; production, 219
capitalism, 58, 117, 119
Chacoan society: collapse of, 466
chain-reaction model, 35–9
chain-reaction process, 379
Champy, James, 400
China, 389
Chomsky, Noam, 274
Christianity, 81, 322–3, 328, 449;

liberal, 126; politics, 288; revelation, 322–3; work, 121
Chrysler, W.P., 261
CIGNA: re-engineering, 405
civilization: collapse of, 466–7
Cobb, John, 50
Colombian valley, 10–11
commoditization: culture, 50; mass
society, 372–4
common law, 61–2
Commoner, Barry, 258
communication, 91–2; children, 88
compensation, 307, 464–5
computer and information revolution, 376–90
computers, 386–8, 401–2, 407–9; art,
479; result of, 435; use in education, 350–1
Comte, Auguste, 205
consequences, 306
consumer society, 383
context: use of, 300–2
control: corporation, 253
corporate planning system, 270
corporation: control of, 253; diversification, 256; goals, 254–5; legal
person, 467–8; psychopathic, 468;
uncertainties, 263–4
creativity; culture, 136
cultural cycle, 18, 80
cultural unity, 94, 100, 102–103, 105,
114, 117, 127–8, 279–80, 455–6; art
and literature, 128–9
culture, 45, 47, 82–4, 98, 102, 108,
200–1, 280, 300, 325–7; advertising,
268–9; change in, 199, 312–14, 316–
17, 466–7; education, 349–50; experience, 90; language, 186; logic,
330–1; loss of, 383–4; market
transactions, 48–9; Middle Ages,
198; myths, 98; science and technology, 205; symbolization, 187;
technique, 334; technology, 163;
tradition, 53
culture-based approach to life,
304–5
culture-based connectedness, 5–9,
12, 21, 23–9, 39–40, 46–9, 59–60,
63–7, 71–5, 82–3, 110, 173–4, 182,
226–7, 250–1, 313, 361, 364, 368–9,
424–5, 460; characteristics, 432–3;
culture, 434–5; information, 377;
progress, 124; technology, 133;
weakening, 107
culture-based expert living, 316
culture-based mediation, 438
Czechoslovakia, 389

daily-life experience: culture, 209
Daly, Herman, 50
Davenport, Tom, 400
Debord, Guy, 383
demand-control model, 413–17
Denning, W.E., 400
Descartes, René, 323
design, 231–2; technology, 230
design exemplar, 232
desymbolization, 107, 456; technique, 445–6
developing nations: industrialization, 69; culture, 69
Diderot, Denis: encyclopedia, 152
differentiation, 178–9
Dreyfus, Hubert, 316, 322; artificial
intelligence, 331; computers, 388;
Internet, 442–3; skill acquisition,
181
Dreyfus, Stuart, 316; artificial intelligence, 331; computers, 388; skill
acquisition, 176–9

Du Pont, Pierre, 261
Durkheim, Emile, 7, 29

Earth Summit: South Africa, 369
ecological footprint, 9, 12, 311
economic world view, 373
education, 276–7, 343–4; computer
    in, 350; modern, 349–51
11 september 2001, 68
Ellul, Jacques, 7, 13, 101–2, 273, 323,
    344, 474, 479; art, 470, 472–3, 475–
    7, 480, 482; education in France,
    345; flows of goods, 33; informa-
    tion, 379; integration propaganda,
    286; mass media, 294–5; propa-
    ganda, 296–8; sport, 340; technical
    bluff, 423–5; technique, 327, 372,
    427, 432
emigration: agriculture, 358
end of pipe, 307
energy, 312; analysis, 38; flows, 38;
    network of flows of, 5–6, 37–8
England, 66, 154, 196; culture, 156;
    Industrial Revolution, 69–70;
    industrialization, 156; technologi-
    cal development, 155; war, 352;
    work, 290
enterprise: integration, 405–6;
    resource planning, 405–6
enterprise-wide systems, 405–6
environmental crisis: association,
    258–9
expertise: universal knowledge, 316
externalization, 18

Feuerbach, Ludwig, 126–7, 447
flow wastes, 22
food-web, 22
Ford, 261; Edsel, 256; Model T,
    238–9; Mustang, 238, 240–3, 261;

re-engineering, 404–5; Rouge
    complex, 392
Fordist-Taylorist system, 391, 394–8
Foucault, Michel; epistemé, 336
France, 111–13, 457; Industrial
    Revolution, 152; protests, 456
free trade, 373–5
Frege, G.: logic, 329
French Revolution: social order,
    277
Freud, Sigmund, 106, 277, 447–9
Friedman, Iona, 478

Galbraith, J.K., 238, 267, 272; corpo-
    ration, 255; industry, 245–6;
    knowledge separated from experi-
    ence, 237; technology, 240–1;
    technostructure, 251–2
Galileo (Galileo Galilei), 201, 323
General Motors, 253, 306; Corvair,
    256; vertical integration, 262
Germany, 66, 270; technical know-
    ing and doing, 175; war, 352
Gandhi, M.K., 12
Giedion, Siegfried, 437
globalization, 373–5
goals: corporation, 254–5
Gödel, Kurt, 333
goods, flows of, 32–3; traditional,
    34–5; mechanization, 35–8; net-
    work of, 36–7
government, 289–90; advertising,
    298; tax, 291–2; work, 291
Greece: culture, 317–22; Hellenistic,
    322; Homeric, 317–18, 322; phi-
    losophy, 328; sport, 340; universal
    knowledge, 315, 324
Green Revolution, 359–60
Gross Domestic Product (GDP), 272,
    301

Guggenheim, Meyer, 261
Gusdorf, G., 204

Hammer, Michael, 400
happiness: myth, 120
Hegel, G.W.F., 323, 447
Heidegger, Martin, 318
high technology, 384–5
homo economicus, 41, 45, 118, 266, 389;
    mass society, 283
homo informaticus, 4, 13, 204, 220, 331,
    350, 386, 389–90
homo logos, 4, 18
homo societas, 4, 13
horizontal division of labour, 391
human factors engineering, 409
human values, 436

IBM: financing, 402–3
idolatry, 122
industrial-military complex, 271–3,
    381
Industrial Revolution, 17; social
    order, 277; war, 352
industrial societies, 130; characteris-
    tics, 131–2
industrial working class; poverty,
    114
industrialization, 18–19, 28, 57–8, 64,
    66, 70, 110–11, 152, 325, 371;
    adaptation, 343; agriculture, 358;
    beginning, 147–8; capital, 382;
    chain-reaction model, 35–9;
    consumption, 354; information,
    389; reverse process of, 67–8; third
    world, 158
information, 377–8; corporation, 378
information machines, 378
information society, 376–8, 390
information theory, 378

input-output methods, 260–3
inputs/outputs, 327; culture, 301–2;
    means of success, 301–2
integration, 178–9
integration propaganda, 286, 295–7;
    culture, 454, 458
intellectual assembly line, 402, 410;
    IBM, 403
internalization, 18
Internet, 390; culture, 442–3
Inuit culture, 440
invention, 135–8
invisible colleges, 215, 218
invisible elbow, 49–50, 375
invisible hand, 48, 50, 59, 266

Japan, 272; work, 392, 397; workers,
    395
Judaism, 81, 322; politics, 288; work,
    121

Kandinsky, Wassily, 474, 478
Karasek, Robert, 226, 413–15
Keynesianism, 270–2, 299, 373
Kierkegaard, Søren, 449, 475
knowing and doing, 346, 363;
    separation of, 347, 362
knowing 'inside out,' 206, 220
knowing 'outside in,' 206
knowledge embedded in experience,
    190, 226, 228–9; corporation, 398,
    400; Ford Model T, 238–9; Ger-
    many, 196; metaconscious knowl-
    edge, 221–5; technological cycle,
    250
knowledge separated from experi-
    ence, 190, 212, 221; corporation,
    245–6, 398, 400; economy, 270;
    England, 196; factor of produc-
    tion, 273; mass society, 283;

mathematics, 202; Mustang, 238, 241–3; politics, 288; social inequality, 278; society, 275; technological cycle, 250, 252; technostructure, 254, 256

Kuhn, Thomas, 94, 197, 216; disciplinary matrix, 336

Kusterer, Ken, 225, 227

labour: intellectual division of, 219; network of flows of, 52; social division of, 23; vertical division of, 391

labour inputs, 51

labyrinth of technique, 379

language, 389; television, 383

laws, 61; changes in, 60; culture, 62–3

Landes, David, 147; flows of goods, 33–4

Le Corbusier, 478; cities, 471

lean production system, 392–7

legislation: government, 55

Lenin, V.I., 59, 282

Leonardo da Vinci: *Mona Lisa*, 160, 199

life-milieu, 82; culture, 444–5; technique, 460, 474–5; technological means, 82–3

logic, 329–30; communication, 331; limitations, 332–3

Lovins, Amory, 68

Mao Tse-tung, 282

management, 398, 400

Marcuse, Herbert, 383–4

Market, the, 45–9, 59–60, 118, 256, 371–2; biosphere, 48; globalization, 374; role of, 260–7

market externalities, 45–6

marketing: design, 249; media, 355

markets: natural capital, 50

Marx, Karl, 7, 29, 58–9, 63, 101–2, 106, 118, 175, 184, 273, 287, 290, 389, 447–9, 475

Marxism, 226

mass education, 343, 349

mass media, 293–6; culture, 383

mass society, 284–5, 292, 437; development of, 281–3

mathematics, 201–5

matter: network of flows of, 5–6, 37–8

Mayan society: collapse, 466

mechanization, 28, 35–7, 54–5; culture, 52

media, 435

medium: dimension of, 108; economic dimension of, 260

megamachine society, 385–6

memory: human, 85; machine, 85

mental map, 92–3, 149; culture, 150

meta-experience, 80

metaconscious, 90, 99, 116; advertising, 269; language, 186

metaconscious knowledge, 85, 89, 96, 115, 188–9, 459; children, 86

metalanguage, 383, 389; culture, 439–41

Middle Ages, 151; Christian, 150; happiness, 124–5

Mies van der Rohe, Ludwig, 478

military, 272–3

Mill, John Stuart, 266

Mondrian, Piet, 478

monocultures, 358–9

Mumford, Lewis, 41

Mutual Benefit Life: re-engineering, 404

myth-information, 295

myth-symbols, 295
myth of progress, 124; of science,
 213–14
myths, 96–100, 103, 116, 118; culture,
 445–7; Greek, 318; progress, 121–2;
 technique, 453

Napoleon, 204
negative feedback, 116, 465
net wealth, 274; GDP, 272
news, 292–4
Newton, Sir Isaac, 201, 220
Nietzche, Friedrich, 106, 318, 447–9
non-renewable resources, 22
non-sense, 302, 304
North America, 66; industrialization,
 157–8

objectivization, 18
Ohno, Taiichi, 392–3, production
operations research, 353
organizations, 356–7
ownership; corporation, 252–3

Pascal, Blaise, 323
Peloponnesian War; Hellenic
 Greeks, 318
people changing technology, 19–20
performance values, 231, 314
Perrow, Charles, 68
personality types: inner directed,
 279; other directed, 279; tradition
 directed, 279
Piaget, Jean, 204
planned obsolescence, 255–6
planning; corporation, 256
Plato, 7, 318–24
Polanyi, Michael; tacit knowledge,
 336

political framework, 56; emergence
 of, 57; pre-industrial societies, 56
political revised sequence, 298
politics, 287–8
population: distribution, 54–5
postmodernism: technique, 446
production, factor of, 41; agricultural
 societies, 43
professional and intellectual fundamentalism, 265
Prozac, 373, 431
public opinion, 284–5, 292–6
public relations, 294
public-sector services, 374–5

quality movement, 400

rational knowledge, 316; *anamnesis*,
 320; Aristotle, 321
rationalism, 151, 333
rationality, 325
reality, 94–7
reality as it is known, 94–7
Reisman, David, 279
religion: England, 155
renewable resources, 22
revised sequence, 267
Ricardo, David, 266
Richta, R., 237
risk: management of, 256–7
Rockefeller, John D., 261
Roman society: collapse, 466
Roman law, 322
Russell, Bertrand, 330

sacralization; capital, 118–19;
 humanity, 439
sacred, 100–1, 103, 122; Jewish,
 101–2; Christian, 101–2, 125–6

secularization: Christian, 126
separation of knowledge from
    experience, 174
service sector, 379–80; society and
    economy, 379–81
Schoenberg, Arnold, 475
school-based scientific knowledge,
    186–8
science, 125–6, 197–8
scientific growth of knowledge,
    380
scientific knowing, 11, 13
Second World War, 352
service economy: energy, 436
Shain, Martin, 418–20
Shannon, Claude, 378
shareholders, 252–3
Silicon Valley: technostructure, 252
Simmel, George, 199, 277
Simon, Herbert: computers, 388
Simondon, G., 204
skill acquisition, 176–9, 212, 315–16;
    artificial intelligence, 332
Smith, Adam, 23, 25, 40, 50, 199, 266,
    290
Smith, A.M., 206
soccer, development of, 341
social mobility, 275
social roles: technology, 168
socialism, 119
socialization, 209–13; primary and
    secondary, 208, 215–16
society, 286, 312; economy, 381–2;
    global village, 383; post-capital-
    ism, 381; post-industrial, 379–81;
    structure, 279; wealth, 307
society of risk, 384, 465–6
socio-cultural wholes, 110
socio-epidemiology, 26, 307

Socrates, 7, 318–21, 324
solar energy, 22
Solzhenitsyn, Alexander, 130
Soviet Union, 256, 389
Spartans, 352
specialists, 301, 306
spectator society, 383–4
sport, 340–2; nationalism, 342
state, 57–60, 288
statistical morality, 284
stock wastes, 22
strategic planning, 399; develop-
    ment of, 398
structure of experience, 79–80, 200;
    Middle Ages, 198; traditional, 109
subconscious, 90
survey methodology, 347–9
symbolic link, 439
symbolization, 5–7, 73, 75, 78–82,
    90–1, 96, 209, 268, 423, 425; cul-
    ture, 443–5; culture-based, 84;
    children, 89; humanity, 439–44;
    technology, 208

Tainter, J., 465–7
taxes, 291
technical approach: context compat-
    ibility, 363–4; environmental
    crisis, 369–70; external regulation,
    366–9; integrality, 363; reification,
    365–6; selective force, 370–2; self-
    regulation, 362–3; separation of
    knowing and doing, 361–2
technical approach to life, 303–6,
    315, 338–9, 381–2, 385, 429, 436;
    consequences of, 361–70; educa-
    tion, 343, 348–9; information, 376–
    7; sport, 340–2
technical development; law, 61

technical division of labour, 23, 25, 26; England, 32

technical knowing and doing, 11, 13, 176, 180–3, 312; separation of, 263

technical knowledge, 176, 181

technical order: society, 313

technique, 327–8, 334, 336–7, 346, 370, 372, 375, 385, 390, 422–7, 441–3; art, 471–82; characteristics, 428–32; collective person, 467–9; consciousness, 438–4; culture, 434–8, 450–9, 461–2; education, 344; human life and society, 463–70; influence of, 445–6; information, 376–7, 381–3; labyrinth of, 379; life-milieu, 434–8, 474; mediation, 434–8; non-sense, 462; organizations, 356–7; phenomenon, 426–33; possession, 444–60; production, 398; result of, 431; risk, 384; system, 461–7, 481; war, 352–3

technique as mediation: culture, 437

technique-based connectedness, 339, 372–3, 377–8, 385, 390, 411, 422–4, 427–9, 445, 448, 450, 460, 468–9; characteristics, 432–3; culture, 386

technological bluff, 286

technological cycle, 134–5, 137, 241, 261–2; mass production, 249–50

technological development, 141–2

technological 'hardware,' 257

technological knowledge, 189–90, 225

technological 'software,' 257

technology, 58, 143–5, 259, 326, 384; culture, 133–4, 139–40; evolution, 140; hardware/software, 134; medieval, 153–4; society, 138–9, 167, 257–8, 271–2

technology as system, 257

technology-based connectedness, 5, 12, 21, 23–30, 39–41, 49–50, 54, 59–60, 64–7, 71, 110, 173–4, 182, 226–7, 250–1, 313, 325–7, 361, 364, 368–71; labour, 52; progress, 124; technology, 133

technology changing people, 19–20, 66–7

technostructure, 246–7, 250–5, 270, 288; design process, 248–9

television, 355, 383

Theorell, Töres, 226, 413–15

third megaproject, 4–5, 9, 12–14, 106, 302–3, 324, 482

Tocqueville, Alexis de, 277

totemic societies, 159–62; life-milieu, 161

Toynbee, Arnold, 7, 218, 466–7; culture, 439

Toyoda, Kiichiro, 392

Toyota, 396; lean production, 392; work strain, 417

Toyota, Eiji, 392

traditional society, 30

traditional technologies, 163

Uddevalla: Volvo, 397

ultimate good, 123

umbrella design exemplar, 232; Ford Mustang, 242

underdevelopment, 143–4

United States, 67, 253, 261, 272, 457; agribusiness, 358–9; defence policy, 353; democracy, 458–9; economy, 258; energy use, 359; work, 417

universal knowledge, 316, 324

universal technology, 236, 238

Vacca, Roberto, 67–9
vanity, 104
Viagra, 373
visualization: infant, 88
Volvo: work, 397

war, 352; technology, 353
weapons: technology, 353
Weber, Max, 7, 13, 29, 78, 273, 326–7,
    421; information, 379; rationality,
    372; technique, 427, 432

Weizenbaum, Joseph, 387–8
Wittgenstein, Ludwig: language, 335
Woodstock, 477
work: factory, 44, 53–5, 401; automa-
    tion, 416; Fordist-Taylorist, 391;
    negative impact of, 412–13; stress,
    417–21

Zuboff, Shoshana, 190, 408–9, 411,
    416; computer systems, 407